Quantitative Zoology
Revised Edition

George Gaylord Simpson
Anne Roe
Richard C. Lewontin

DOVER PUBLICATIONS, INC.
Mineola, New York

Copyright

Copyright © 2003 by Dover Publications, Inc.
All rights reserved.

Bibliographical Note

This Dover edition, first published in 2003, is an unabridged republication of the revised edition published by Harcourt, Brace and Company, Inc., New York, 1960. The work was originally published by McGraw-Hill, New York, in 1939.

Library of Congress Cataloging-in-Publication Data

Simpson, George Gaylord, 1902-
   Quantitative zoology / George Gaylord Simpson, Anne Roe, Richard C. Lewontin.
      p. cm.
   Originally published: Rev. ed. New York : Harcourt, Brace, 1960.
   Includes bibliographical references (p.) and index.
   ISBN 0-486-43275-0 (pbk.)
   1. Zoology. 2. Biomathematics. I. Roe, Anne, 1904- II. Lewontin, Richard C., 1929- III. Title.

QL58.S55 2003
590'.1'51—dc21
                                                                                    2003060372

Manufactured in the United States of America
Dover Publications, Inc., 31 East 2nd Street, Mineola, N.Y. 11501

Contents

	PREFACE	iv
	PREFACE TO THE DOVER EDITION	ix
1	Types and Properties of Numerical Data	1
2	Mensuration	20
3	Frequency Distributions and Grouping	31
4	Patterns of Frequency Distributions	48
5	Measures of Central Tendency	65
6	Measures of Dispersion and Variability	78
7	Populations and Samples	96
8	Probability and Probability Distributions	117
9	Confidence Intervals	148
10	Comparisons of Samples	172
11	Correlation and Regression	213
12	The Analysis of Variance	258
13	Tests on Frequencies	306
14	Graphic Methods	339
15	Growth	373
	APPENDIX TABLES	421
	SYMBOLS	427
	BIBLIOGRAPHY	430
	INDEX	435

Preface

Zoology, for our purposes, is a systematic branch of biology, distinct from the primarily experimental branches. The primary subject of this book, then, is the gathering, handling, and interpretation of numerical data from zoological investigations in this stricter sense. The basic statistical techniques included are those with explicit application in this field, and these suffice for most of the problems likely to arise in current zoological research.

The concepts and principles underlying quantitative and especially statistical methods are explained at some length, because we feel that a knowledge of the philosophy of statistical inference is essential to proper practice. Together with the basic techniques, this knowledge will provide a foundation for exploring and understanding more specialized or advanced biometrics if the need arises.

The exposition of the material proceeds simply, step by step, with numerous accompanying examples, the working of which is fully explained. No knowledge of mathematics beyond the most elementary algebra and use of simple logarithms is assumed. Specifically, we do not suppose the reader to know anything about statistics, nor is the calculus ever employed in our discussions.

In respect of the purpose and scope of the book and the audience to which it is directed, there are no really essential changes from the first edition, but there have been some important alterations of approach and attitude.

The use of quantitative methods in zoology has changed considerably since the first edition of this book was written (mostly in 1937) and published (1939).[1] Then the application of any but extremely elementary numerical techniques was quite unusual in this field. Students were almost never given explicit training in handling quantitative data. Practicing zoologists were not only, as a rule, profoundly ignorant of the principles

---

[1] *Quantitative Zoology*, George Gaylord Simpson and Anne Roe. New York, McGraw-Hill, 1939.

of statistics but also, in many cases, outspokenly antagonistic to any statistical approach to their problems. It was that situation and our dissatisfaction with it that led the original authors to write this book.

A change was then in the air, especially as regards systematics which among all the ramifications of zoology necessarily remains its basic discipline. The practice of systematics was still usually typological in 1939, but a change to population systematics was incipient. *The New Systematics*,[2] which as much as any one work both signalized and stimulated the change, appeared in 1940, the year after *Quantitative Zoology*. The satisfactory treatment of populations absolutely requires the application of statistical concepts and the use of some, even if only the simplest, statistical methods.

There are still some typological systematists, but the population approach has now become usual in systematics and has spread into all branches of zoology. It is today widely admitted that all zoological problems without exception relate to populations and that their valid study always involves inference from samples to populations. Such inference is, by definition, statistical. Now advanced students of zoology are commonly required to take some basic training in quantitative, including statistical, procedures. Professional zoologists can no longer consider themselves competent unless they have at least elementary notions of this aspect of zoological methodology.

The first edition of *Quantitative Zoology* was frankly addressed to zoologists not only lacking in mathematical background but also quite skeptical as to the applicability and utility of mathematical treatment of their data. Later developments have made some change of approach possible and advisable. It is, surely, no longer necessary to argue with the few remaining skeptics. It is possible to assume that zoologists, especially those early in their careers, want to learn the fundamentals of quantitative and statistical treatment and will take some pains to do so. Nevertheless, in preparing the second edition we have retained these elements of the original approach: clarity and simplicity of exposition, the requirement of only minimal mathematical ability, and specific applicability to the most frequent kinds of zoological problems.

Biological statistics, or biometrics, has become a profession in itself, and an intricate and difficult one. It would be fine if all zoologists were also master biometricians. Some are, or will become so. Others do not have the time for a thorough study of biometrics or perhaps, although fully competent in their own field of zoology, lack the special and different interests and abilities required for professional biometrics. Nor is it necessary that an expert zoologist also be an expert biometrician. What is necessary is that he be able to use basic techniques in his own field and that he have sufficient grasp of statistical principles to understand the general signifi-

[2] Edited by Julian S. Huxley, Oxford, Clarendon Press, 1940.

cance of more advanced biometric techniques. With that equipment he is furthermore in a position to utilize specialized biometric assistance in instances (relatively few in usual zoological research) where basic techniques are insufficient. Since basic techniques and general principles come first in any case, he is also in a position himself to proceed further in the study of technical statistics if the need and occasion arise.

There are excellent books on biometry, statistics, and quantitative methods in general, many more than when this book was first published. But even now, none seems to meet the demand that *Quantitative Zoology* tries to fill. The most nearly similar treatments are strongly oriented around experimental design and interpretation, especially with agricultural applications. They contain a great deal of material unlikely to be useful to the zoologist, strictly speaking, and they omit much that is essential to him. Others, while including almost the whole field of quantitative zoology, are also exhaustive beyond the needs of the average zoologist and presuppose a level of mathematical and statistical sophistication considerably higher than he is likely to have reached. Increased interest in the subject and the fact that this particular need is not supplied by other books have been evidenced in the continuing demand for this book, a demand that seemed to increase rather than decrease after the first edition went out of print. That is the reason why we finally accepted the task of preparing a second, completely revised edition. That task has been lightened for the original authors and its outcome has been improved by adding a third author (Lewontin) who has specialized in both zoology and biometrics.

In revising the book, a somewhat different general approach to statistics has been adopted. The parameters of theoretical distributions and the estimates derived from samples have been more clearly and consistently distinguished. The precise operational meanings of hypothesis testing and of probability have been stressed. The approach to statistical significance has been changed from the first edition, where the distribution with infinite frequency was taken as the general case, and the sample with finite, especially with small, frequency was taken as a special case. Now the distributions of finite samples, however small or large, are considered as the general case and the special case arises when the sample is large enough to be treated, for practical purposes, as infinite in frequency. This approach, using confidence intervals, is not only more logical and consistent in modern statistical theory but also more appropriate for practical applications in zoology.

The only major addition is the insertion of a wholly new chapter on the analysis of variance. This must now be considered an essential and basic technique in quantitative zoology. It also rounds out the concepts and principles really fundamental in statistics. Many procedures that are or could be occasionally used in zoology are necessarily still omitted, because they are highly specialized, are not really fundamental, or are considered inferior to similar methods that are included. This treatment is introductory and it cannot at the same time be exhaustive.

In the first edition stress was placed on hand calculation, with some special hand formulae and with detailed instructions for calculation with various aids. The calculating formulae now given can be readily adapted to hand or slide-rule calculation, but in general they assume that a machine is available, and no special instructions for purely arithmetical calculations are given. It is now reasonable to believe that even the beginner in quantitative zoology can handle such simple operations or can better learn them elsewhere.

There are no important deletions of topics, but some have been rearranged and some chapters reorganized. Errors have, of course, been eliminated as far as possible, and the whole text has been clarified and brought up to date. Much the greater part of the text has been completely rewritten, although still within the framework of the original intention.

As with any book, the preparation of this second edition has benefited from the efforts of many hands and heads besides those of the authors. Dr. C. C. Cockerham has helped us greatly by his views on some of the more debatable statistical questions, although we have not always chosen to take his sound advice. Dr. John Imbrie and Dr. James C. King have very carefully reviewed the manuscript and their suggestions have been of great help. Dr. William Hassler and Dr. Edward Lowry have been very generous in allowing us to use their unpublished data for some of the examples. The tedious job of typing the several versions of the manuscript has been done by Mrs. Frances Ingram and the equally tedious task of making the index by Mrs. M. J. Lewontin. The illustrations have all been redrawn and a number of new ones added by Mr. Jefferson D. Brooks III.

We are indebted to Professor Sir Ronald A. Fisher, Cambridge, to Dr. Frank Yates, Rothamsted, and to Messrs. Oliver and Boyd Ltd., Edinburgh, for permission to reprint Tables III and VI from their book, *Statistical Tables for Biological, Agricultural, and Medical Research*.

During most of the work on this revision, Simpson was at the American Museum of Natural History and Columbia University, Roe at New York University, and Lewontin at North Carolina State College. Prior to publication Simpson and Roe moved to Harvard University and Lewontin to the University of Rochester.

G. G. S.
A. R.
R. C. L.

Preface to the Dover Edition

Biology has long had a very varied relationship to quantitative concepts and methods, especially to statistics and probability. On the one hand, the science of genetics was founded by Mendel on the basis of inference from the numerical proportions of individuals of different types among the progeny of a cross. Population and evolutionary genetics, population ecology and studies of species diversity are quantitative sciences based in mathematical and statistical structures and their methods of inference are essentially statistical. Indeed, R.A. Fisher, one of the founders of theoretical evolutionary genetics and the genetics of continuously varying characters, was also a founder of modern statistics. On the other hand, those aspects of biology concerned with a description of the gears and levers of biological processes, molecular and developmental biology are singularly non-quantitative and one can, at present, make a perfectly successful lifework of studying the molecular basis of development while remaining in complete ignorance of even the most elementary notions of probability and statistics.

What distinguishes those aspects of biology that use statistics and probability from those that do not is the issue of whether individual variability is central to the questions being asked. Mendel discovered the principles of formal genetics precisely by concentrating on the differences between individuals within a progeny group. It was not the *similarity* of offspring to parents, but the pattern of *deviation* of offspring from their parents that provided the key to inferring the principles of heredity. Population genetics and population ecology are studies of the temporal changes and spatial divergence that occur in the proportions of different forms within populations. Questions about individual variation form the problematic of these fields. In contrast, the molecular biologist assumes that the forms of molecules and processes in which they participate are the same in all cells of a particular type. Developmental biologists want to know why all vertebrates have similar body segments and why lions are different in their development from lambs, but they never ask why lions are different from each other or, for that matter, why the fingerprints on my left hand were different from those on my right hand from the day I was born. The variability among individuals is simply not part of the set of questions that agitate molecular or developmental biology, although no doubt it will be at some future time.

A paradoxical feature in the history of biology is that until the Second World War, the study of the growth and morphology of whole organisms was as disinterested in individual variation as is present-day molecular biology. This lack of interest in individual variation was paradoxical, because these morphological studies were largely devoted to inferring the evolutionary relationships between species and tracing the lines of descent of living forms from their long-extinct ancestors. But the processes of evolution depend critically on the existence of *variation* between individuals within a species. The Darwinian explanation of evolution is precisely that heritable variation between individuals within populations is converted into differences between populations and species in space and time. It seems odd that the study of systematics could have remained so long obdurately typological. I remember the frustration I felt as a graduate student when, trying to identify a species of *Drosophila* using the standard key, I came across the choice "costal index (the ratio of the lengths of two wing veins) greater than 4.0 or less than 4.0" when my specimens, all the offspring of a single fertilized female, varied in their costal indices from 3.9 to 4.1.

The first edition of Simpson and Roe's *Quantitative Zoology*, written in 1939, was an attempt to introduce statistical and graphic methods into the study of growth and form and, in particular, into the practice of systematics. In a sense, *Quantitative Zoology* was more a propaganda instrument than a fulfillment of a widely perceived need. George Gaylord Simpson was already well known as a leader in vertebrate paleontology, and the later publication (in 1944) of his Columbia University Jessup Lectures as *Tempo and Mode in Evolution* simply confirmed his preeminence in the field. His authorship of a book on numerical and statistical methods in zoology put the weight of his growing prestige behind the drive to quantify systematic practice. He had not used statistical methods to any great degree in his own earlier work. In the Preface to the first edition he writes of

> "the experience of the senior author of
> the present text some ten years ago when

he set out to work toward a conscious,
rational numerical methodology, the
absence of which was increasingly apparent
to him in his own work and that of almost
all his colleagues."

Anne Roe, while not a biologist, did research in education, a field in which statistical and numerical methods had long played an important role, and her joint authorship was intended to provide a professional expertise that Simpson lacked. The title of the book they produced, "Quantitative Zoology," is to be taken seriously. It was not their intention to produce a textbook that could serve in a standard course in statistics for natural scientists, and which would be as applicable to the design and analysis of experiments as it would be to descriptive biology. Their purpose was to introduce zoologists to the elements of statistical and numerical reasoning (the same methods could, of course, be used by botanists), not to make them into generally competent users of statistics. Their view, expressed in the Preface, was that many of the concepts and methods of statistics were too difficult to be absorbed by the unsophisticated zoologist, who needed to be led gently into the realms of quantification. Thus, there were chapters on mensuration, growth, and graphic methods. They explicitly refused to discriminate in their notation between observed sampled values of quantities and the parameter values in the unobservable universes from which the samples were taken. The notion that a sample value is an estimator of a parameter of some unobservable population, an estimator that has a well-defined probability distribution, is fundamental to statistics, but was deliberately rejected by Simpson and Roe as too difficult for zoologists to handle. Thus, in testing whether two samples had means that were significantly different from each other, they employed the concept of the standard error of a difference, but were unable to tell the reader how to find the probability that two samples from the same population would differ by any particular multiple of the standard deviation. A standard statistical text would have introduced Student's t-test for the difference between two sample means, but this was regarded by the authors as too advanced for their readers.

A major change in the education of biologists followed the Second World War as a result of the prestige that had accrued to physicists and mathematicians from their war-related research. Biologists were now expected to know a minimum of mathematics and even the systematic zoologist could be expected to comprehend and apply the standard methodology of statistics. Simpson and Roe decided that a major revision of *Quantitative Zoology* was possible, bringing it more into line with modern statistical theory and they asked me, as someone with recent professional training and published research in both zoology and mathematical statistics, to undertake the work of *aggiornamento*. The flattery was more than I could resist, and the result was the second, revised edition of *Quantitative Zoology*, now reproduced in this Dover edition. Had I been left to my own devices I would have omitted the chapters on mensuration and graphic methods as too elementary, and left the chapter on growth unchanged through my own ignorance of the subject. When Simpson and Roe saw what I had produced, however, they wisely insisted that this was still a book on quantitative zoology and not simply on statistics, so those chapters were updated and included. Nevertheless, the Revised Edition is much closer to a standard text of statistics. The fundamental distinction between sample quantities and the parameters they are meant to estimate is central. The model of statistical testing is the erection of a null hypothesis; for example, that the means of two populations are the same, followed by the comparison of the observed test statistic calculated from the samples with the probability distribution of that statistic under the null hypothesis. If the probability is smaller than some agreed-upon small value, say .05, the null hypothesis is rejected. The main body of the text was expanded and completely rewritten to include such standard statistical methods as the Students t-test on means, the F-test for the equality of variances and the analysis of variance. In addition, an appendix was added containing the tabulated distributions of the commonly used descriptive and test statistics. While it is very far from a comprehensive text of biostatistics such as Sokal and Rohlf's *Biometry*, a zoologist marooned on a desert island with nothing but his or her specimens and the Revised Edition of *Quantitative Zoology*, could produce a perfectly acceptable manuscript for publication.

In the forty-five years that have passed since the publication of the second edition of *Quantitative Zoology* and the appearance of the Dover edition, immense changes have occurred in statistics and in systematics—both as a consequence of the development of high-speed computers. The standard run of statistical tests depends on probability distributions of test statistics under the null hypothesis. These probabilities are calculated from explicit mathematical equations derived from the null hypothesis. But the problem is that, in addition to the null hypothesis of interest, say, that the means are equal in the populations from which two samples have been taken, it is necessary to make a lot of assumptions about the populations; for example, that their variances are equal and that the variable in the population has the normal distribution. That is, any statistical test is a test of a complex null hypothesis, but we are interested in testing only one element of that complex hypothesis so we pretend that the other elements are true and we sweep them into a category called "assumptions," leaving behind as the "hypothesis" to be tested, the one question we care about. But suppose our "assumptions" are not met. Suppose the means of the two populations are really equal, but the variances are not. How sensitive is Student's t-test for the equality of two means to a difference in variance, rather than a difference in means? How sensitive is it to deviations of the underlying distribution of the variable from the normal distribution? If the Student's t-test were very sensitive to these departures from the assumptions, then many of our conclusions about the facts of nature based on doing t-tests on means would be wrong! For most of the history of statistics there was no way of knowing how sensitive our tests were to faulty assumptions. High-speed computation has changed all that. Using computers it has been possible to simulate populations that differ from each other in different ways and then to apply the standard statistical tests to tens of thousands of random samples from these simulated populations in order to determine how sensitive the tests actually are to different kinds of deviations from the complex null hypothesis. The response to many such "Monte Carlo" sampling experiments, as they are called, has been a collective sigh of relief on the part of statisticians. It turns out that Students t-test is insensitive to unequal variances and non-normality. The F-test as used in the analysis of variance is also insensitive to deviations from the normal distribution. Alas, however, the F-test for the equality of two sample variances turns out to be rather sensitive to deviations from the normal distribution, so an unknown number of conclusions about the variability of different populations, based on that test, are wrong.

Equally as important as the use of computers to check up on standard statistical tests has been the possibility of inventing entirely new statistical procedures for specific problems where no mathematical work has been possible. We can now simply invent new statistical tests of very complex hypotheses or new methods of estimating quantities of biological interest, and then see how these methods behave in samples by producing huge numbers of simulated samples in the computer. We can, for example, ask whether there is any evidence that natural selection has played a role in causing the divergence between the nucleotide sequence of a gene in two different species. Unfortunately, to investigate this particular problem free of any assumptions about populations requires an immense amount of computing—more than has been deemed worthwhile up to the present—so most such tests have included untested assumptions. But the importance of those assumptions can be tested by the same computer simulation techniques if anyone cares to do it.

The practice of systematics has also been revolutionized by high-speed computational techniques. There are two common methods currently in use for reconstructing the tree of relationships among organisms. One, the cladistic methodology, makes a few assumptions about the nature of character evolution; for example, that a given character change is irreversible in evolution and that the same character change does not occur more than once in the history of a related group. The problem then is to find a phylogenetic tree involving a large number of related species and a large number of characters that has the minimum number of exceptions to these basic assumptions. Such a tree is said to show "maximum parsimony." The number of possible trees that can be drawn for even a moderate number of species is immense, so only with computational techniques can the space of all possible trees be examined to find the one with the highest parsimony. In fact, it often turns out there are a number of equally parsimonious trees, as well as a large number that are close to that maximally parsimonious configuration. The other methodology is an adaptation of R.A. Fisher's method of "maximum likelihood" to estimate the parameters of distributions. It makes a complex quantitative model of the evolutionary process that involves the probabilities of various events and then searches for the phylogeny that has the maximum likelihood under this probability model. Again, very large amounts of calculation are required and the difference in likelihood among a number of the best alternative phylogenies may be quite small. At the present time virtually all systematic reconstruction for organisms of all kinds uses one or the other of these systems. Thus systematics has, in its everyday practice, become far more quantitative and probabilistic than could have ever been dreamed of by the original authors of *Quantitative Zoology*.

When the 1960 revised edition of this book was published, George Gaylord Simpson and Anne Roe were both at Harvard, and the present author was at the University of Rochester.

> Richard Lewontin
> Harvard University
> Cambridge, Massachusetts
> August, 2003

CHAPTER ONE

Types and Properties of Numerical Data

Variables in Zoology

Zoology is concerned with the study of things, of whatever sort, that vary in nature and that are in any way related to animal morphology, physiology, or behavior. Thus, when a zoologist sets out to describe or discuss any animal, he almost inevitably finds that he is using some numbers. Usually, measurements of the dimensions of individual animals are given; the proportions of the different parts of the animal are considered; different animals are compared as to size and proportions; abundance or scarcity of a species may be mentioned; the number of teeth, scales, fin rays, vertebrae, and the like are recorded. In many other ways, essentially numerical facts and deductions enter into the work. Commonly these observations are expressed by actual numbers, but not infrequently they may be expressed in words, without the use of figures. When it is said that one species is larger than another, that a given animal is abundant in a certain area, or that a certain mammal lacks canine teeth, for instance, this is only a verbal expression of a numerical idea. If such observations can be reduced to concrete figures, the expressions will usually be more accurate and more succinct. Even if they cannot well be expressed except in words, the essentially numerical nature of the concepts demands recognition and requires knowledge of the properties of numbers and of the ways in which they should be used and understood.

Because variables are not alike in their properties, a clear distinction must be made among the sorts of quantities with which the zoologist deals, in order that they may be treated intelligently.

The basic distinction to be made is that between *continuous* variables and *discontinuous*, or *discrete*, variables. While other categorizations of variables are possible, it is this dichotomy that is logically and operationally of greatest significance.

Continuous variables are those which can take any value in a given interval. The three basic physical units—time, length, and mass—are

1

clearly continuous variables. Moreover, all continuous variables in zoology, or indeed in any descriptions of the real world, are expressible in one of these units or in some combination of them. It is obvious that no matter how close together two points in time are, there is some point which lies between them. There is, in fact, an infinity of values between any two points in a continuum, and it is this property which distinguishes continuous variables.

Discrete (or discontinuous) variables, on the other hand, take only certain values, so that it is possible to find two points in a discrete series between which no other value exists. The integers form such a discrete series: there is no integer, for example, between 5 and 6 or between 107 and 108. This immediately suggests that a common discrete variable in zoology is an enumeration of the number of objects in a given situation. Thus 4, 5, and 6 eggs in a clutch are values in a discrete series since, presumably, 4.367 or 5.237 are not allowable values. Nearly all discrete variables in zoology take on integral values only, since for the most part they are counts of objects; but they are not exclusively so. Thus, degrees of genetic relationship among members of a family take values such as 1/2, 1/4, 1/8, 1/16, etc., with no intermediate steps. What is important in considering discrete variables is not whether they actually are restricted to integral values, but that there are no intermediate values between any two consecutive steps.

The distinction between kinds of variables has been made in terms of numerical values, but there are also variables in zoology, such as color, shape, and behavior, that are not numerically expressed, either because it is inconvenient and unnecessary to do so or because suitable techniques are not available. Such variables are *by their nature discrete*, although they may *represent* series which are, in fact, continuous. Thus, the division of animals into "large," "medium," and "small" or "dark" and "light" represents the conversion of continuous variables into discontinuous ones by including a range of values within one class. Furthermore, there are some nonnumerical variables which cannot be considered as representations of numerical ones. For example, an aperture may be described as "round," "triangular," or "square"; the coiling of a gastropod shell may be either "dextral" or "sinistral"; a structure may be "present" or "absent." Such descriptive variables which do not have values falling in some logical order from smallest to largest, as do numbers, are termed "attributes." Although they are nonnumerical, they share with discrete numerical variables the property that the various classes may be assigned arbitrary integral values —that is, they may be enumerated. Variables, whether continuous or discontinuous, which take on numerical values are termed "variates" to distinguish them from "attributes."

The existence of nonnumerical variables poses a constant problem in zoology, or indeed in any science. To begin with, there is a tendency to reduce numerical observations to nonnumerical valuations. The substitution of description such as "larger than," "heavier than," "older than" for actual measurement is, for the most part, unwarranted. Only in special cases should this qualitative representation be used in place of actual quantities.

Second, there is the possibility of assigning arbitrary numerical values or "scores" to variables which are not directly measurable. Unless the particular situation imposes some obvious order on these nonnumerical variables, the assignment of scores is not advisable. Very few operations which are performable on scored data cannot just as easily be applied to the primary class designations. It must be remembered that assigning numerical scores to nonnumerical classes is simply a renaming of these classes; it cannot create numerical accuracy where none exists.

There are, however, some problems which are most easily treated by numerical scores. A case of this type is discussed on page 14. Our intention is not to dismiss scores as totally useless, but rather to caution against their indiscriminate use.

Finally, there is the possibility of accurately quantifying variables which have formerly been given only qualitative treatment. Colors, which are usually considered to be nonnumerical attributes, can be described in terms of their wave length and intensities in a perfectly rigorous numerical way by the use of appropriate measuring devices. While such precision is not always necessary, it is generally desirable to treat basically numerical concepts as numerical rather than to sacrifice information and precision.

There is nothing to be gained, however, in quantifying a variable such as color when there is a clear and unambiguous distinction among the various observed classes. The distinction between black and white or between red and green is sufficiently obvious to require no further numerical precision. The distinction among various shades of grey, on the other hand, especially if there is an imperceptible gradation of shade, does require some numerical specification to make the character a useful one. Dice (1933), for example, has made extensive use of the tint photometer for the study of pelage coloration in *Peromyscus*. With this technique it is possible to assign numerical values to the intensity of red, yellow, green, blue, and violet coloration in the pelages of each specimen.

While it is true that variables may be continuous or discrete, *in practice all measurements are discrete variables*. This is so because of the real limitations of accuracy inherent in any measuring device. Electronic instruments exist which will measure time in millionths of a second (microseconds), but there is still an infinity of intervals between 1 and 2 microseconds, let us say, which are unmeasurable. No length can be measured with perfect accuracy, nor can any mass.

The degree to which measurements may approximate continuity depends, of course, upon the fineness with which the units of measurement can be subdivided. The distinction between continuous and discrete variables is useful in practice only to the extent that truly continuous variables, like length, are measured with sufficient fineness to give to the observed values some semblance of continuity. If organisms vary between 1 and 3 inches in length, a measurement to the nearest inch cannot be regarded as a value of a continuous variate, for it would provide only three distinguishable classes.

The Meaning of Numbers in Zoology

When it is said that a bird lays clutches of 4 eggs each, or that its eggs are 4 centimeters long, the number 4 is being used in two quite different and not interchangeable ways. In the first instance the number 4 is a count of discrete objects; it means that there were 4 such objects, neither more nor less. It is exactly accurate. Saying that an object is 4 cm. long, however, is only an inaccurate representation of the object's true length, which is, of course, impossible to measure exactly. It is necessary, then, that some convention be established regarding the range of values of a continuous variate which is implied in a given measurement. The convention which is universally accepted, although often misused or misunderstood, is:

1. The observed value is the *midpoint* of the implied range of this measurement of the variate.
2. The range is equal in length to the *smallest unit specified in the measurement*.

Thus an observed measurement of 4 cm. means that the true value of the variate lies between the limits 3.5000 . . . and 4.4999 . . . . Notice that the upper limit of the implied range is *not* 4.5000 . . . , for if this were the case, a true value of 4.5000 . . . could be signified by 4 or 5, the implied ranges of both of these numbers having the point 4.5000 in common. The correct definition of the range avoids such ambiguity.

In the same way, a measurement of 4.0 cm. implies a range of true values between 3.95000 . . . and 4.04999 . . . . The numbers 4 and 4.0 are not equivalent in meaning, the addition of a zero in the first decimal place indicating a refinement of accuracy. Example 1 shows a series of progressively more accurate measurements of a true value assumed to be 2.3074999 . . . , together with the implied range of each and the length of that range.

In dealing with discrete variates, the problem of implied range does not exist. To say that there are 4 eggs in a clutch means precisely that. To be consistent in the treatment of measurements of continuous variates, it would be necessary to write 4.000 . . . eggs, indicating that this number is accurate to an infinite number of decimal places. Such a usage would be cumbersome and is never observed. Nevertheless, the expression "4 cm." with its implied range of 3.5000... – 4.4999... and the expression "4 eggs" which implies 4.000... must always be distinguished in practice.

EXAMPLE 1. Increasingly accurate measurements of a true value of 2.3074999... with their implied ranges.

MEASUREMENT	IMPLIED RANGES	LENGTH OF RANGE
2	1.5000... – 2.4999...	1
2.3	2.25000... – 2.34999...	.1
2.31	2.305000... – 2.314999...	.01
2.307	2.3065000... – 2.3074999...	.001
2.3075	2.30745000... – 2.30754999...	.0001

Significant Figures

In the record of any numerical value, significant figures are, strictly speaking, those digits that are accurate—that is, the last digit is correct within a half step and thus implies a range within which the exact value lies. In a broader sense, one digit beyond these may be considered significant if it is nearer to the exact value than would be the limit of the range implied by the preceding digit.

As we have shown, the number actually recorded as the value of a continuous variate is only a symbol implying a range in which the actual value is known to lie. If, for example, a variate obtained by measurement or calculation is known to be within the range 1.5000... – 2.4999..., it is wrong to record the measurement as 2.0, for that symbol implies a range 1.95000... – 2.04999.... The information at hand does not show the true value to lie within this much shorter range (although, in fact, it may). The problem of significant figures, then, is one of assuring that the recorded symbol implies the range actually established from the available information. Thus, in the preceding example, the correct symbol is 2 and not 2.0. The 2 is a significant figure, but the .0 is not.

The first consideration in the quantitative treatment of variables is that of the number of significant figures used in recording observations.

Accuracy of measurement depends on the nature of the material, the equipment used in measuring, and personal factors such as bias and consistency (reliability), which are discussed in a following section. The degree of accuracy obtained under given circumstances can be determined experimentally. For example, one of the present authors made a measurement of tooth length, as is routine in vertebrate paleontology, using a low-power binocular microscope and a caliper calibrated to .1 mm. He repeated the operation on five consecutive days. The results were as follows:

  13.0 mm. 13.3 13.2 13.1 13.0 12.9

Expressed in integral millimeters, these measurements are all 13, while in tenths of a millimeter they range from 12.9 to 13.3, averaging 13.1. From the distribution of these measurements and other criteria extraneous here, it was certain that the exact value was somewhere in the range of 13.0–13.2. All the measurements are thus accurate to two figures (13), for that implied range 12.5000 ... – 13.4999 ... certainly includes the true value. They are not accurate to three figures (one decimal place), for no single one of these more refined figures certainly includes the true value, and two of them (12.9, 13.3) certainly do not. This is nevertheless a case in which records to three figures, one inaccurate, are preferable to the accurate two-figure measurements. All the three-place figures, even the most divergent, are closer to the exact value than are the limits 12.5000 ...– 13.4999 ... implied by the two-place figure 13. As a general criterion, inaccurate figures are useful if their range of error is less than the implied range of the accurate figures available.

The smallest of six measurements made in this experiment is certainly 98 per cent or more of the exact value and the largest 102 per cent or less. It is thus certain that any one measurement was within 2 per cent of the real value of the dimension measured. Supposing, as other experiments show to be highly probable, that this represents the degree of accuracy generally obtainable with such equipment and with little or no personal bias, it is possible to work out a schedule such as the following:

  Between .2 and 2, use two decimal places (.20–1.99)
  Between 2 and 20, use one decimal place (2.0–19.9)
  Between 20 and 200, use units (20–199)
  Etc.

Another expression of the same rule is: under the given or similar conditions of material and technique, record three digits if the first is 1, and otherwise record only two. In practice this means that a record of a tooth as being 15 mm. in length is, for practical purposes, absolutely accurate, and a record of 15.8 is a better approximation for most purposes although not absolutely accurate, but a record as 15.82 is in no respect better than 15.8. If the measurements are large, it is advisable to change the unit so that no number larger than 199 need be used. Thus, under these conditions, 390 mm. should be recorded as 39 cm., for the figure 390 implies a range of 389.500 ... – 390.4999 ..., whereas the range really intended is 385–395 mm. expressed by 39 cm.—i.e. 38.5000 ... – 39.4999 ... cm.

Such a rule, naturally, is valid only for the given conditions, but there is no difficulty in applying similar methods to any sort of measurement. If the degree of accuracy obtained proves to be insufficient for the purposes in mind, a refinement of technique and increase of accuracy are usually possible. Considerable inaccuracy is inseparable from the nature of some material, and in such cases refinement of technique is useless and only problems soluble by the relatively inaccurate data can be usefully attacked. In most paleontological work the degree of accuracy shown by the preceding example is quite adequate for the purposes involved, and in many cases a markedly higher degree of accuracy is impossible. In some other fields such measurements would be grossly inadequate, and accurate four- or five-digit measurements may be both possible and desirable.

In the absence of any other criterion, it is proper to record as many digits as are accurate or are found to be useful approximations by tests like that just described. When refinement can be increased indefinitely by changes in technique, there nevertheless comes a point beyond which it is useless to go, and for the determination of this point, statistical methods provide the best criterion. If a series of specimens is to be measured, the most useful rule is to measure the largest and smallest specimens and then to adopt as a minimum unit of measurement one that is contained at least 16 and up to 24 times in the range. If an adequate series is not available, a much rougher but still useful rule applicable to most linear dimensions is simply to record three digits. (If fewer than 16 steps are used, the approximation of the measured values to a continuous series becomes poor, and the methods which will be developed in later sections for the treatment of continuous variates are inapplicable.)

In practice this means that measurements of a variate ranging, say, from 10 to 12 mm. should be taken to .1 mm. This would give 20 steps within the range, which sufficiently meets the first rule. If the range were 75–95 mm., no decimal places need be recorded, for there are 20 integral steps in the range. The first example conforms also to the second rule. The second does not, but the rougher rule would result only in making measurements somewhat more refined than necessary. Except in a few special cases, it is useless to exceed greatly the requirements of either rule and unnecessary work can thus be avoided. For instance, with a range 10–12 mm., measurements to .01 mm.—giving 200 steps within the range—even though entirely accurate, would generally serve no useful purpose; and the refinements of technique and added labor involved in making such minute measurements would simply be wasted.

In the great majority of cases, these rules ensure data that will provide a maximum of useful information, enough for efficient statistical or any other usual zoological purpose. It is not true, however, that the rules *must* be met in order to provide useful data. When measurements of the optimum refinement are not practicable, the substandard data developed may still be highly useful and no less accurate. They are merely less efficient.

The significance of figures resulting from calculation is equally important.

8 QUANTITATIVE ZOOLOGY

Neither simple nor obvious, this is a subject which requires further consideration. The number of significant figures resulting from an arithmetical operation on observed numbers can be determined by performing the operation on the implied range of numbers. For example, the sum of 2 (a continuous number) and 2 (another continous number) has no strictly significant place. The implied range of 2 is from 1.5000 ... – 2.4999 .... Now adding 1.5000 ... and 1.5000 ... yields 3.0000 as the lower limit of the implied range of the sum, while in a like manner 2.4999 ... added to 2.4999 ... gives 4.999 ... as the upper limit of the sum. Thus the sum of 2 and 2 has an implied range of between 3.000 ... and 4.999 .... To say that the sum is 4 is to imply that the range is 3.5000 ... – 4.4999 ..., which is too small. A generalization of this result is that the sum of continuous numbers will have one less strictly significant place than the numbers themselves. In the unfortunate case where the numbers have only one strictly significant figure, there are no significant figures in the sum, although the result is considered broadly significant. A similar procedure may be followed with other operations with the following results.

1. In any operation involving *only* discrete numbers, all the resulting figures are significant.
2. The sum of a discrete number and a continuous number has as many strictly significant figures as does the continuous number.
3. The sum or difference of continuous numbers has one less strictly significant figure than there are in the number with the fewest significant figures.
4. The product of a continuous and a discrete number has one less strictly significant figure than does the continuous number.
5. The division of a continuous by a discrete number yields a quotient with as many strictly significant figures as the continuous number.
6. The product or quotient of two continuous numbers has one less strictly significant figure than the number with the fewer significant figures.
7. The square root of a continuous number has the same number of significant figures as does the number.
8. All the above operations yield one more *broadly* significant figure than *strictly* significant figures.

The reader may easily verify these rules by applying the method outlined of operating on the extremes of the ranges, writing down the number which correctly symbolizes the resulting range, and then comparing it with the result obtained by operating on the original numbers themselves.

The rules discussed above leave the impression that calculations made from observations usually have fewer significant figures than do the observations themselves. As a matter of fact, the reverse may be true if the calculated figure is derived from many observations. The true values of some of the observations will lie at the lower end of their implied ranges while others will lie at the upper end. The net result of adding, let us say, 100 observations will be that these variations will cancel each other out, the resulting sum being even closer to the true value than the original numbers themselves. This has been experimentally verified. A conservative general rule which may be used is that a calculation involving the summation of numerous implied ranges will have as many significant figures as the observations themselves. A more rigorous rule can be made based upon probability considerations, but the one given above is sufficiently refined for any practical purpose.

A corollary to the discussion of significant figures is that it is a waste of time to make some measurements more accurately than others in the same series, since for any calculation made from these observations it is the one with the least accuracy which governs the number of significant figures in the result.

Rounding Figures

When a number is recorded or calculated with more figures than are significant, it is necessary to reduce the number of figures. This cannot be done simply by dropping the nonsignificant places since by so doing a bias would be introduced. Thus the number 2.376 cannot be rounded to 2.37 since it is obviously much closer to 2.38. The usual rule is to round the number "down" if the first nonsignificant figure is less than 5 and to round the number "up" if the first nonsignificant place is greater than 5. In this way 2.34 is rounded to 2.3, but 2.36 is rounded to 2.4. The problem then arises as to the disposition of a number whose first nonsignificant figure is 5. The only completely accurate method of rounding such numbers is to make the measurement again (if it is a measurement which is being rounded) with greater accuracy. For example, an observation recorded as 2.35 might prove to be 2.347, in which case it will be rounded to 2.3; or it might prove to be 2.353, which will be rounded to 2.4. Because it is often impractical or impossible to take a measurement again with greater refinement, some convention must be adopted. One common convention which gives satisfactory results where a fairly large number of observations is involved is to round "up" when the figure before the 5 is odd and "down" when it is even. In the long run, about an equal number of such observations will be rounded up as are rounded down, so that no bias will be introduced.

Data from Direct Observation

The raw data for the numerical analysis and synthesis of zoological materials must be derived from direct observation. In starting work, for instance, on an unstudied group of specimens, observations are in most cases lists of the specimens with simple measurements suspected of being significant and verbal notes of qualitative differences. As study progresses, some of these first observations will, in all probability, prove to be unimportant for the object in view and will therefore be discarded, while new observations of the same sort but of different variates or attributes may prove to be desirable. When the work has progressed to the point of recognizing particular groupings, whether qualitative or quantitative, it becomes possible to compile numerical values of a different category—frequencies— that is, counts of the numbers of observations belonging, in a given respect, to one of the categories recognized. This operation often derives numerical data from observations that are not numerical in character. Thus the presence or absence of a keel on a given tooth cusp would not be expressed primarily by a number, but if it appears that the presence or absence of a keel has some significance for the work being done, it becomes subject to numerical analysis and statistical study when the number of specimens with the keel and the number without it are counted. Or, in taxonomic work, after all the specimens have been identified, the number of individuals in the collection belonging to each species gives numerical data involving biological conclusions not themselves of a numerical character. From the point of view of basing inferences of a higher order on the data and particularly of using statistics as a basis for such inferences, all of these direct numerical observations are primary observations, or raw data, even though, as in the last example given, they can be made only after many secondary observations (necessary for recognizing the groups involved) are made.

There is an almost unlimited variety of types of primary numerical data possible under the broad categories of continuous variates and frequencies—about as many sorts as there are different zoological problems to be solved. In the field of animal morphology and taxonomy, the greater number of useful continuous variates are linear dimensions. Areas have some significance—for instance, the area of grinding teeth in mammals is important in considering food habits, the area of the caudal fin in fishes is essential in studying their locomotion. Areas have, however, a serious disadvantage: they cannot be directly measured but must be calculated from linear dimensions or indirectly measured from drawings, projections, or photographs. This calculation, often difficult, may introduce errors or inaccuracies and involves certain obscure peculiarities analogous to those of ratios (discussed on a later page). For these reasons, it is usually preferable wherever possible to avoid using area and to use instead the more directly measurable dimensions from which area would have been calculated, i.e. its linear components. Volume, to even greater degree, is open to objection on the same grounds, and if it must be calculated from linear dimensions, it should generally be used only if the problem cannot be attacked efficiently in any other way. Volume can, however, also be measured directly, as by displacement of liquids or by filling cavities with a measured volume of fine shot or similar substances. Such measurements may be both reliable and useful. In mammals, cranial capacity is an important character properly recorded in this way.

Angles measure an important category of animal characters not measurable in any other way. The numerical results are continuous variates subject to much the same sort of comparison and analysis as are linear dimensions. Angles are usually measured in degrees, minutes, and seconds, which though not decimal units may be converted to decimal fractions by the use of *radians*. The radian equivalent of an angle is the length of the arc cut off by that angle in a circle of unit radius. There are thus 2π radians in 360 degrees, π in 180 degrees, $\pi/2$ in 90 degrees, and so on. Adequate tables of this conversion are readily available in the *C.R.C. Standard Mathematical Tables* (1957). Even without the tables, an angle is easily converted to its equivalent radian measure by the following relations:

1 radian = 57.2958 degrees
1 degree =   .0174533 radians

Angles record biologically and taxonomically important characters such as cranio-facial flexion, limb angulation, or axial rotation of skeletal processes. The exact measurement of angles in zoological material is difficult but can be adequately achieved by methods of graphic projection.

Temperature typifies a class of physiological characters—to which basal metabolism, blood pressure, pulse rate, and numerous others also belong—that are essentially continuous variates and may be treated as such mathematically. Although it is generally impractical to use them in that way, they clearly can be related to taxonomy. Principally, they are involved in physiological problems in which they are of the greatest importance.

The measurement of periods of time delimited by some animal activity is also important in physiological and ecological studies, and these are also continuous variates. Among the many essential time measurements involved in zoological research are pulse and respiration rates, which may be expressed as the periods between pulsations and respirations, periods of incubation and gestation, length of life, time of hibernation, and length of oestrous cycle. All of these are time-period variates.

Discrete variates, although not always recognized as such, are almost as abundant as continuous ones in zoological data. They are of major importance in taxonomy because they often have more limited individual and specific range and variability than do continuous variates and hence may characterize genera or higher groups. Their character and significance are more often obvious on inspection and without analysis, although this is not always true. Dental, vertebral, and phalangeal formulae often characterize superspecific categories and usually are of obvious significance. Cuspule or striation counts on mammal teeth, fin-ray counts on fishes, feather or egg counts of birds, blood cell counts for any vertebrate, and many others are discontinuous variates, commonly highly variable and demanding some formal analysis for their successful interpretation. Any serial or repetitive structures are discontinuous variates whatever the scope of the taxonomic or other category within which they vary, and all may, if desirable, be treated as such statistically by methods discussed on later pages.

Frequencies are simply counts of individuals belonging to any selected category. The categories may be based on any measurements or counts of variates, either as observed or as gathered secondarily into groups. The categories may, furthermore, be based on any logical consideration, even one wholly nonnumerical or fundamentally subjective. Thus frequencies may be based on simple attributes such as the presence or absence of a vestigial tooth or differences in geological or geographical origin. They may be counts of the individuals of each species in a certain collection, counts of the number of known species in each of several genera, counts of the species of a given fauna grouped by their probable habits of life, and so forth, each of these and the innumerable other possibilities having a definite bearing on some type of zoological research. All observations involve frequencies, even if the frequency be 1 (or 0, the characteristic sought not being found in any case), and in many cases these frequencies are at least as essential to consideration of the problem in hand as are other types of data.

Since a continuous variate may theoretically take any of an infinite series of values, it follows that absolutely accurate measurements of any two values of such a variate would never be the same and consequently that the frequency of any one value would always be 1 and the concept of frequency useless. In fact, it has been pointed out that such absolutely accurate measurements are not possible (or desirable) and that the measured and recorded value of the continuous variate is in practice only a conventional means of defining a greater or smaller span on the continuous scale within which the real or absolute value is known to lie. The record 3.1 mm., for instance, can include various different exact values of a continuous variate between 3.05000 ... and 3.1499 ... mm., and recorded values of continuous variates can and do have frequencies greater than 1 in practice. The groups of values thus brought together can be made larger or smaller at will, and a similar sort of grouping may be applied to discontinuous variates, so that the frequencies can be manipulated into the form most advantageous for the problem in hand, a subject discussed in detail in Chapter 3.

Ratios and Indices

Ratios, products, indices, and other numbers obtained by the combination in various ways of two or more numbers are themselves raw numerical data from a statistical point of view, but they are secondary, not derived from direct observation, and they have properties unlike those of numbers obtained by direct observation.

From the standpoint of any particular problem, the purpose is to find a figure, all the elements of which are related to the problem, which has some property not possessed by its primary elements, such as being more constant, or varying in some definite and ascertainable way with respect to a different variate, to function, and so forth. General possibilities to bear in mind aside from ratios (quotients) are modules (arithmetic means), areas (and other products) and many secondary or tertiary figures such as powers of ratios or of deviations and quotients of modules. Such figures should appear in the final work, however, only if they really prove to express characters or have useful properties other than those of the original measurements.

Among secondary numbers the most important are ratios, which express in a single number the relative sizes of two other numbers. The most widely used ratios are the quotients of two numbers that express observations of the same sort, (e.g., linear dimensions), and that are in the same unit (e.g., millimeters). The resulting ratio is independent of the absolute size of the original figures as well as of the orginal unit of measurement. Thus, 5:10 is the same ratio as 500:1000 and 5 mm.:10 mm. is the same ratio as 5 years:10 years. The result, ordinarily expressed as .5 for all these examples, is a pure number divorced from any particular system of mensuration. It should be pointed out, however, that some commonly used ratios are not dimensionless. One of the most usual is the "surface-volume ratio," an important characteristic of animals in relation to their heat and water balance. The ratio of surface to volume is a ratio of square inches to cubic inches or square centimeters to cubic centimeters. Such a ratio will have the dimension of 1/inches or 1/centimeters; that dimension measured in inches will be 2.54 times as large as that measured in centimeters. Because of this, such ratios must always be accompanied by the units in which they have been measured. Otherwise comparison between ratios becomes impossible.

The word "index" is used in a variety of ways which are not consistent with each other. In the most general sense, an index is any arithmetical combination of different measurements, more often than not of nonhomologous dimensions. Despite the difficulty of giving it a direct biological meaning, such an index may be quite useful in comparing populations of animals. An example is "Reed's wing index," used for distinguishing sibling species of *Drosophila*. This index is obtained by multiplying the wing area (in square millimeters) by the cubed wing length (in cubic millimeters). The "discriminant function" of R. A. Fisher, in which measurements of different parts of an organism are multiplied by certain calculated constants and then added together, is an example of an index which is widely used by physical anthropologists.

A third example is the "hybrid index," which up to now has been little employed in zoology. An excellent illustration of its use is the work of Sibley (1954) on hybridization in red-eyed towhees. *Pipilo erythrophthalmus* differs from *P. ocai* in the following six color characters: pileum color, presence or absence of wing and back spots, back color, throat color, flank color, presence of tail spots. Sibley noted that for each character there were five distinguishable grades in the hybrids, which he numbered 0, 1, 2, 3, and 4. A score of 0 on a given character indicates an expression of pure *P. ocai*, while a score of 4 indicates an expression of pure *P. erythrophthalmus*. With six characters scored in this way, *P. ocai* has an index of 0 (6 $\times$ 0), *P. erythrophthalmus* has an index of 24 (6 $\times$ 4), and hybrids fall somewhere between these two values. This "hybrid index" cannot be given a direct biological interpretation. It does not, for example, give the exact degree of genetic relationship. Nevertheless, it does characterize by a single number the degree of resemblance between the hybrids and the parental species, and this resemblance could be equated to relationship if enough were known about the genetic determination of the characters.

In a stricter sense the term "index" is used for a figure obtained by dividing a given dimension by some larger dimension of the same anatomical element and then multiplying it by 100 (or expressing it as a percentage). Unless the dimensions are otherwise specified, it is generally understood that they are minimum and maximum dimensions of the anatomical unit. In contradistinction the term "ratio" usually (although not always) refers to proportions between dimensions of different anatomical elements.

Ratios of two continuous variates are in proper and widespread use in zoology, and they express characters that are of fundamental importance. They have, however, certain peculiar and generally ignored properties that must be kept in mind and may in some cases make conclusions based on them inaccurate or even invalid. Ratios are themselves continuous variates, and the numbers in which they are written are of the indefinite kind that express approximate position in a continuous series; but the accuracy and limits implied are not the same as for the direct measurements on which the ratios are based.

Ratios frequently vary more than do the dimensions on which they are based. Thus if the lengths of a given sample of homologous teeth vary from, say, 0.9 to 1.1 mm. and the widths also vary from 0.9 to 1.1 mm., the possible length–width ratios vary from 0.8 to 1.2, a markedly greater range. The relative variabilities of ratios and of their constituent dimensions are tied up in an intricate way with the correlation between the latter (see Chapters 9 and 11).

The most confusing characteristic of ratios is that they are grouped in a peculiar way not determinable by simple inspection of the figures and that this may be a source of error in basing deductions on them. A length recorded as 1.0 mm. is known to be somewhere between .95000 . . . and 1.04999 . . . on the continuous scale, a simple and obvious relationship, but this is not true of a ratio recorded as 1.0. For instance, a length–width ratio of 1.1:1.1 mm. would be recorded as 1.0, but its real value may be anywhere between .92 and 1.09, or, in round figures, from .9 to 1.1. Furthermore, this peculiarity may result in writing two really different ratios as the same or two really identical ratios as different. It has been shown that the ratio 1.1:1.1 may really be anywhere from approximately .9 to 1.1; the ratio 1.0:1.1 may really lie anywhere from .8 to 1.0, a range widely overlapping that of the other and apparently different ratio. Again, the real value of the ratio .9:.9 is somewhere in the range .90–1.11 and that of 1.9:1.9 somewhere in the range .95–1.05—a considerable difference in accuracy; but written as a single figure (i.e., 1.0), according to usual practice, these ratios are given as identical.

These difficulties are far outweighed by the usefulness of ratios, but they must be understood, and it should not be supposed that a figure representing a ratio is necessarily as accurate as those on which it was based. If minute differences are important and the status of ratios is doubtful, it may occasionally be advisable to abandon the ratios and deal with the problem directly from the original measurements.

Ratios may also be usefully based on discontinuous variates and on frequencies. The ratio of dorsal to lumbar vertebral counts, for instance, may express an important character in the clearest way, or, as another example, the ratio of number of individuals (frequency) with skulls longer than a selected standard to the number of those with skulls shorter than the standard may be a valuable means of characterizing the group as a whole. Ratios based on such data are themselves discontinuous variates. They do not have the disadvantages of ratios that are based on continuous variates, but they have an extraordinary peculiarity of their own: although discontinuous, they are usually fractional and sometimes indeterminate.

16 QUANTITATIVE ZOOLOGY

An example will make this clear. Suppose that each of two discontinuous variates can take the value 1, 2, 3, or 4. Ratios between these two can take the values shown in Example 2. This series of eleven possible values is irregular and follows no obvious system; seven of the values are fractional and three are infinite repeating decimals. Nevertheless, they are the possible values of a discontinuous variate. Each value is definite and exact, not an approximation or group symbol as we would have for a continuous variate. Under the postulated conditions these are the only values that the variate can take, intermediates between them being impossible. It is also noteworthy that more combinations of the original dimensions result in a ratio of 1 than any other figure, a peculiarity that also may strongly affect conclusions based on such ratios.

EXAMPLE 2. Ratios between two discontinuous variates, each with values of 1 to 4. (Hypothetical data)

1:4	= 0.25
1:3	= 0.333 ...
1:2, 2:4	= 0.5
2:3	= 0.666 ...
3:4	= 0.75
1:1, 2:2, 3:3, 4:4	= 1
4:3	= 1.333 ...
3:2	= 1.5
2:1, 4:2	= 2
3:1	= 3
4:1	= 4

The useful broad categories of ratios based on dimensions appear to be as follows:
1. The ratio (or index) of two different dimensions of one anatomical element of one individual.
2. The ratio of two analogous dimensions, one for each of two anatomical elements of one individual.
3. The ratio of homologous dimensions of homologous anatomical elements of two individuals.

There are, of course, many other comparisons possible, but in the main the parts compared are not related to each other in some simple, orderly, and significant way; or, differently expressed, they obscure significant dependence by introducing variables too numerous or essentially independent. A ratio is useful only insofar as it makes apparent some biological characteristic not apparent in the original measurement. No rules can be formulated for the appropriate use of ratios. As in all mathematical procedures relating to zoology, there are no substitutes for zoological understanding and intelligence.

The three useful categories of ratios of dimensions express different sorts of characters or concepts, and the inferences based on them are of different kinds. Indices or ratios in (1) above are essentially unit characters not markedly unlike linear dimensions in the concept involved. The index (breadth × 100)/length of a given tooth is a simple character for that tooth, as are its breadth and its length taken individually. Such indices are sometimes designated by their supposed or actual correlation with some other function or character; e.g., the index (length × 100)/breadth of a limb bone has been called the "speed index" because it is advanced as a hypothesis or supported as a theory that the larger the value of this index the more rapid, in general, the locomotion of the animal. Even aside from the fact that this is not a constant relationship (and even that the exact opposite can be demonstrated to be true in some cases), this naming of a ratio by the inference that is expected to be drawn from it is unsound. The conclusions that may be drawn from numerical data should not be confused with the data themselves.

Ratios in (2) above express a different sort of character, for they are descriptive of a larger anatomical unit than that measured by either of the primary figures from which the ratio is derived. Thus if teeth are used as examples again, the ratio length of trigonid/length of talonid belongs to this category, and it expresses numerically a character of the tooth as a whole, whereas neither of the direct measurements applies to the whole tooth. Similarly, length of humerus/length of radius is a character of the forelimb and length of humerus/length of femur a character of the locomotive apparatus as a whole. By "analogous dimensions" we mean length against length, and so forth. There may well be some relationship between the length of one element and, say, the width of another, but this is a somewhat confusing concept and one of little practical use.

There may be some confusion as to the meaning of "analogous" when dimensions are thought of in the sense of common usage. The "length" of a structure as opposed to its "width" is, in common parlance, the largest linear dimension. Curiously enough, this is not always the case in zoology, where a structure may in fact be "wider than it is long." To define "analogous dimensions" in zoology, the points of reference must be the various anatomical axes of the organism. "Dorsal" and "ventral," "anterior" and "posterior," "abaxial" and "adaxial" are unambiguously defined for any group of animals and thus provide suitable reference points for the determination of analogous measurements.

Ratios listed in (3) are by far the most common in zoological work and in some form or other are almost universally employed. The statement that one species is larger than another is merely a crude expression of a ratio of this sort. On the other hand, a statement that one species is, for example, 20 per cent larger than another is a gross misstatement of the actual facts. What this usually means is that some given dimension of one specimen of one species is 20 per cent larger than the same dimension of one specimen of another species. That all the dimensions of all the individuals of one species should be 20 per cent larger than the corresponding dimensions of all the individuals of another species is impossible. It is preferable to say what is really meant. This example illustrates the usefulness of defining species, whenever possible, by the statistical constants of their several variates, rather than by individual values of these variates, and of always specifying the particular variate involved.

There is a higher, derived category of ratios which are not usually formally recognized but are often implied and may be useful, i.e., the ratio of two ratios. Thus the ratio of the cephalic index of one specimen to that of another is a ratio of two ratios which can be written in this way:

$$\frac{\text{Breadth of skull A} \times 100}{\text{Length of skull A}} : \frac{\text{breadth of skull B} \times 100}{\text{length of skull B}}$$

This is a means of comparison as logical as the ratio of linear dimensions, for instance:

$$\frac{\text{Breadth of skull A}}{\text{Breadth of skull B}}$$

However, the ratio of two ratios suffers in an exaggerated degree from the peculiar and disadvantageous properties of ratios in general and should be used with the greatest caution.

Ratios and indices may be expressed numerically in several different ways:

1. as the unreduced ratio of the actual measurements, e.g., 5:10 mm. or 5:10
2. as a fraction, e.g., 5/10 or 1/2
3. as a quotient, e.g., 0.5
4. as a percentage, e.g., 50 per cent
5. as a quotient multiplied by a constant, e.g., (using 100 as the constant—the usual form), 50

For the purposes of inference or of analysis, ratios are still raw data. Their only essential difference from the numbers on which they are based is that they express a different sort of character. In general, the further study of ratios follows the same lines as for any other raw numerical data. As with any other data, however, their morphological meaning and arithmetical derivation must be kept clearly in mind. For instance, in considering variation and variability, the fact that linear dimensions may or do vary within related groups relative to the mean of each group does not warrant the assumption, a priori, that ratios based on these dimensions will vary in the same or a similar way, since the ratios are not dimensions but are usually pure numbers derived from, but independent of, the mean dimensions (see also coefficient of variation, Chapter 6).

There are several other types of calculated but essentially raw data that are analogous to ratios in expressing by one number some relationship between two measurements but involving distinct concepts and operations. Few of these are in use in zoology; we know of none likely to be of really general value, though some may be useful in certain special problems. Such a figure is, for instance, (length + width)/2, sometimes called a "module." This may be a useful concept where there is no marked functional difference between the two dimensions and they tend to vary about the same or approximate means. In cases in which one dimension tends to increase as the other decreases, this module will generally vary less than does either dimension (whereas the ratio will vary more than either) and may be useful on this account. If the dimensions vary together so that an increase in one is accompanied by an increase in the other, then, of course, the ratio is less variable than either module or original dimension. Measures of area (length × width) are in a sense analogous and have the same property of tending to be less variable than either length or width if these have an inverse relation to each other. There are numerous cases in nature (e.g., the surfaces of grinding teeth) in which the functional character is the module, or area, rather than the linear dimensions. In such cases the character is better described in terms of a module or similar figure than in terms of the original linear dimensions.

Various limb modules, such as

$$\frac{\text{Length of humerus} + \text{length of radius}}{2}$$

or

$$\frac{\text{Length of tarsus} + \text{length of metatarsus}}{2}$$

are also logical concepts that may serve to bring out relationships not immediately visible from the original measurements, and many similar formulae will suggest themselves in the course of special investigations.

CHAPTER TWO

Mensuration

Requirements of Good Measurement

An infinite number of numerical observations may be made on any one zoological specimen, and each observation may be made in many different ways. The first step is to decide what is to be measured and how. The most important criteria of good numerical observations are that they should be logical, related to a definite problem, adequate, well delimited, and comparable and standardized.

a. Measurements should be logical. Paleontologists seem to use illogical and nonunit measurements more often than do neozoologists. They may, for instance, measure the length from the second premolar through the first molar in a mammal. This measurement has no natural unity, measures no biologically important single character, and is poor for comparisons (the only purpose for taking it) because such a measurement is not likely to be available in the literature on other specimens, and on many specimens otherwise comparable it may be impossible to make. Measurements of each single tooth should be given, and measurements of groups of teeth should be of natural groups—of the whole cheek series, of all the premolars, or of all the molars.

A general principle of measurement, involved in several of the criteria listed below and violated in the example of illogical measurement just given, is that those measurements are usually best that permit the greatest number of valid comparisons. In paleontology violations of this principle such as that in the example cited are generally caused by incompleteness of the material. It is possible, of course, to measure only what is preserved, but this is hardly worthwhile at all unless natural units can be measured. Instead of measurements of individual teeth, such an odd and relatively useless dimension as length P^2—M^1 is probably given on the premise that the percentage of error will be less for a large measurement than for a small one. This argument, however, merely indicates that the technique used should produce accuracy at the desirable degree of refinement, whatever the size of the measurement. In fact, in paleontology this premise is often fallacious, for a longer measurement is more likely to be affected by distortion than a shorter one. Its accuracy, therefore, as an estimate of what the dimension was in the living animal may be as low as for a smaller dimension, or even considerably lower. This is particularly true in dealing with teeth or similar series in which the individual elements are usually little distorted but the series as a whole is frequently seriously distorted.

b. Measurements should be related to a definite problem. This requirement is so obvious and so rarely transgressed that it is necessary only to point out that the relationship should be as direct and as simple as possible, and that the problems of other workers should be kept in mind to some extent. Brain growth, for instance, can be studied from the skull dimensions or endocranial capacity, and in some cases must be so studied because other data are unobtainable; but neither factor is related directly and simply to the question of brain growth, for which the best measurement is naturally that of the weight or volume of the brain itself. It is, however, pertinent to give measurements that will be useful to others working on related problems, even though they may not be necessary for the purpose of the immediate enquiry. In taxonomy many standardized dimensions may be quite unnecessary to define a species or subspecies and yet should be included as a regular practice to facilitate future work.

It is always better in assembling raw data to take too many measurements than too few. At this stage in research, it is commonly inadvisable to adhere too rigidly to a criterion of direct relationship and preferable to measure any variates that have a conceivable bearing on a problem, for in this way important and unsuspected relationships are often discovered. Such data require, in any case, careful analysis. Certain of them will probably turn out to be unnecessary to demonstrate the point at issue. In that case (except for standard dimensions that will surely be useful to others immediately or in the future), they should be discarded, no matter how much work has been involved in obtaining and evaluating them. Zoological literature is replete with long tables of measurements that prove nothing and the publication of which is unnecessary, expensive, and really a discourtesy to other students. In this respect the methods, largely statistical, which are discussed in succeeding chapters are invaluable. They provide definite tests as to whether measurements really are germane, thus facilitating the selection of essential data and the rejection of nonessential data. They also assist in reducing raw data to the most compact and useful form.

c. Measurements should be adequate. Equally common and perhaps still more open to criticism is the gathering and publication of inadequate numerical data. In discussing a species from a taxonomic point of view, it is usually unnecessary to give all the pertinent dimensions for each of a large series of specimens; but at least this practice does make the data available and is preferable to the practice of using only the dimensions of the type or of giving only the mean dimensions of the whole series. Many studies that purport to deal with variation give only the maximum and minimum dimensions observed, or sometimes these plus the mean. Occasionally the number of specimens involved is also given, but the frequency of omission of this absolutely essential datum is remarkable. For a real study of variation and indeed for most purposes of valid comparison, such data, even if the means and the number of specimens are recorded, are a little better than nothing—but not much better. Far more important are data on the way in which the observations were distributed about the mean, on the probable relationship of the observed extremes and the mean to those of the whole population, on differences in ranges and means, and on similar questions. Measurements and other observations are inadequate if they do not permit the calculation of such data, and the publication of results is inadequate if such data are not obtained and recorded.

d. Measurements should be well delimited. Some measurements are useless or nearly so because they have no well-defined limits and hence cannot approach an adequate standard of accuracy and refinement. For instance, an attempt has been made to use the distance of the narrowest point on a slender limb bone from the proximal end of that bone as a numerical character of animals. The sides of a limb bone are nearly parallel, and hence its narrowest point is so vaguely defined that any reasonable degree of accuracy is impossible and the character, although real, is generally useless because it is not well delimited.

e. Measurements should be comparable and standardized. To be comparable, all measurements must be taken in the same way. This requirement is largely mechanical and depends on adequacy of equipment, practice, and experimentation to produce sufficiently consistent results. Absolute consistency is impossible, but assurance is necessary that it is approached closely enough not to distort the results derived from data. Specification demands mention not only of exactly what measurement was taken but also of exactly how it was taken, unless both are obvious or understood by the reader addressed. In taxonomic work on fishes, which is largely based on proportions (that is, on ratios), proportions are often obtained by using dividers which are set at the smaller dimension; the integers of the proportional value of the larger dimension are then stepped out with the dividers and the fractional excess is estimated by eye. Such grossly inaccurate methods are not to be condemned on that score alone if the low accuracy is really adequate for the purposes intended; but clearly they are not comparable with more refined methods and their use should be specified. Similarly, mammalogists usually measure the longer dimensions with a ruler and the shorter dimensions with calipers; but some use ruler and some calipers for both, some use simple dividers read against a ruler, and some use other methods, such as measuring with the short end of proportional dividers and reading the long end against a ruler. The refinement of each method is different and may need specification, even though in this case all the methods mentioned may be sufficiently refined for the usual purposes.

The condition of the material and the way in which it is held for measurement also affect accuracy and comparability and may require specification. Living animals, dead unprepared animals, animals in preservatives, and skins all differ to some extent in dimensions, and the different preservatives and methods of preparation may also have effects so different as to render measurements incomparable. Sumner (1927) has shown, for instance, that the mean total length of 10 mice (*Peromyscus maniculatus gambelii*) was 166.65 mm. at the time of death and 164.10 mm. two hours later. Differences between freshly killed animals and skins as customarily preserved in collections are usually still greater. Measurements of one specimen held free, one lying flat, and one stretched out may differ considerably for some types of material, especially live or freshly killed.

Specification of the thing measured is equally important, and current practices are still more varied and confusing. Checking over some of the literature, the dimension given simply as "length" for mammal teeth was found to have been applied in at least six different ways:

1. Greatest distance between planes tangential to the margin of the crown and at right angles to the longitudinal skull axis.
2. Distance between planes tangential to the crown margin, parallel to each other, and approximately parallel to the anterior and posterior edges of the crown.
3. Greatest horizontal distance along the outer or inner face of a tooth (e.g., along the ectoloph).
4. Distance from anterior to posterior borders along the midline of a tooth.
5. Greatest diameter of the tooth crown (sometimes longitudinal, sometimes transverse, sometimes vertical, and generally somewhat oblique).
6. Distance from tip of crown to tip of root.

Probably other usages are also current. Obviously "length of tooth" is a meaningless designation unless some further specification is made or distinctly understood.

It is, however, most usual for a dimension taken to be the maximum distance between parallel planes tangential to the designated anatomical element. For length, the planes are usually considered to be oriented vertically to the axis of the body through the axial anatomical divisions and their parts—teeth, skull, vertebrae, and so forth—and vertically to the proximodistal axis for nonaxial elements—ribs, limbs, and so forth. Width is the dimension at right angles to the length and most nearly in a horizontal plane, and depth or height is the dimension at right angles to these two and nearly in a vertical plane. These definitions apparently conform to a consensus at present and although not recognized as rules, might well be made so. In some groups, specialists understand other conventional designations without specification, but in general any departure from the general definitions just given should be specified.

Systems of Mensuration

Experience in ascertaining the most useful measurements, the irksomeness of fully specifying a dimension each time it is used, and the need to make the work of different observers as comparable as possible have led to some standard systems of mensuration more or less generally used within the various zoological groups and for various types of zoological problems. There is not and cannot be a single standardized system for zoology in general. Even the vertebrates differ too much in structure, their dimensions differ too much in significance, and the variety of problems that arise is too great for such an end to be practicable or desirable. Systems already in use are so numerous that they cannot be usefully summarized by the few examples briefly mentioned in a general work like this. The student must first become familiar with the systems employed in his own field through its special literature and then adopt them or replace them by one specifically suited to his own problems.

Measurement of linear dimensions of animals is most suited for reduction to a standard system, supplemented in most instances by some counts of discontinuous variates. In most cases the principal purpose of such a system is taxonomic, and it usually concentrates on external characters.

For fishes the standard linear dimensions currently in use by ichthyologists are given by Hubbs and Lagler (1947). This system is not always adhered to, however, so that there is some variation in usage. Wildlife management workers especially do not follow the standard system.

In the case of lizards and snakes few linear dimensions, except total length and tail length, are commonly used in taxonomic work, which is based mainly on discontinuous variates such as tooth counts, scale counts (on rather elaborate systems), and counts of elements in repetitive color patterns. An excellent exemplification of such a system is given by Blanchard (1921). For turtles the simple linear dimensions of carapace, plastron, tail, and so forth, are the usual numerical data. Kälin (1933) has given a complicated system of numerical study of the crocodile skull involving numerous linear dimensions and twelve indices.

The measurement of birds for taxonomic purposes is more nearly standardized than for most lower groups, Ridgway's system being employed Because it is so widely accepted, its standard measurements—all linear dimensions—are listed below as an example of a standardized system (from Ridgway, 1904).

> LENGTH. From tip of bill to tip of tail. (This may differ greatly in recently killed birds and prepared skins and may also be difficult to measure accurately.)
>
> WING. From the anterior side of the carpal bend to the tip of the longest primary (feather).
>
> TAIL. From between the shafts of the middle pair of rectrices at the base, pressed as far forward as possible without splitting the skin, to the extremity of the longest rectrix.
>
> CULMEN. From the tip of the bill to the edge of the feathers on the dorsal side. (This is sometimes called "bill" if it extends to the true base of the bill and "exposed culmen" if the base is partly covered by feathers.)
>
> DEPTH OF BILL AT BASE. From the lower edge of the mandibular rami to the highest portion of the culmen.
>
> WIDTH OF BILL AT BASE. Across the chin between the outside of the gnathidea at their base.
>
> TARSUS. From the tibiotarsal joint on the outer side to distal end of the tarsus.
>
> MIDDLE TOE. From the distal end of the tarsus to the base of the claw, not including the claw unless so stated.
>
> GRADUATION OF TAIL. From the end of the outermost rectrix to that of the middle or longest, the tail being closed.

As proposed, the length was to be taken with tape or ruler, the other measurements with dividers (then read against a ruler). As with all such systems, the whole series of measurements is not invariably made—frequently only of wing, tail, and culmen, and in some groups other measurements may be needed. Except for the total length, these dimensions are nearly the same on skins as on the living birds.

A more recent and widely used system is that of Baldwin, Oberholser, and Worley (1931), which attempts to standardize over 200 measurements in birds, most of which are clearly shown by diagrams. In the main, Ridgway's standards have been used by these authors, though they also use a large number of dimensions not considered by him.

For mammals, the standard external measurements are given by Anthony (1925) or Sumner (1927).

It is customary to take the longer dimensions with a ruler and the shorter with calipers or dividers. Except for foot length, all these measurements may be significantly different in the living animal and in prepared specimens, so that they are generally taken on the freshly killed animal. Any deviation from this practice must be specified.

Taxonomists tend, for practical reasons, to concentrate on external characters like those given above for birds, especially when they are interested in smaller groups such as species and subspecies. These characters are superficial, both literally and figuratively, and so are not very reliable for the taxonomy of higher groups. They are usually not available in fossils, which, with unimportant exceptions, can be studied only by osteology and dentition. To compare these internal characters with those of living animals also requires study of the hard parts in the latter. Numerical and other characters derived from the teeth and skeleton are of great value and are widely used in mammalogy, both recent and fossil. As regards the skeleton, these characters are of equal value among the lower vertebrates but have as yet been less used for recent animals.

Paleontological mensuration differs little from that of the hard parts of recent animals. Fossil material is almost invariably less complete, so that a standardized system of a few measurements is less practical and requirements must in each case be adjusted to possibilities. Fossil bones are also commonly distorted so that their measurements are generally less reliable than are those of recent animals. This may make some measurements, especially those of proportions, unusable. Some groups of extinct animals are so unlike any living forms that they present a different problem in mensuration. All of these factors also militate against systematization of paleontological data, but they do not make it impossible. In earlier paleontological publications, aside from a few obvious measurements, the numerical data were frequently inadequate or seemed to have been selected at random and without rational criteria. Recent work has gone far toward correcting this fault.

Perhaps the most detailed system of osteological mensuration for mammals is that of Duerst (1926), who also gives references to and synonymy with the practices of other workers. Osborn's elaborate studies (especially those of 1912 and 1929) on the osteometry and craniometry of perissodactyls, although based on a single group of mammals, also repay close study by anyone engaged in gathering numerical zoological data. Within the limitless field of special problems, only two strikingly different and suggestive examples will be mentioned. Zeuner (1934) has used a system of cranial angles as a basis for biological inferences regarding rhinoceroses, and Soergel (1925) has employed numerical and mathematical procedures in studying footprints and inferring from them the sort of animal that made them.

Aside from dimensions and counts like those mentioned above, color is a very important character in the study of recent animals. Usually this is roughly described in the vernacular, or an attempt, much better but still inexact, is made to match the color against a standard chart, of which Ridgway's (1912) is the most widely used. The most precise method of analyzing color is by photometric spectroscopic analysis, but this is such an elaborate and exacting process that it is impractical in most zoological work. Numerical data on color can be obtained more simply with a color top or a tint photometer. A color top (see Collins, 1923) is a device containing adjustable segments of white, black, and a set of standard colors, usually complementary and primary. When the top is spun, the colors blend into a single shade, which can be made to match the color being measured by adjusting the size of the segments. Adjustment of the segments, which must be done by trial and error, is a long process, and the matching is subjective and does not give very consistent results. In a tint photometer (see Sumner, 1927), reflected light from a white surface and from the colored object to be measured are viewed simultaneously through a color filter, and the light from the white surface is cut down by a diaphragm until it matches in intensity that from the object. This process gives us a relative measure of the amount of light, i.e., of those wavelengths passed by the screen, reflected by the object. The percentage of closure of the diaphragm is read from a scale and recorded numerically. If several screens are used and a reading is taken for each, a good numerical measure of color can be obtained. The procedure is reasonably rapid and simple, and the estimate of relative intensity of light is easier and involves less subjective inconsistency than does the matching of colors. This method also has drawbacks, especially its requirement of a complex apparatus and the fact that it does not measure the whole color but only certain components of it—namely, the color bands passed by the filters. Without the use of an impracticably large number of filters, the color cannot be reproduced exactly from data gathered in this way. This is, however, the most practical valid method for reducing color to exact numerical terms that has yet been devised.

Bias and Consistency

One of the most troublesome difficulties in using numerical data is bias, a tendency to favor some hypothesis or to lean toward a numerical result which is not purely objective. In this sense, bias is assumed to be unconscious and to have no flavor of disingenuousness. It usually arises either in sampling, which is discussed in Chapter 7, or in measurement.

Bias in measurement is subjective and personal. It usually takes such forms as tendency to overrun or underrun the accurate figure for the measurement in question, tendency toward or away from integral or some other certain values, or tendency to favor or oppose a given hypothesis. The existence of a tendency to overrun or underrun measurements can usually be detected by having two workers independently make a large series of measurements of the same objects in the same way. If the average result obtained by one worker is significantly smaller than that obtained by the other, the existence of bias may be assumed and further tests of a similar nature may be made to determine whose the bias is, its direction, and its amount. The same sort of bias may often be both detected and corrected by taking measurements in duplicate in two different directions; for instance, by opening calipers to the dimension sought and then closing them to it, and taking the mean if the two measurements differ.

There is also a tendency when taking a series of homologous or numerically closely similar measurements to make them more nearly similar than is correct. This tendency, almost universal if attention is not paid to it, may be largely eliminated (1) by deliberately ignoring preceding readings and (2), when using calipers, by throwing them far off the last measurement before bringing them to the next. This precaution is an essential feature of good measuring technique.

If not forewarned, many students have a bias toward integral values; and if detected, this may be overcompensated by bias away from them. Such bias with respect to particular numbers can usually be detected by checking over a large series of measurements of many different sorts and determining whether any one final digit occurs oftener than would be likely by chance. Care must be taken that the data are not such as would really tend to be concentrated about any one number in the last place.

Tendency to favor a hypothesis is perhaps the most obscure bias of all and the most difficult to detect or to avoid. If there is any real possibility of such bias, measurements may be made by a worker not acquainted with the hypothesis in question.

In addition to the forms of bias mentioned, there are also biases of procedure, of instruments, and of materials. Some systems of dealing with specimens consistently make them appear longer or shorter than others. Biased instruments, such as an inaccurately calibrated ruler or an instrument that does not return precisely to zero when closed, naturally produce biased results. Measurements of shrunken or swollen skins and other specimens are biased with respect to fresh materials. Inexact or incorrect specification of the dimension measured also produces an effect analogous to bias. The correctives for all these are fairly obvious.

The possibility of bias can generally be reduced to insignificance by duplication of measurement (perhaps varying the direction), by maintenance of an objective attitude, by carefully standardized procedure, by the use of highly refined instruments, by recording exactly what the measuring instrument says, by ignoring the purpose of the measurements as far as possible while they are being made, and by recording the results in smaller units than are to be used in ensuing calculation or publication. For trained observers some of these precautions are automatic and others are unnecessary; but the complete elimination of bias is very difficult.

The distinctive feature of bias is some degree of consistency, a tendency to deviate from the ideal more often in some particular direction than in others. Since the usual purpose of measurements is to make comparisons, such deviations may have little or no effect on the conclusions drawn. Thus, a form of bias such as the almost unavoidable shrinkage of dead materials may be of no importance if it is sufficiently consistent, and the deviations from live measurements are hardly to be considered as bias so long as comparisons are made only between specimens comparably preserved. Similarly, a worker may have a marked bias and yet it may not affect his comparisons so long as he is highly consistent and uses only measurements made by himself. It is a well-recognized fact in zoology that measurements made by one observer compare more closely than those made by two or more different observers. Here there is not only the element of bias as it has hitherto been defined but also the related element of personal idiosyncrasies regarding the exact definition and orientation of measurements, which even the most rigidly standardized systems of mensuration do not wholly eliminate.

The factor of consistency is, strictly speaking, at least as important as that of bias. Both factors are visible in examples such as that given by Sumner (1927) in recording the means obtained by each of three different observers measuring the same sample of ten specimens on two successive days. The figures, which are for tail length in a sample of *Peromyscus maniculatus gambelii*, are given in Example 3.

EXAMPLE 3. Mean measurements of tail length of the deer mouse *Peromyscus maniculatus gambelii* taken by three observers on two successive days. (Data from Sumner, 1927)

	FIRST DAY	SECOND DAY
Sumner	74.9 mm.	74.4 mm.
Second observer	70.9 mm.	72.2 mm.
Third observer	70.2 mm.	71.1 mm.

The second and third observers working on this experiment were clearly biased with respect to Sumner, or he with respect to them, for his mean on both days is considerably larger than theirs. The consistency involved is of two sorts: that of the figures of a single observer and that of those given by different observers. Each observer is reasonably consistent with himself,

Sumner more so than the other two. The figures of the second and third observers are fairly consistent, but those of Sumner are not consistent with theirs. In fact, these figures strongly suggest that the second and third observers used the same technique in nearly the same way and that Sumner used a different technique. Judging from the data, it does not necessarily follow that Sumner's technique was more accurate or more refined than the techniques of the observers, although this also is hinted. However, such was the case. Sumner measured the specimens on a special measuring frame with calipers calibrated to .1 mm., and the other two measured the loose specimens with a ruler. Incidentally, the figures clearly show that measurement to .1 mm. was here unduly refined, even for Sumner's more precise methods, and that the last digit is not in any case either accurate or useful.

CHAPTER THREE

Frequency Distributions and Grouping

Frequency Distributions

The first step in reducing original observations to more compact form and in preparing to draw any sort of conclusions from them is to tabulate them in the form of a frequency distribution. A frequency is the number of observations that fall into any one defined category, and a frequency distribution is a list of these categories showing the frequency of each. Such distributions are the basis for almost all important numerical operations in zoology, and the use of numerical data depends on the definition of the categories or groups in which the data are to be placed.

In constructing a frequency distribution there are two essential criteria for defining the classes. First, the classes or groups must be *mutually exclusive*. That is, it must be absolutely clear into which class each observation falls. For example, 1.5–2.5 and 2.5–3.5 are not valid group limits because there is ambiguity as to where the measurement 2.5 lies. Second, the groups must be *exhaustive*. In other words, every measurement must belong to some class. As an example, classes such as 1.5–2.5 and 3.5–4.5 are insufficient because a measurement such as 3.2 does not lie in either of them.

In qualitative grouping, whether on numerical or other bases, the principles of exhaustiveness and exclusion sometimes are more obscure than for numerical variables, so that failures in this respect are common in the literature. It is frequently stated that a given character is present in a certain number of cases (i.e., that it has a certain frequency) and absent or indeterminate in so many others. This grouping is invalid because the second group may or does include among the indeterminate cases some that had the character and hence belong in the first group. This twofold grouping is thus not mutually exclusive, and there are really three groups: present, absent, and indeterminate. But since it is presumably the presence or absence of the character that is being studied, the indeterminate specimens have nothing to contribute to the problem and should not be included in the data. This simple logic is so often contravened that it apparently is not obvious and requires statement. Commonly, the form of the error is to say that 50 per cent of the specimens have the character and that in the other 50 per cent it is absent or indeterminate; or, slightly better but still wrong in most cases, that 50 per cent have the character, 30 per cent do not, and 20 per cent are indeterminate. The correct expression of these facts is that of the determinable specimens 62.5 per cent have the character and 37.5 per cent do not.

Attributes

A grouping need not be and often is not in itself numerical. A common zoological grouping is that of the taxonomic system, the group being a subspecies, species, genus, or larger category in the hierarchy. Example 4

EXAMPLE 4. Specimens of Diptera trapped on Squaw Peak, Montana, in the summer of 1952. (Data from Chapman, 1954)

	BY SPECIES	BY GENUS	BY FAMILY
Syrphidae			106
Arctophila flagrans	10	10	
Chrysotoxum ventricosum	18	18	
Cynorhina armillata	14 ⎫	22	
Cynorhina robusta	8 ⎭		
Eristalis tenax	37	37	
Sphecomyia pattoni	19	19	
Tabanidae			55
Hybomitra rhombica var. *osburni*	9 ⎫	47	
Hybomitra rupestris	12 ⎬		
Hybomitra zonalis	26 ⎭		
Tabanus aegrotus	6 ⎫	8	
Tabanus sequax	2 ⎭		
Tachinidae			73
Fabriciella nitida	32	32	
Gonia porca	4	4	
Mochlosoma sp.	9	9	
Peleteria conjuncta	7 ⎫	28	
Peleteria iterans	21 ⎭		

Example 4 illustrates how the primary observations, themselves forming a frequency distribution, may be grouped secondarily by genus and finally by family. Secondary and tertiary groupings will often bring out relationships not obvious from the primary observations. The use of secondary groupings is especially important for continuous variates and will be discussed in some detail under that heading.

is a frequency distribution of this sort. Frequencies are employed when it becomes necessary to count the number of individuals within a given taxonomic unit observed under certain conditions: the number observed in traversing a defined area, the number caught by fishing operations, etc.

In other studies the groups may be defined ecologically, and the frequencies may be either of individuals or of species or genera observed within certain limits. Thus for the Bridger (Middle Eocene of Wyoming) mammalian fauna as known to Matthew (1909), a frequency distribution can be compiled as in Example 5.

The groups may be geographic or based on habits and activities or on nonnumerical anatomical characters. Examples 6 and 7 will suggest the wide range of possibilities of this sort.

EXAMPLE 5. Distribution of Bridger mammalian fauna by habitat type. (Data from Matthew, 1909)

HABITAT	NUMBER OF KNOWN GENERA	NUMBER OF SPECIMENS IN AMERICAN MUSEUM COLLECTION
Aerial	0	0
Arboreal:		
surely	13	184
probably	11	485
Terrestrial	17	314
Fossorial	3	8
Amphibious or aquatic	1	12

EXAMPLE 6. Physiological condition of specimens of the eastern chipmunk *Tamias striatus* taken in July. (Data from Schooley, 1934)

CONDITION	FREQUENCY
With embryos	11
Has ovulated recently	14
Not gravid, no recent ovulation	22

EXAMPLE 7. Bill color in the European starling *Sturnus vulgaris vulgaris* in February. (Data from Hicks, 1934)

COLOR	FREQUENCY
Yellow (more than 85% yellow)	33
Partial (20 – 85% yellow)	20
Dark (less than 20% yellow)	29

Discontinuous Variates

As with attributes, the raw observations of discontinuous variates give rise directly to frequency distributions. The values of discontinuous variates are the *names* of *classes*, just like the species in Example 4, so that in this sense there is no difference between a frequency distribution of attributes and of discrete variates. A difference which does exist between these two types of variables, however, is that there is a logical order in which the discrete variates fall, a property which attributes do not always have. Thus, in Example 8A–D the values of the variate are placed in ascending order. Because attributes generally have no logical ascending or descending order, they may usually be placed in any order in constructing a frequency distribution.

EXAMPLE 8. Distributions of discontinuous variates.

A. Discontinuously variable physiological function. Number of breaths taken in a single breathing period by a young Florida manatee. (Data from Parker, 1922)

BREATHS TAKEN	NUMBER OF TIMES OBSERVED
1	16
2	13
3	2
4	2

B. Discontinuously variable reproduction. Number of young in nests of tree swallow *Iridoprocne bicolor*. (Data from Low, 1933)

NUMBER OF YOUNG	NUMBER OF NESTS
1	1
2	4
3	7
4	31
5	56
6	17
7	4

C. Discontinuously variable anatomical character. Number of serrations on the last lower premolars of specimens of the extinct mammal *Ptilodus montanus*. (Original data)

NUMBER OF SERRATIONS	NUMBER OF SPECIMENS
13	8
14	19
15	2

EXAMPLE 8. *continued*

D. Discontinuously variable anatomical character. Number of caudal scutes in the king snake *Lampropeltis getulus getulus*. (Data from Blanchard, 1921)

NUMBER OF CAUDAL SCUTES	NUMBER OF SPECIMENS
40	3
41	2
42	4
43	4
44	4
45	7
46	6
47	3
48	9
49	2
50	5
51	2
52	1
53	2
54	3
55	0
56	1
57	0
58	1

For a discontinuous variate, the raw data are not grouped but are absolute values. The 6 specimens of *Lampropeltis getulus getulus* with recorded vertebral counts of 46 do not have from 45.500 . . . to 46.4999 . . . caudals; each has exactly 46. Occasionally there is some doubt in the observations; it may be, for instance, that a count could be equally well called 5 or 6. The best procedure is to make some arbitrary convention as to whether a structure will be considered absent or present and be consistent in its use. Sometimes the existence of intermediate cases is in itself an important observation. For example, it may indicate the breakdown of a mosaic developmental pattern. It is best to mention specifically the number of cases in which unusual or indeterminate observations occurred as a separate item of information along with an explanation of the convention used in calculation.

Just as in the case of attributes, larger classes of discontinuous variates may be constructed from the raw observational classes by grouping. In such grouping, the counts defining a group are midpoints and not limits, and the midpoints of the groups are halfway between the original counts. Among numerous other ways, Example 8D could be grouped as shown in Example 9.

In the grouping with interval 4 (or any other even number), the midpoints

EXAMPLE 9. Two forms of secondary groupings of data from Example 8D.

A. INTERVAL 3			B. INTERVAL 4		
GROUP	MIDPOINT	FREQUENCY	GROUP	MIDPOINT	FREQUENCY
40–42	41	9	40–43	41.5	13
43–45	44	15	44–47	45.5	20
46–48	47	18	48–51	49.5	18
49–51	50	9	52–55	53.5	6
52–54	53	6	56–59	57.5	2
55–57	56	1			
58–60	59	1			

are fractional and are not values that the variate can actually take. This introduces a series of imaginary values if the midpoints are used in calculation; the conception is difficult and (some believe) logically objectionable. For this reason it may be preferred to use only groupings with intervals of odd numbers so that the midpoints will be integers and will be real values of the variate. In fact, however, the use of imaginary values as midpoints is not mathematically invalid. The results based on them are just as accurate as those based on real values as midpoints, and there is no reason why they should not be used if they are more convenient in other respects.

For purposes of calculation, the ungrouped data are usually used for discontinuous variates. This is more accurate, and the number of steps is seldom so great as to make the calculation unduly laborious. The form of distribution is, however, usually clearer if some grouping is employed, especially with the small samples usual in zoology.

Continuous Variates

It was shown in the preceding chapter that the values of a continuous variate are grouped as they are originally observed and recorded. Thus 9.2 mm. is not an absolute measurement but the designation of the midpoint of a group of measurements, the absolute values of which may be anything greater than or equal to 9.15 and less than 9.25. A figure such as 9.2 in this usage is thus simply a conventional way of representing a group with the limits 9.15000 . . . and 9.24999 . . . .

When continuous variates are designated by a single figure, this figure is almost always the midpoint of the implied group, There are, however, some exceptions, and these may lead to serious numerical errors if not detected and adjusted to the usual convention. Thus ages are usually represented by the lower limit of the group. A child 2 years old is not between the ages of 1 year 6 months (1.5 years) and 2 years 6 months (2.5), but between the ages of 2 and 3. In all statistical operations on such data this convention has a strong influence, as the following hypothetical distribution shows:

Recorded age	Frequency
2	6
3	20
4	5

Calculated on these data in the ordinary way (which is more fully expounded in Chapter 5), the mean or average age of this group of infants would appear to be 3.0 years. The calculation is, however, invalid unless the records are adjusted to represent group midpoints, thus:

Midpoint of age group	Frequency
2.5	6
3.5	20
4.5	5

The mean age is now correctly found to be 3.5 years, a decided difference.

Some other age records are even more confusing. For instance, horse breeders advance the nominal age of all horses, regardless of when they were foaled, on January 1, so that a "1-year-old" horse may in reality be anything from just over 0 to just under 2 years in age, a "2-year-old" between 1 and 3 years, etc.

Almost all numerical procedures are based on the convention that the figure recorded is the midpoint of the group, and if this is not true of a given set of data, an adjustment must be made.

Some workers take measurements in units that are not decimal and yet write them in the ordinary way, e.g., measure only to half millimeters but record these as decimals. This practice is confusing and indefensible in the face of the universal convention as to limits in decimal measurements. Such an author will record 2.3 mm. as 2.5 because it is nearer to that than 2.0. By 2.5 he means a group 2.25000 . . . – 2.74999 . . . , but his reader can only infer that, according to convention, 2.5 stands for the group 2.45000 . . . – 2.54999 . . . , a group to which the measurement does not really belong. It would be preferable to write the measurement as 2 1/2 mm., thus showing that the unit of measurement was 1/2 mm. and that the group implication is that the dimension is nearer 2 1/2 than any other multiple of 1/2, i.e., that the class limits are 2 1/4–2 3/4. Such a record, however, has the serious drawback that the integer, such as 2 in this case, does not indicate the unit of measurement. This difficulty could be overcome only by writing all measurements as fractions, multiples of the unit used, i.e., writing 2 as 4/2 and 2 1/2 as 5/2 if the unit were 1/2 mm. Even

38 QUANTITATIVE ZOOLOGY

this is clumsy and makes subsequent calculation based on the measurements difficult. Still worse are cases in which nondecimal fractional measurements are used but the fractional unit is not the same for the different measurements to be compared; for instance, one measurement may be recorded as 3 1/3 and another to be compared with this may be recorded as 3 1/8, etc. It is practically impossible to base valid frequency distributions and make accurate comparisons and calculations on such data. The general solution of these difficulties is to make measurements in decimal units whenever possible and, when this is not possible or for some special reason is undesirable, to make records by class limits, not by class midpoints. Thus 2 1/2 mm. should be recorded decimally as 2.3–2.7 or 2.25000 . . . – 2.74999 . . . , not as 2.5.

Secondary Grouping

Measurements are recorded to the nearest unit, which may be at any point on the decimal scale, and the implied grouping is of the sort just discussed, with the record understood to be the midpoint of a group extending one-half unit below and above this point. In compiling frequency distributions, it is often advisable to expand the group limits (secondary grouping), thus giving fewer groups and higher group frequencies.

In secondary grouping a requirement is that intervals should be of equal size, so that within a single distribution groups such as 10.5–11.4 and 11.5–11.9 should never be used. Exceptions to this rule are instances in which the class zero (exactly) is an important and qualitatively distinct class—for example, in fertility records, in which case complete infertility (0 eggs or offspring) is qualitatively distinct from any other value.

The relationship between the original measurements, the so-called group limits, the real limits of the groups so designated, and the midpoints of the groups is somewhat confusing. If original measurements are taken to .1 mm., then the classes of their distribution are designated by a series of single figures each .1 mm. larger than the last, as in Example 10. If these figures were translated into the real group limits of the measurements—a form of record unnecessarily complex and never employed, although many errors might have been avoided by using it—they would read as in Example 11.

If, now, it is decided to gather these measurements into larger groups, these new groups are usually designated by the smallest and the largest original measurements placed in them:

Group	Frequency
9.1–9.3	16
9.4–9.6	12

These figures, 9.1–9.3 and 9.4–9.6, are what are called the group or class limits, but obviously they are not real limits. The real limits of the implied range are 9.05000 ... – 9.34999 ... and 9.35000 ... – 9.64999 .... However, since the original measurements were taken only to the nearest .1 mm. there is absolutely no ambiguity in stating the limits of the secondary groups as 9.1–9.3 and 9.4–9.6. That is, our criteria of *mutual exclusion* and *exhaustiveness* of classes (see page 31) are perfectly fulfilled, for all measurements fall in one or the other of these groups and no measurement falls in both. Any confusion which may arise here is due to an inadequate separation of the idea of a measurement and the range of true values of which it is a symbol.

EXAMPLE 10. Distribution of measurements as usually given. (Hypothetical data)

MEASUREMENT	FREQUENCY
9.1	1
9.2	5
9.3	10
9.4	7
9.5	3
9.6	2

EXAMPLE 11. Distributions of measurements by real group limits

IMPLIED LIMITS	FREQUENCY
9.05000 ... – 9.14999 ...	1
9.15000 ... – 9.24999 ...	5
9.25000 ... – 9.34999 ...	10
9.35000 ... – 9.44999 ...	7
9.45000 ... – 9.54999 ...	3
9.55000 ... – 9.64999 ...	2

The midpoints of the new groups thus formed are respectively 9.2 and 9.5. In calculations based on grouped data, these midpoints are taken to approximate the average value of all the measurements placed in the corresponding groups; and calculations are based on the midpoints, the fixing of the true value of which is thus important. For purposes of calculation, the midpoint is often added in tabulating a frequency distribution of this sort:

Group	Midpoint	Frequency
9.1–9.3	9.2	16
9.4–9.6	9.5	12

Some workers have assumed that the lower figure in the secondary group designation is, in fact, the true lower limit of the variate, so that the group 9.1–9.3 does not include any value of the variate below 9.1 (not even 9.0999 . . .) and does include all values between 9.1000 . . . and the lower limit of the next group, 9.4. That is, they assume that 9.1–9.3 symbolizes a true range of 9.1000 . . . to 9.3999 . . . . This is obviously at variance with our stated convention of what range of values a given number is meant to symbolize. If this false assumption were correct, the midpoint of the group 9.1–9.3 would be 9.25, not 9.2.

In designating secondary numerical groups, it must be clear whether the numerical designation is the midpoint, lower limit, upper limit, or both limits, and whether the limit is absolute or is in terms of the original measurements. It is assumed that a single number designates a midpoint unless the contrary is explicitly stated. If only the lower limit or only the upper limit is given, this usage must be specified. If two figures separated by a dash are given, these are the two limits. It may usually be assumed that these are given in terms of the original measurements and hence that they are midpoints of the smaller groups of observation from which the larger groups have been derived. If the figures are intended as absolute limits, they are generally and should always be distinguished either in words or by added decimal points on the second figure. Thus 20–22, designating a group for a continuous variate, will be assumed to be in terms of original measurement and hence to have the true limits 19.5–22.5 and midpoint 21; but 20–21.99 is assumed to represent absolute limits, not 19.5–22.5 but 20–22, the midpoint still being 21.

The relationships between recorded measurements, conventionally stated class limits, real limits and midpoints, and the false limits and midpoints sometimes used are clearly shown in the diagram on page 42 (Fig. 1). The magnitude of the groups formed is designated by the class interval, which is the distance from any point within a group, such as the lower stated limit or the midpoint, to the corresponding point in the next higher or lower group. The class interval is .3 in the example just discussed. Although it is usual and preferable for most purposes to designate secondary groups by their conventional limits, a distribution may also be given by midpoints alone, even though the grouping is larger than that of the original measurements. If the classes are designated by one number and the difference between successive designations is not a single unit, it may be understood that the numbers are midpoints of enlarged groups and not measurements.

Example 12 shows a frequency distribution in terms of the original measurement to .1 mm. and with three different secondary groupings, two with class interval .3 mm. but with the limits at different points on the scale and one with class interval .5 mm.

EXAMPLE 12. Frequency distributions. Length of P_4 in a sample of the extinct mammal *Ptilodus montanus*, from the Gidley Quarry. (Original data)

A.		B.		
ORIGINAL MEASUREMENT, MM. (CLASS INTERVAL .1 MM.)	FREQUENCY	CLASS LIMITS (INTERVAL .3 MM.)	MIDPOINTS	FREQUENCY
7.1	1	7.1–7.3	7.2	2
7.2	1	7.4–7.6	7.5	8
		7.7–7.9	7.8	17
7.3	0	8.0–8.2	8.1	20
		8.3–8.5	8.4	21
7.4	0	8.6–8.8	8.7	6
		8.9–9.1	9.0	1
7.5	6			
7.6	2			

		C.		
7.7	5	CLASS LIMITS (INTERVAL .3 MM.)	MIDPOINTS	FREQUENCY
7.8	8			
		6.9–7.1	7.0	1
7.9	4	7.2–7.4	7.3	1
		7.5–7.7	7.6	13
8.0	4	7.8–8.0	7.9	16
		8.1–8.3	8.2	22
8.1	8	8.4–8.6	8.5	21
		8.7–8.9	8.8	0
8.2	8	9.0–9.2	9.1	1
8.3	6			
8.4	8			

		D.		
8.5	7	CLASS LIMITS (INTERVAL .5 MM.)	MIDPOINTS	FREQUENCY
8.6	6			
		6.8–7.2	7.0	2
8.7	0	7.3–7.7	7.5	13
		7.8–8.2	8.0	32
8.8	0	8.3–8.7	8.5	27
		8.8–9.2	9.0	1
8.9	0			
9.0	0			
9.1	1			

To compile such frequency distributions, it is first necessary to make the measurements to as fine a point as will be required for any desirable grouping. These records will be irregularly scattered, for it is not practical to make them in the order of their magnitudes. The next procedures are to write down all the steps from the smallest to largest in the unit of measurement (to .1 mm. in the example), to tally against this the original measurements, and then to reduce the tally marks to numbers. This results in the first form of distribution given in Example 12A. If a larger unit of secondary grouping is to be employed, the interval to be used and the point at which to start (or positions of the midpoints, as determined by this) are decided and the frequencies are taken from the distribution of the measurements. This may be done as in Example 13, using data from sections of distributions in Example 12A and B.

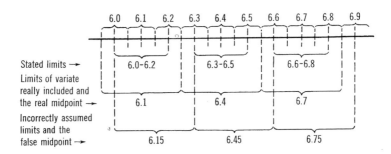

FIGURE 1. Midpoints and limits in primary and secondary grouping. The horizontal line represents the scale of all possible measurements of a continuous variate. The numbers above this line are original measurements, to .1 mm., which are in fact the midpoints of primary groups, the ranges of which are shown by the brackets beneath the recorded measurements. Below the line is indicated secondary grouping with interval .3 mm.

This also facilitates the selection of the best secondary grouping, discussed in a later section.

It is customary to speak of the distribution in terms of the original measurements (i.e., with the class interval equal to the smallest unit of measurement) as "ungrouped" and of a distribution with a larger class interval as "grouped", but we shall avoid this. Especially in conjunction with the record of measurements by midpoints rather than by limits, this practice obscures the fact that the measurements (if a continuous variate) are really grouped, a fact that should always be kept in mind.

EXAMPLE 13. Secondary grouping, or decreasing the number of groups in a frequency distribution, using data from Example 12.

ORIGINAL MEASUREMENTS (MM.)	FREQUENCY	FREQUENCY (INTERVAL .3 MM.)	LIMITS (INTERVAL .3 MM.)	MIDPOINTS
7.7	5			
7.8	8	17	7.7–7.9	7.8
7.9	4			
8.0	4			
8.1	8	20	8.0–8.2	8.1
8.2	8			
8.3	6			
8.4	8	21	8.3–8.5	8.4
8.5	7			

Numerical Qualitative Grouping

In the distributions of variates discussed so far, the categories in which the observations are grouped and for which frequencies are recorded are themselves quantitative concepts. It is also possible and often highly useful to employ categories that are defined by numerical data but are conceptually qualitative, the consideration and analysis of which should be from the viewpoint and with the methods of the study of attributes rather than of variates. Because such categories are defined numerically, they are easily confused with truly quantitative distributions, and it is important to recognize the distinction.

One of the commonest of such arrangements of data, especially useful in studying association (see Chapter 13), is the division of a frequency distribution into two groups—one in which the values of the variate exceed and one in which they are less than a given value. The value selected may be the midpoint of the distribution or may be at a break in the distribution or at any other point suggested by the problem in hand. In any case, the resulting two-fold grouping, although literally quantitative, is in effect qualitative. It is a division not into a series of equal, quantitative steps, but into larger and smaller qualitative groups. Such a division, from data for a continuous variate, using a break in the distribution as the division point, is given in Example 14.

Such a grouping might be made, for instance, to see whether larger size and smaller size, as attributes, can be associated with greater age and lesser age, with occurrence in two different regions, or with any other factors.

EXAMPLE 14. Distribution of total length in a sample of 34 females of the king snake *Lampropeltis elapsoides elapsoides*. (Data from Blanchard, 1921)

A. QUANTITATIVE GROUPING (INTERVAL 25 MM.)		B. QUALITATIVE GROUPING	
LENGTH	FREQUENCY	LENGTH (MM.)	FREQUENCY
150–174	1	Less than 300	9
175–199	2	Greater than 300	25
200–224	3		
225–249	2		
250–274	0		
275–299	1		
300–324	0		
325–349	2		
350–374	2		
375–399	5		
400–424	4		
425–449	4		
450–474	3		
475–499	2		
500–524	3		

Sometimes a multiple grouping, as in Example 15, suggests a quantitative frequency distribution and may be mistaken for one or may even be supposed by its author to be one. If it were so intended, it would be a very careless and unsound grouping, and conclusions based on it would be inaccurate and probably invalid. As a qualitative grouping, however, it is valid and useful. In Example 15 these are actually qualitative age groups necessarily defined indirectly in terms of size and shown to be associated with various growth phenomena, although the evidence for this is not given in the example.

EXAMPLE 15. Distribution of total length in a sample of the recent fish *Caranx melampygus*. (Data from Nichols, 1935)

QUALITATIVE LENGTH GROUPS (IN MM.)	FREQUENCY
104–115	4
120–125	7
128–136	5
147–169	2
195–208	2

Criteria for Secondary Groups

Decision as to what secondary grouping is to be made depends on the uses to which the groups are to be put. These uses are discussed in detail in the following chapters, and the purposes and procedures of grouping will be clear when these chapters have been read. In general, the purpose of secondary grouping is to simplify calculation and to bring out formal characteristics of the distribution. Frequently, especially with small samples, the same grouping will not serve well for both purposes.

Grouping is defined by the class interval and by the position of any one limit or midpoint. The class interval together with the total range of the observations to be grouped determines how many classes or steps there will be in the grouped distribution. This, in turn, determines the concentration or dispersion of frequencies. Since the total frequency is fixed, if there are fewer classes each will tend to have a higher frequency, and if there are more classes, each will tend to have a lower frequency. In grouping for calculation the number of classes should generally be between 15 and 25, if the original data cover only 25 or fewer steps, calculation should be based on these data and not on further grouping.

As we have pointed out earlier, if measurements representing continuous variates are themselves to be treated as continuous variates, it is important that the entire range in which the measurements lie be finely enough subdivided to give a semblance of continuity.

In calculating from secondary grouped data, the class midpoints are taken to represent all the observations included in the class. It therefore follows that in grouping for this purpose that arrangement is best which produces groups in which the midpoint of each class most nearly corresponds to the mean of the individual values included in the class, or in other words, in which the individual values in each class are most symmetrically distributed around the midpoint of the class. If the secondary grouping is done from a frequency distribution of the individual values (original measurements), the degree to which this ideal is approached and the position of the classes that best corresponds with it may easily be determined by inspection and trial. Thus, in Example 16, an extract from a (hypothetical) distribution, arrangement B is clearly better than arrangement A although the interval is the same.

In general, even if the secondary grouped distribution is not to be used in calculation it is well to follow this criterion as much as possible. The more nearly these conditions can be fulfilled, the more proper it is to reduce the number of classes or to increase the class interval.

A good method of grouping is to choose the most frequent class as the midpoint of a group (if the number of steps in a group is odd). This then

EXAMPLE 16. Two arrangements of secondary grouping with the same interval.

ORIGINAL MEASUREMENT	FREQUENCY	A. GROUPED FREQUENCY (INTERVAL .3)	B. SAME, IN DIFFERENT POSITION
8.7	0		
8.8	0	1	
8.9	1		1
9.0	0		
9.1	8	18	
9.2	10		25
9.3	7		
9.4	5	14	
9.5	2		14
9.6	7		
9.7	0	7	
9.8	0		

A. CLASS LIMITS	MIDPOINTS	AVERAGE OF INDIVIDUAL MEASUREMENTS INCLUDED	B. CLASS LIMITS	MIDPOINTS	AVERAGE OF INDIVIDUAL MEASUREMENTS INCLUDED
8.7–8.9	8.8	8.90	8.8–9.0	8.9	8.90
9.0–9.2	9.1	9.15	9.1–9.3	9.2	9.20
9.3–9.5	9.4	9.36	9.4–9.6	9.5	9.51
9.6–9.8	9.7	9.60			

In the first grouping A, the frequencies are irregularly distributed within each of the classes shown. As a result, the class midpoints differ by .10, .05, .04, and .10 from the means of the individual measurements which, in calculation from such grouped data, they are taken to represent. In the second grouping B, the frequencies are almost symmetrically distributed in each class, and the class midpoints agree almost exactly with the means of the individual measurements.

---

defines one group, all the others being fixed by this choice. This is, in fact, what has been done in Example 16B.

If the frequencies in each class are low and the whole sample small, secondary grouping for calculation may be unjustified even when the number of classes in the original data is greater than 25. In such cases, secondary grouping into 15 or more classes should produce a marked concentration of frequency in some classes and should tend to eliminate frequencies of zero or one except towards the ends of the distribution.

Grouping that is satisfactory for calculation is also satisfactory for the further purpose of reducing the bulk of the data for publication. All that is required for publication is that satisfactory results be derivable from the data published; hence, a compact table on which accurate calculations can be based is just as good as a much longer and more diffuse table of the raw measurements.

The way in which secondary groupings can be used to bring out the form of the distribution is well exemplified by the figures for caudal scutes of *Lampropeltis getulus getulus* (Examples 8D and 9). The frequency distribution of the original data is long and irregular, and it is difficult to detect any pattern in it. When these are grouped with interval 3, giving seven classes, a very definite pattern emerges. When they are grouped with interval 4, giving five classes, a similar pattern is evident but it is now so compressed as to be less clear. Evidently, for these data a secondary grouping with interval 3 reveals the distribution pattern more readily than the raw data or than any other grouping. That secondary grouping is best for this purpose which most clearly and smoothly brings out such a pattern, a criterion that will be more easily applied when the sorts of patterns involved have been considered in the next and subsequent chapters.

Secondary grouping for the purpose of bringing out the form of a distribution generally requires fewer classes than are advisable for calculation, and this is particularly true with the small samples usual in zoology. The number of classes should usually be less than 16 and more than 4, while for calculation they should, if possible, be more than 15. In grouping for pattern, it is often an advantage to have an odd number of classes, for this will give a middle class, an important point in most zoological distributions. This should tend to smooth out any small random fluctuations in the frequencies so that they tend to rise or fall steadily through several successive classes. In Example 9, with interval 3, they rise through the first three and fall through the last five classes in an orderly way, and in the raw data (Example 8D) they reverse direction eleven times. The grouping should tend to eliminate frequencies of 0 within the distribution and also any very low frequencies, except toward the ends. In Example 8D, the raw data have two internal zeros and several low frequencies of 1 to 3 far from the ends, while the grouped data (Example 9) have no internal zeros and have relatively low frequencies only in the last two classes, where they may be expected to occur in any case.

No matter what system of grouping is used, a certain amount of subjectivity cannot be avoided. While it is desirable that grouping "bring out" a pattern, it is probably more accurate to say that a pattern is *created* by the grouping, different groupings creating somewhat different pictures. As in other cases, intuition and experience with his material will to some extent govern the results of the zoologist's endeavors.

CHAPTER FOUR

Patterns of Frequency Distributions

Graphic Representation

A frequency distribution has characteristics of its own, not seen in the isolated observations, and these are properties of the data as a whole on which the most important deductions and comparisons can be based. The essential characteristic of a distribution is a pattern formed by the rise and fall of the values of the frequencies as the values of the variate increase. This pattern is shown by the distribution in numerical form, but it almost always stands out more clearly if it is made into a diagram or picture, and such graphic distributions may convey much of the information in the most simple, rapid, and concise way.

In all graphs of frequency distributions, the values of the variate are laid down on a horizontal line, the X-axis (or abscissa), starting at the lower left-hand corner of the diagram, and frequencies are measured from the same point upward along the vertical Y-axis (or ordinate). For purposes of such plotting, the axes may be called the X- and f-axes, X being a conventional symbol for the value of a variate and f for its frequency which in these cases takes the place of the conventional Y in mathematical curve plotting.

Aside from a few exceptional cases, the initial value of the f-axis should be 0. It would be preferable also to begin the X-axis scale at 0; but if the lowest X of the distribution is a large number, as it often is, this means that a large blank space will occur to the left of the diagram. In such cases it is usually advisable to begin the X-scale at an arbitrary number shortly below the lowest observed value of X.

The simplest way to construct such a diagram is to place dots at points defined by the pairs of corresponding X- and f-values. These are not very satisfactory because the scattered dots do not readily suggest a pattern and the magnitudes involved are not readily grasped (see Figs. 2A and 4A). They are also liable to confusion with a scatter diagram, which is quite different from a frequency distribution (see page 218).

A dot diagram of this sort is changed into a frequency polygon by drawing a line from each dot to the next (Figs. 2B, 2C, and 4B). The line is preferably joined to the edge of the diagram by including on each side a

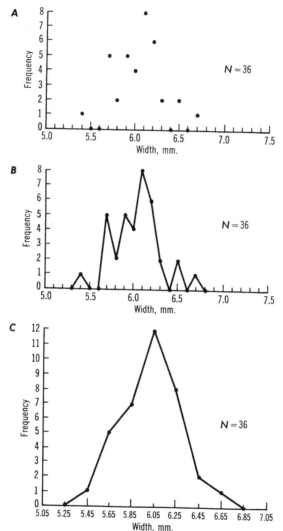

FIGURE 2. Graphic representation of a continuous frequency distribution. Width of the last upper molar in the fossil mammal *Acropithecus rigidus* (data of Example 17A). *A:* the raw data plotted by dots. *B:* the raw data as a frequency polygon. *C:* frequency polygon of the data regrouped to interval .2 mm.

value of the variate for which the frequency is 0. In a frequency polygon (if both ends have 0 frequency), the whole area is proportional to the total frequency, and the distances from the points (usually angles of the polygon) to the X-axis are proportional to the class frequencies. This type of diagram has the disadvantages that the verticals to the X-axis are proportional to frequencies only at these points, where the frequency is supposed to be concentrated, and that the areas above the X-axis for the given classes—magnitudes generally clearer to the eye than linear distances—are not proportional to the frequencies. The principal advantage of the frequency polygon is that it nearly resembles a curve, the theoretical form to which the angular pattern is to be related. This advantage is generally not great for anyone accustomed to the use and characters of distributions, and

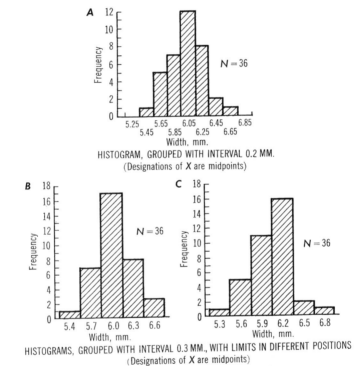

FIGURE 3. Histograms of a continuous frequency distribution (same data as in Fig. 2). *A:* regrouped to interval .2 mm., corresponding to the polygon of Fig. 2C. *B:* regrouped to interval .3 mm., showing change of form by broadening of class intervals. *C:* regrouped to interval .3 mm., with the midpoints taken at different values.

frequency polygons are not very commonly used. They should particularly be avoided if there are abrupt changes of slope, which tend to make the polygon misleading.

The commonest and for most purposes the best graphic representation of a frequency distribution is by a histogram (Figs. 3 and 4c). To make a histogram, a vertical line is erected at each class limit, and these are connected across their tops by horizontal lines at a height equal (on the f-scale) to the frequency of the class.

If raw measurements are used, it should be remembered that these are really midpoints of an implied range of true values. The vertical line should be erected at the limits of the implied range, with the measurement itself shown as the midpoint of this range.

In the same way, if secondarily grouped measurements are used, the divisions between the classes should be at the limits of the implied true range of this class. Thus, in Example 17 if the measurements were grouped in classes of .2 mm., the first grouped class would be called 5.4–5.5. The midpoint of this class is 5.45 and the true limits are 5.35000... –5.54999....

EXAMPLE **17**. Frequency distributions.

A. Widths of last upper molars of the extinct mammal *Acropithecus rigidus*. (Original data)

MEASUREMENT (MM.)	FREQUENCY	MEASUREMENT (MM.)	FREQUENCY
5.4	1	6.1	8
5.5	0	6.2	6
5.6	0	6.3	2
5.7	5	6.4	0
5.8	2	6.5	2
5.9	5	6.6	0
6.0	4	6.7	1

B. Size of sets of song sparrow *Melospiza melodia beata*. (Data from Nice, 1933)

NUMBER OF EGGS (X)	NUMBER OF NESTS (f)	NUMBER OF EGGS (X)	NUMBER OF NESTS (f)
1	1	4	25
2	2	5	14
3	19		

Sometimes in the final drawing, vertical lines within the diagram are omitted and only the external boundaries of the histogram drawn, but such a diagram is usually harder to read.

The f-scale is marked to the left of the diagram either in units or in convenient multiples, e.g., by fives or tens. The unit of the X-scale should be the class interval, and designations should be either at (true) limits or at midpoints. The latter is usually preferable, and in either case the numbers should be so placed as to leave no doubt as to their positions in the classes. In frequency polygons and in graphs of discontinuous variates the designations of the X-scale must represent midpoints.

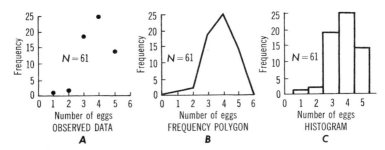

FIGURE 4. Graphic representation of a discontinuous frequency distribution. Number of eggs in nests of the bird *Melospiza melodia beata* (data of Example 17B). *A:* the raw data plotted by dots. *B:* the same plotted as a frequency polygon. *C:* the same plotted as a histogram.

In a histogram each class is represented by a rectangle. The horizontal widths of these are all the same, and their heights are proportional to the class frequencies. The areas are therefore also proportional to the class frequencies, the great advantage of this sort of diagram.

The same distribution may be represented by histograms of markedly different superficial aspect, depending on where the classes are placed and on their magnitude. If the class interval is increased, the frequencies of many or of all classes will also be increased. The histogram with larger class intervals may therefore rise higher (Figs. 3A, B, and C). This is characteristic only of the secondary grouping and not of the distribution so that it is necessary to recognize essentially the same types of curves with different groupings and also to employ, as far as possible, the same class intervals for two distributions that are to be compared. In placing the classes on the scale, the position that gives the most symmetrical result is usually preferable.

The Meaning of Distribution Patterns

If a frequency polygon of measurements were based on a series of observations that could be multiplied indefinitely and at the same time made more refined at will, it would be possible to decrease the class interval and at the same time to increase the number of observations so that the class frequencies remained reasonably large. Continuing this process, a condition would be reached when the dots, the angles of the polygon, were so close together that they became indistinguishable, for the horizontal distance between any two successive dots is equal to the class interval and this is made indefinitely small. The polygon would then cease to have visible corners and angles and would become a smooth curve. The same procedure applied to a histogram would produce the same result, since the horizontal lines forming the tops of the rectangles would become shorter and shorter with decrease of the class interval, to which they are equal, until eventually they would appear only as points which would coalesce and form a curve.

This curve that is approached as a limit when the class intervals are decreased and the total frequency increased is the ideal pattern of the corresponding frequency distribution. In practice the curve cannot be obtained in this way; for no method of measurement is sufficiently refined for the indefinite reduction of the class interval, nor can the number of observations ever be really increased indefinitely. The true ideal curve would, indeed, only be reached when the class interval reached zero and the total frequency infinity, an obvious impossibility in practice. Yet the approach of the distribution to this purely theoretical limit is a real phenomenon, and the theoretical curve is the best possible representation of the distribution as a whole. The study of a frequency distribution thus commonly involves setting up a hypothesis as to the curve represented by the data of the actual observations and estimating the mathematical constants by which the curve can best be defined.

This concept of a theoretical curve which is approximated by the observed frequency distribution applies, obviously, only to continuous variates. Discrete variates have an irreducible class limit, so that no matter how accurate the observations, the theoretical picture is always a series of points and never a smooth curve. In a mathematical sense, the theoretical frequency curve for continuous variates is itself continuous (it can be represented by a smooth curve), while for a discrete variate the theoretical function is defined only for distinct values of the independent variable. Therefore it can be represented correctly only as a series of unconnected points.

General Types of Distribution Patterns

In the great majority of cases the characters—anatomical, physiological, psychological, or other—with which a zoologist deals are distributed in such a way that certain classes of these variates are more frequently observed than others and that the frequency becomes progressively less as the classes are farther in either direction from these most common values. This fact, so often seen in dealing with zoological data that it becomes a basic assumption of the science, is often called Quételet's law or, better, Quételet's principle. As with most of the so-called laws of biology and zoology, there are some exceptions; but these are rare and usually belong to certain distinctive classes of data so that zoological variates may generally be assumed to fall into a pattern approximately specified by Quételet's law.

A large number of specific types of curves have been observed in frequency distributions of zoological variates. The distinction and specification of many of these require such extensive data and such intricate mathematical procedures that they are of little or no use to the zoologist, and even if not entirely beyond his powers, such work would be a waste of time and effort. Moreover, many of these curves—most of those commonly involved in zoological work—approach a few standard types so closely that they are most usefully studied as approximations of these standard curves and specified in terms of the latter with, if necessary, estimates of deviation from them.

All such curves can be classed in four general groups:
1. Those high at the midpoint and sloping away nearly symmetrically on each side of this.
2. Those with a high point not at the midpoint of the distribution and sloping away from this with moderate asymmetry.
3. Those with the high point near or at one end of the distribution and strong asymmetry.
4. Those with a low point within the distribution and rising at both ends.

These are not absolutely clear-cut categories: 1 grades into 2, 2 into 3, and 3 into 4; but a given distribution can usually be referred to one of these general types.

Absolute symmetry almost never occurs in a limited set of observations, indeed so rarely that its appearance may be viewed with suspicion. Distributions nearly enough symmetrical to be considered as essentially so are, however, common. This is the ideal form of most animal characters that follow Quételet's law. Numerous examples appear in the pages of this work, and Example 18, given in graphic form in Fig. 5, serves to illustrate the type here.

Moderately asymmetrical curves are spoken of as being moderately "skewed" and may be loosely defined as those in which the highest frequency is definitely not near the middle or near the ends of the distribution. Skewed curves in which the right-hand limb tapers off more gradually than the left-hand limb, hence in which the class with highest frequency is below the middle of the distribution, are said to be positively skewed, or skewed to the right. Similarly those with the left-hand limb longer and the class with highest frequency above the middle are negatively skewed or

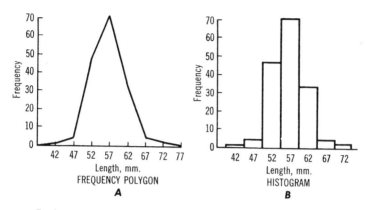

FIGURE 5. An essentially symmetrical frequency distribution following Quételet's law. Lengths of the fish *Pomolobus aestivalis* (data of Example 18). *A:* as a frequency polygon. *B:* as a histogram. The degree of asymmetry observed is usual in samples of essentially symmetrical populations.

EXAMPLE 18. Frequency distribution. Lengths of specimens of the glut herring *Pomolobus aestivalis*, caught in Chesapeake Bay, October 16-31. (Data from Hildebrand and Schroeder, 1927)

LENGTH (MM.)	FREQUENCY
40–44	1
45–49	4
50–54	47
55–59	71
60–64	33
65–69	4
70–74	2

This is a slightly asymmetrical distribution, since classes at corresponding distances on each side of the middle do not have the same frequencies, and the total frequency of 52 below the middle class is greater than that above it, which is 39. It is, however, an approximation to symmetry, about as close as is commonly expected unless the total frequency is very large. The highest frequency, 71, is in the middle class, and the distribution does taper off steadily for equal distances on both sides of this.

skewed to the left. Interesting examples such as Fig. 6 of the two types are furnished by the data in Example 19 on samples of the same subspecies of fish caught at different times in the year.

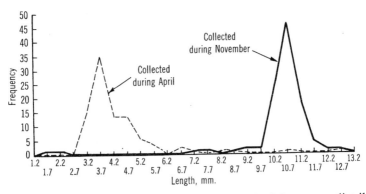

FIGURE 6. Moderately but significantly asymmetrical frequency distributions. Lengths of the fish *Parexocoetus brachypterus hillianus* (data of Example 19). The polygon in continuous outline represents the November catch and is skewed to the left. The polygon in broken outline represents the April catch and is skewed to the right.

The distribution for November is skewed to the left, or negatively, since the class with highest frequency is well above the middle class; it is the 19th of 23 classes. The distribution for April is skewed to the right, or positively, since the class with highest frequency is far below the middle class; it is 2nd of 20 classes. The skewing in this instance is so well marked that it might, especially for the April sample, be considered an example of extreme rather than of moderate skew.

The biological significance of the skewing and its reversal at different seasons in this example are clearly related to the existence of a restricted spawning season and to changing growth rates. If it were possible to gather a sample of these fishes all of the same age, the curve would almost surely be symmetrical. As in many cases of marked asymmetry, the asymmetry in this example is probably due to heterogeneity of the sample. It is not a characteristic of length distribution in specimens essentially similar in everything but length.

Every gradation from no skewing to extreme skewing may be encountered. Indeed, as will be shown in Chapter 6, a slight degree of skew, usually to the right, is to be expected with many zoological variates and may usually be ignored. A large skew, however, demands recognition

EXAMPLE 19. Frequency distributions. Standard lengths of samples of the flying fish *Parexocoetus brachypterus hillianus*, collected in the Atlantic during two different months. (Data from Bruun, 1935)

STANDARD LENGTHS (CM.)	FREQUENCIES A. NOVEMBER	B. APRIL
1.5– 1.9	1	0
2.0– 2.4	1	0
2.5– 2.9	0	0
3.0– 3.4	0	15
3.5– 3.9	0	35
4.0– 4.4	0	13
4.5– 4.9	0	13
5.0– 5.4	0	5
5.5– 5.9	0	3
6.0– 6.4	0	0
6.5– 6.9	0	2
7.0– 7.4	1	1
7.5– 7.9	1	0
8.0– 8.4	0	1
8.5– 8.9	1	1
9.0– 9.4	2	0
9.5– 9.9	2	0
10.0–10.4	21	0
10.5–10.9	46	1
11.0–11.4	19	0
11.5–11.9	4	0
12.0–12.4	0	0
12.5–12.9	0	1

and explanation either as a character of the variate or a result of peculiarities in the sample.

The most extreme form of skew is one in which a terminal class (in practice, usually the lowest) has the highest frequency. Such a curve, with its high point at one end and hence with the frequencies dropping at first rapidly and later increasingly slowly to zero, is usually called a J-shaped distribution; it does not follow Quételet's law. With most characters of animals, especially with continuous variates, it will be found that an apparently J-shaped distribution is simply a humped, very asymmetrical distribution with the class intervals too large. Thus, if the April distribution of Example 19 were grouped with interval 2.0 instead of .5, it would appear as in Example 20.

In zoology, true J-shaped distributions are usually of discontinuous variates, anatomical or otherwise, although this is not the commonest

EXAMPLE 20. Regrouping of Example 19B, with interval increased from .5 to 2.0.

LENGTH (CM.)	FREQUENCY
3.0– 4.9	76
5.0– 6.9	10
7.0– 8.9	3
9.0–10.9	1
11.0–12.9	1

This appears to be a J-shaped distribution but is really not, since splitting up the first class into much smaller classes, as in Example 19, would show the frequencies falling steadily to 0 at this end of the series, as they do, although more slowly, at the other end.

---

form even for discontinuous variates. Simply from the logical and biological aspects of the problem, it is usually obvious when a terminal class may be expected to have the highest frequency and hence when a J-shaped curve is the probable pattern of the frequency distribution. Such variates are either structures, events, etc., not normally occurring but occasionally observed in varying number or, more rarely, variates in which the usual value is never exceeded or is always reached or exceeded. Example 21 gives two such distributions (see also Figs. 7 and 14).

U-shaped distributions, rising to a peak at both ends, are of rare occurrence with any type of data and are almost nonexistent for variates used in zoology. Some of the distributions of the special sort mentioned at the end of this section may approximate the U-shape, however. In other instances, an apparently U-shaped distribution may result from incorrect sampling or similar errors in technique, resulting in the fusion of two curves strongly skewed in opposite directions. Thus, the combined data for the whole sample of *Parexocoetus brachypterus hillianus* from which the two skewed distributions in Example 19 were drawn could be presented in this form:

Standard length (mm.)	Frequency
1.5– 4.4	66
4.5– 7.4	28
7.5–10.4	39
10.5–13.4	83

As it stands, this is U-shaped; but this is because most of the specimens were collected in two months, November and April, when their distributions were strongly and opposingly skewed. If the specimens had been

EXAMPLE 21. J-shaped distributions.

A. Number of times individual female snowshoe hares were live-trapped. (Data from Aldous, 1937)

NUMBER OF TIMES TRAPPED	NUMBER OF HARES
1	365
2	163
3	103
4	58
5	33
6	14
7	6
8	4
9	1
10	0
11	0
12	0
13	1

B. Number of dorsal soft fin rays in the fish *Caranx melampygus*. (Data from Nichols, 1935)

NUMBER OF FIN RAYS	NUMBER OF FISHES
20	1
21	2
22	6
23	11

Obviously more individuals will be trapped once only under ordinary conditions than will be trapped two or more times, so that a J-shaped distribution, as actually occurs in A, is to be expected. This cannot be made into a moderately skewed distribution by splitting the classes, since an animal cannot be trapped a fractional number of times.

As it stands, B is a J-shaped distribution skewed to the left. The species usually has 23 such rays, and as far as these data show, it never has more but may have less. It is probable, in this and in most analogous cases, that the J-shape is illusory, however, and is only a chance result in a small sample. It is highly probable that further search would result in finding some individuals with more than 23 rays, for most distributions of this sort are only moderately skewed and there is no obvious reason why this should be extremely skewed. Most J-shaped distributions, in which the class with highest frequency is not 0 (or 1 in cases like example A), would probably lose the J-shape if a very large total frequency were available; and this pattern in such a case is distinguished from the sort of asymmetrical distributions, previously discussed, only by being still more skewed.

collected in about equal numbers at all times of the year or if only those collected at one time were counted, the distribution would not be U-shaped. Moreover, if the class intervals were made smaller, as they should be, it would be obvious that this is not a U-shaped curve but two moderately skewed curves. An apparently U-shaped distribution of zoological variates is usually an indication of faulty procedure or of heterogeneity of the material included.

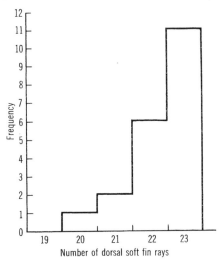

FIGURE 7. An extremely asymmetrical or J-shaped distribution. Number of dorsal soft fin rays in the fish *Caranx melampygus* (data of Example 21B). Such a left-skewed distribution is less common than one skewed to the right (e.g., Fig. 14).

There are a few zoological variates that tend to fall into a curve more complex than those already mentioned. Generally, the presence of two high points on a curve is a sign that the sample is heterogeneous and that the curve is really composed of two or more curves that should, if possible, be separated. An exception to this rule is the possibility that the variate may naturally take only low or high values, a rarity with zoological materials. Thus the Patagonian rhea frequently lays one or a few eggs in several isolated spots but otherwise tends to concentrate a large number of eggs in one spot, a crude nest. Figures on this do not seem to be available; but the observed habit suggests a hypothetical distribution of this general form, as in Example 22.

The question of whether a truly homogeneous population can have two high points of frequency is really a logical rather than a biological one. By a homogeneous population showing two frequency peaks with respect to a given measurement, we mean that the individuals clustered around one peak differ, on the average, from those clustered around the other peak *only* with respect to the measured variable and no other. As, in practice, it is impossible to demonstrate that no other difference can ever be found no matter how hard one looks, the problem is insoluble. When confronted with such a frequency distribution, the zoologist ought to suspect variation in some other factor and, within reason, attempt to uncover it. There is, however, no guarantee that it ever will be found.

EXAMPLE 22. A U-shaped curve. Hypothetical data on sets of rhea eggs.

NUMBER OF EGGS	NUMBER OF SETS
1–5	20
6–10	5
11–15	10
16–20	15
21–25	20
26–30	10
31–35	5

The curve begins high, falls, then rises to a second apex, then falls again. Even such cases, however, may properly and most conveniently be considered as composed of two separate curves—in the above example, a J-shaped curve of sets not in nests (1-5 eggs) and an approximately symmetrical curve of sets in nests (more than 5 eggs).

In a broader sense, a population showing a bimodal frequency distribution is *always heterogeneous* because there are two fairly distinct subpopulations with respect to the measured character itself.

Cumulative Distributions

In the distributions previously discussed in this chapter, the frequency within each class is given. It is sometimes more convenient, and may be more directly related to a problem in hand, to give the total frequency below (or occasionally above) each class. Such distributions are called *cumulative*. The construction of such a distribution from the frequency distribution itself is quite simple. The cumulative frequency of a given class (usually symbolized by "C.F." to distinguish it from the usual frequency) is the sum of the frequencies of all the classes up to and including the class in question. If the frequency distribution starts with a class not represented in the sample measurements, the cumulative distribution will have an initial value of 0. It will then rise until, in the last class represented in the sample, it reaches a frequency equal to the total number of measurements, or sample size.

EXAMPLE 23. Ordinary and cumulative distributions. Data from Example 18.

ORDINARY DISTRIBUTION		CUMULATIVE DISTRIBUTION			
		A.		B.	
LENGTH (MM.)	FREQUENCY	LENGTH	C.F.	LENGTH	C.F.
40–44	1	Below 44.5	1	Above 39.5	162
45–49	4	,, 49.5	5	,, 44.5	161
50–54	47	,, 54.5	52	,, 49.5	157
55–59	71	,, 59.5	123	,, 54.5	110
60–64	33	,, 64.5	156	,, 59.5	39
65–69	4	,, 69.5	160	,, 64.5	6
70–74	2	,, 74.5	162	,, 69.5	2

Example 23 shows an ordinary frequency distribution and the two cumulative distributions, one cumulative from below, the other from above, which can be constructed from it. The two cumulative distributions are approximately mirror images of each other and would be exactly so were the frequency distribution perfectly symmetrical. The figure given for the terminal class (the last class in cumulative distributions of type A and the first in those of type B) is always the total frequency. Note that classes are designated by limits and that care should be taken to use the real limits and not, as is usually done, the conventional limits. Cumulative distribution A in the example would usually have the classes listed as "below 45," "below 50," etc.; but this is wrong.

Figure 8 is a graphical representation of these two cumulative distributions.

Relative Frequencies

Up to this point the word "frequency" has signified the absolute number of observations in a given class. In the future discussion of statistical concepts, an extremely important consideration will be not the absolute but the *relative frequency*; that is, the proportion of the observations falling in a given class. Clearly, the relative frequency of a class is the number in that class divided by the total number of observations. Such proportions are usually given not as percentages but as a decimal fraction. Thus, the relative frequency of a class may range from 0 (none of the observations lie in the class) up to 1 (all of the observations lie in the class). The decimal fraction thus obtained is a pure number accurate to an infinite number of places since it is the ratio of two integers.

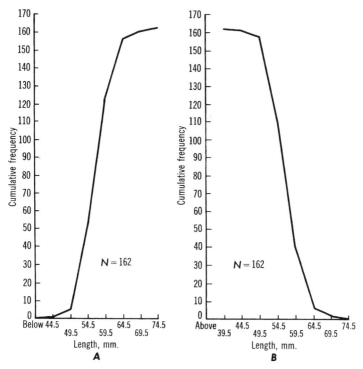

FIGURE 8. Graphs and cumulative distributions. Lengths of the fish *Pomolobus aestivalis* (data of Example 18, as rearranged in Example 23). *A:* frequencies cumulative from below. *B:* frequencies cumulative from above.

One obvious advantage of relative frequencies is that they make any two distributions directly comparable, even though these are based on different numbers of observations. There are numerous other advantages which will become apparent later, the most important of which is the direct relation between probability and relative frequency. In order to avoid any ambiguity the following notation will be adopted:

f will signify the absolute frequency (number) in a given class.

N will signify the total frequency (total number) of observations.

$\dfrac{f}{N}$ will signify the relative frequency for any given class.

From this definition it follows that the sum of all the relative frequencies in a frequency distribution is 1, since the sum of all the absolute frequencies is equal to N, the total frequency. Further, the cumulative distribution of relative frequencies will then express the *proportion* of all the observations below (or above) a given class and will rise from a relative frequency of 0 in the initial class to 1 in the final class. This property of a cumulative distribution, that of giving the proportion of observations falling below or above a certain value, will be extremely important in the discussion of probability and statistical testing.

CHAPTER FIVE

Measures of Central Tendency

Arithmetic mean

Most zoological variates are so distributed as to be more frequent near some one value and to become less and less frequent in departing from this value in either direction, a fact of experience summed up in Quételet's principle (page 54). Clearly the most important things to observe and to measure in such a distribution are (1) the point around which the observations tend to cluster and (2) the extent to which they are concentrated around this point. The most widely used measure of central tendency is the arithmetic mean, usually called simply the "mean."

This is by far the most common statistic, and everyone who uses numerical data at all has at some time calculated a sample mean. The sample mean is an average obtained by adding together all the observed values and dividing by the number of observations. Its general formula is

$$\bar{X} = \frac{\Sigma X}{N}$$

where $\bar{X}$ = the mean (arithmetic only)
Σ = a sign of operation, indicating that all the data represented by the symbol or symbols following it are to be added together
X = any given value of the variate
N = the number of observations made (total frequency).

These symbols and a few others explained as they appear are used consistently throughout the present book, and learning this shorthand notation greatly simplifies not only the explanation of these processes but also their use. An instance of the simplest possible calculation of a mean is given in Example 24.

A word about calculating formulae is required here. Often the formula given as defining a statistic is not the most convenient one for calculation. The best form for calculation will depend upon whether a desk calculator capable of adding, subtracting, multiplying, and dividing is available. In every case we have given both the defining expression and either a special machine formula or special hand-calculating formula, if one or the other of these is not identical with the definition. In the case of the arithmetic mean, the formula first given is most convenient for machine work.

	MEASUREMENTS (X)
	3.0 mm.
	2.8
	3.4
	3.2
	3.0
EXAMPLE 24.	2.9
Calculation of the	2.6
arithmetic mean of	3.3
the length of the third	3.1
upper premolar of	2.9
the extinct mammal	2.9
Ptilodus montanus.	3.0
(Original data)	2.8
	2.9
	2.7
	2.9
	3.1
	2.8
	3.0
	3.1
	3.0

$\sum X = 62.4$ (sum of measurements taken)
$N = 21$ (number of measurements taken)
$$\bar{X} = \frac{\sum X}{N} = \frac{62.4}{21} = 2.97$$

If the number of observations is large, say, several hundred, and a calculating machine is not available, it is better to calculate the mean with a modified but arithmetically equivalent formula:

$$\bar{X} = \frac{\sum fX}{N}$$

in which f designates class frequency. The calculation made in this way is shown in Example 25. If the number of *classes* is very large (generally, if it exceeds 20 or 25), it may, however, be advisable to increase the class interval and group the data more broadly. In such a case the operation is carried out by the same formula, remembering that X is the true class midpoint. This is done in Example 26, with the same data as that of the last two examples for the sake of comparison although in ordinary practice the secondary grouping would not be justified in this case.

EXAMPLE 25. Same data as in Example 24, recast as a frequency distribution and mean based on this.

MEASUREMENT (X)	FREQUENCY (f)	FREQUENCY × MEASUREMENT (fX)
2.6	1	2.6
2.7	1	2.7
2.8	3	8.4
2.9	5	14.5
3.0	5	15.0
3.1	3	9.3
3.2	1	3.2
3.3	1	3.3
3.4	1	3.4
$\sum f = N = 21$		$\sum fX = 62.4$

$$\bar{X} = \frac{\sum fX}{N} = \frac{62.4}{21} = 2.97$$

EXAMPLE 26. Same data as in last two examples, grouped with interval .2 mm. for calculation of the mean.

CLASS	CLASS MIDPOINT (X)	CLASS FREQUENCY (f)	FREQUENCY × CLASS MIDPOINT (fX)
2.6–2.7	2.65	2	5.30
2.8–2.9	2.85	8	22.80
3.0–3.1	3.05	8	24.40
3.2–3.3	3.25	2	6.50
3.4–3.5	3.45	1	3.45
		$\sum f = N = 21$	$\sum fX = 62.45$

$$\bar{X} = \frac{\sum fX}{N} = \frac{62.45}{21} = 2.97$$

The total here obtained differs slightly from that based on the original data because of inaccuracies introduced by the grouping.

The calculation of the mean from secondarily grouped data is not advisable unless numerous means are to be calculated each from a large number of classes. If only a few means are to be obtained, the time saved, if any, by secondary grouping may be more than offset by the chance of error involved in incorrect grouping. Grouping adds a complexity to the operation, increasing the opportunity for a mistake.

Calculation of the mean from grouped data depends on the assumption that each class midpoint does not differ significantly from the mean value of the observations that fall in the class. This assumption is more likely to be true with small class intervals than with large, because the midpoint cannot differ from the mean for the class by more than half the class interval. It is also more likely to be true with high class frequencies than with low, because if many observations enter into a single class they are likely to be well scattered in it and hence to have a mean value near its midpoint. Grouping always involves some inaccuracy, even when it is only the grouping of the original measurements of a continuous variate; but if these sources of inaccuracy are kept in mind, its extent seldom is great enough to affect the final result significantly. If there is any question about this, it is invariably true that the mean as calculated from grouped data is within one-half class interval of the true mean, and usually it will be within one-tenth class interval or even less. Despite the unduly large interval of .2 in the example just given, the calculated mean is probably not more than .01 from the true mean.

Many older works on biometry, including the first edition of this book, discussed the use of the so-called "assumed-mean" method of calculation. We have avoided this as it does not simplify calculation but rather makes it more complex, again increasing the chance for errors.

The Mean of Means

It is valid to base a mean on two or more other means; but this requires logical consideration of what is being done and in most cases involves a correction, or weighting, if the means are derived from distributions with different total frequencies. Suppose that a sample of animals is composed of a number of subsamples each of different size. For example, animals may be collected in different localities as in Example 27. Two different grand means can be calculated from these data. First, each subsample mean may be treated as the basic variate and the ordinary formula

$$\bar{\bar{X}} = \frac{\sum \bar{X}}{N}$$

employed. Here $\bar{X}$ is the value of the *mean for each locality* and $\bar{\bar{X}}$ is the grand mean. In a sense, the original observations on which the subsample means are based have been forgotten. In Example 27 this calculation yields a grand mean of 61.25 millimeters. The grand mean obtained by averaging the individual means is not the same logically or arithmetically as taking the mean of all the observations. If one wishes to calculate the average tail length of all the specimens in the three subsamples together, it is necessary to weight each subsample mean by its total frequency.

EXAMPLE 27. Calculation of the mean of means by two methods on tail length in the deer mouse *Peromyscus maniculatus bairdii*. (Data from Dice, 1931)

LOCALITY	SAMPLE SIZE (N)	MEAN TAIL LENGTH (MM.)
Ann Arbor	106	57.20
Alexander	86	60.43
Grafton	78	66.13

Method 1

$$\bar{\bar{X}} = \frac{\sum \bar{X}}{N} = \frac{57.20 + 60.43 + 66.13}{3} = 61.25$$

Method 2

$$\bar{\bar{X}} = \frac{N_1 \bar{X}_1 + N_2 \bar{X}_2 + N_3 \bar{X}_3}{N_1 + N_2 + N_3} = \frac{106\,(57.20) + 86\,(60.43) + 78\,(66.13)}{270}$$
$$\bar{\bar{X}} = 60.81$$

This can be done by multiplying each subsample mean by its corresponding frequency (N), adding the figures thus obtained, and dividing by the sum of the frequencies, which can be expressed in a formula by

$$\bar{\bar{X}} = \frac{N_1 \bar{X}_1 + N_2 \bar{X}_2 + \cdots}{N_1 + N_2 + N_3 + \cdots}$$

Since N for the whole sample is equal to $N_1 + N_2 + N_3 + \cdots$ and since $N_1 \bar{X}_1 = \sum f_1 X_1$ and $\sum fX$ for the whole sample is $\sum f_1 X_1 + \sum f_2 X_2 + \sum f_3 X_3 + \cdots$, it is evident that this formula is merely a convenient form for the special case exactly equivalent to the usual formula

$$\bar{X} = \frac{\sum fX}{N}$$

This weighted formula for Example 27 gives 60.81 as the grand mean which is clearly different from the result of the first calculation.

The first calculation treats each locality as a separate entity and the grand mean thus calculated ignores the frequency distribution of tail length within each locality. The second calculation ignores the differences between localities entirely and treats all the individuals as a single sample.

The zoologist must choose his procedure on the basis of the particular problem in mind.

Median

The median of a sample is the observation which has an equal number of observations below and above it, but this strict definition obviously can apply only to a sample with an odd number of observations. For instance in Example 25 (page 67), N is 21 so that the middle observation is the 11th from either end of the sample distribution, in this case the observation with the value 3.0. If there were, say, 22 observations, the 11th observation would have 10 below it and 11 above it. There is no single middle observation when N is even, so that in a sample containing an even number of observations the sample median, as such, does not exist. For such cases the usual quantity employed is the value half-way between the two middle observations. As an example, assume that there are 100 observations and that after they have been arranged in ascending order, the 50th observation is 61.3 and the 51st 61.7. Then the median lies half-way between these values and is equal to 61.5.

By its definition the sample median cannot be expressed to more significant figures than the observations themselves. In Example 25 the 11th observation from the smallest is 3.0 so that the sample median is 3.0. It is possible, however, to calculate a more refined quantity so that more decimal places result, in the following way:

In Example 25, the middle value is the 11th, which is the lowest of the 5 in the class with midpoint 3.0 and implied true limits 2.95000...–3.04999.... For purposes of calculation, it is assumed that the observations within the group are evenly distributed. Thus with $f = 5$, as here, it may be assumed that the middle one of these 5 is at the class midpoint, that those on each side are at a distance equal to 1/5 class interval above and below the midpoint, and that the last two are at distances of 2/5 class interval above and below the midpoint. This leaves a distance of $1/2 \times 1/5$ or 1/10 class interval between the most divergent observations and the true class limits. Such considerations lead to the following general formula for a more refined estimate of the median:

$$\text{Median} = L_l + \frac{(n - 1/2)i}{f}$$

where L_l = the true lower limit of the class in which the sample median lies
n = the serial number of the desired observation within the class
i = the class interval
f = the absolute frequency of the median class

In Example 25 the median class has the true limits 2.95–3.05 (actually

2.95000 .... −3.04999 ..., but such exactitude is unnecessary). Hence $L_l = 2.95$. The median observation is the 11th, and there are 10 below the median class; so the median observation is the first in that class, hence $n = 1$. Also, $i = .1$ and $f = 5$. The median is then estimated by

$$\text{Median} = 2.95 + \frac{(1 - 1/2) \cdot .1}{5} = 2.95 + .01 = 2.96$$

Calculation from the same data grouped with interval .2 (Example 26) gives

$$L_l = 2.95 \quad n = 1 \quad i = .2 \quad f = 8$$

$$\text{Median} = 2.95 + \frac{(1 - 1/2) \cdot .2}{8} = 2.95 + .01 = 2.96$$

The preceding formula assumes that the median is found by counting up from the lower end of the distribution. There is no particular advantage in doing so, and the same result can be achieved in counting down from the upper end. In this case the formula is

$$\text{Median} = L_u - \frac{(n - 1/2)i}{f}$$

The sample median is usually easier to calculate than the mean, and it has a few other advantageous properties. It is less distorted by and less sensitive to extreme values than is the mean—often advantageous although this may also be a disadvantage. It can be calculated from imperfect data—for instance, when the more divergent observations are grouped as so much "and over" or "and under"—while the mean cannot. (But the mean can also be calculated from data inadequate for a median—for example, from means of subsamples together with the sizes of these subsamples.) Either the mean or the median can be calculated from any properly collected and tabulated data.

An observation selected at random is as likely to be above the median as below. The sum of the deviations (ignoring negative signs) is less about the median than about any other point. These properties make the median an advantageous average in certain cases, but in ordinary practice they are outweighed by disadvantages. It is impossible to base a median on other medians. Medians cannot enter into many important algebraic calculations. They cannot be compared so simply and accurately as can means. Their standard errors (see Chapter 9) are larger than for means. In general, the mean is a far more important and useful average than is the median. One essential use of the median (which also requires the use of the mean) is to approximate the value of the mode, as explained in the next section.

Sometimes the median is a more logical quantity than the mean for a given purpose. For example, the median income of the American population provides a kind of information which the mean cannot. Whereas in the case of the mean a single millionaire will compensate for many unemployed persons, the median will not be thus affected for it is the value below which half the population falls. For this particular case, the mean is always well above the median, giving a somewhat false picture of the economic condition of the population. This same reasoning applies to the size of a prey species. In fish populations especially, when a prey species reaches a certain size, the predator can no longer feed on it. In such a case, the median size will give a better idea of the available food for the predator than will the mean.

In any perfectly symmetrical distribution, the median is equal to the mean. Since actual distributions are rarely completely symmetrical, however, there is usually a small difference between these two averages.

Mode

For distributions following Quételet's principle, it is a fair generalization to say that an average is a value around which observations tend to cluster. The idea of the observations being crowded toward an average value applies well to the median and almost as well to the mean in such cases. In extremely skewed or J-shaped distributions, however, an average like the mean is not really a nucleus or a point of concentration of values. It is still true that the observations are arranged around the mean in such cases, and the calculation of the mean is still an essential part of their study, but the point around which they are really clustered may well be removed from the mean.

For studying distributions in which there is a significant degree of skewness, it is therefore necessary to have another sort of average, one that really designates in all cases a center of clustering or piling up of observations. Such a value is that at which the frequency is greatest, and this average is called the mode.

The sample mode is often very unsatisfactory because sampling fluctuations will affect it greatly. In a frequency distribution so grouped as to approach a fairly smooth curve—that is, with one class of outstanding frequency and with the frequencies of the other classes falling away evenly and definitely from this—the sample mode is a single value obvious on inspection. Thus in Example 17A (page 51), the mode is evidently the class 6.1. In other samples, however, there may be more than one modal class, as in Example 25.

One accurate way to estimate the mode is to fit to a distribution the closest possible ideal mathematical curve and then to calculate the point at which this curve has the highest ordinate. Such close curve fitting is an extremely complex process and requires more extensive data than are commonly available in zoology. The most accurate estimation of the mode, therefore, has no practical value in zoology.

There are, however, several methods of estimating the mode that are useful in zoology and that give values accurate enough for practical purposes. The first and least refined of these is that of grouping a distribution so that it is regular and has one class of outstanding frequency, then taking that class as an estimate of the mode. Such methods of grouping were discussed at length in the last chapter.

A second method takes advantage of the fact that for moderately skewed curves the median lies at about one-third of the distance from the mean to the mode. This empirical rule, found to be closely followed by all but strongly skewed distributions, depends on the fact that the mode is not at all affected by extreme values, the median is somewhat affected, and the mean is most strongly affected, the effect on the last being about one and one-half times as strong as on the median. This relationship gives an estimate of the mode:

$$\text{Mode} = \text{mean} - 3(\text{mean} - \text{median}) = 3\,\text{median} - 2\bar{X}$$

This formula cannot be used for extremely skewed distributions, for which approximation by inspection is the only easy and practical method.

There are several other methods giving still closer approximations; but they are also more complex mathematically, and the two mentioned suffice for any ordinary zoological work.

In the distribution of Example 18 (page 55) the mode is seen to lie in the class 55–59, i.e., within the true limits $54.5000\ldots - 59.4999\ldots$. The mean of this distribution is 56.66 and the median 56.54. Then, since

$$\text{Mode} = \text{mean} - 3(\text{mean} - \text{median})$$

$$\text{Mode} = 56.66 - 3(56.66 - 56.54) = 56.66 - .36 = 56.30$$

This gives a reasonably accurate value which is within the class selected as modal by inspection.

The importance of the mode is that since it is the value taken by the greatest number of observations, it is in that sense the most typical. It can be approximated roughly by simple inspection with no calculation, and it is independent of extreme values. It has the serious disadvantage that a reasonably efficient estimate is practically impossible with limited data and is in any case extremely difficult and that, like the median, its usefulness for further calculations and for comparisons is far less than that of the mean. It may have the further disadvantage that for very small samples, such as are common in zoology, the mode may be quite indeterminate or may even be said, as far as a given concrete sample is concerned, not to exist.

In practice the most important property of the mode and the only usual reason for its use is its being unaffected by extreme values. For instance, in a right skewed curve there is an excess of high values to the right in a graph. These affect the mean so that it lies well to the right of the mode and difference between mean and mode thus provides a measure of skewness (see Chapter 8). Like the median, the mode is equal to the mean in a normal or other perfectly symmetrical distribution. In skewed distributions, the only ones for which its use is worth while, the mode may be a zoologically more important average than any other.

Other Measures of Central Tendency

Several other measures of central tendency have been devised and are in occasional use, but they have relatively little practical value in zoology except in a few special problems, and only four of them will be mentioned here.

Range midpoint. This value is obtained by adding the lowest and highest observed values and dividing by 2. It is thus determined entirely by the extreme values and depends more on chance than on any real characteristic of the distribution. It is mentioned here only to observe that it has no practical use and should not be employed. It is generally avoided but occasionally appears in zoological work, sometimes with the wholly unwarranted assumption that it approximates or is equal to the arithmetic mean.

Geometric mean. The geometric mean is obtained by multiplying all the observed values and taking the Nth root of the product (N being total frequency, as before). In mathematical notation

$$\text{Geometric mean} = \sqrt[N]{X_1 X_2 X_3 \cdots X_N}$$

X being the value of one observation of the variate.

Many zoological variates tend to be asymmetrically distributed showing a right, or positive, skew. For such positively skewed distributions, the *logarithm of the values of the variate* may tend to be more symmetrically distributed around their mean than are the values of the original variate around their mean. The arithmetic mean of the logarithms is

$$\frac{1}{N}(\log X_1 + \log X_2 + \log X_3 + \cdots \log X_N)$$

which may be rewritten as

$$\log \sqrt[N]{X_1 X_2 X_3 \cdots X_N}$$

Thus the arithmetic mean of the logarithms is equivalent to the logarithm of the geometric mean so that in a positively skewed distribution the geometric mean will tend to coincide with the median more nearly than will the arithmetic mean. For this reason the geometric mean may be preferred in such cases.

The geometric mean has many of the advantages of the arithmetic mean, but it is relatively difficult to compute, the concept is neither easily grasped nor used, and it is indeterminate when negative values or zero occur among the observations. Its principal use is in the computation of index numbers as used especially in commercial statistics.

Harmonic mean. The harmonic mean is the reciprocal of the arithmetic mean of the reciprocals of the observed values. It may be written thus:

$$\frac{1}{H} = \frac{1}{N} \sum \frac{1}{X}$$

H being the usual symbol for the harmonic mean.

The harmonic mean is always smaller than the geometric mean based on the same data, and the geometric mean is always smaller than the arithmetic mean. The harmonic mean is used in averaging rates.

Quadratic mean. The quadratic mean is the square root of the arithmetic mean of the squares of the observed values. It may be written thus:

$$\text{Quadratic mean} = \sqrt{\frac{\sum X^2}{N}}$$

It is seldom used as such, but it is involved in some methods of calculating measures of dispersion (see Chapter 6).

The various measures of central tendency are illustrated graphically in Fig. 9.

The Meaning of Average, Typical, and Normal

An average, as defined and used in this chapter, is any constant measuring central tendency in a frequency distribution. It is generally given as a single figure, representing a point or a small class in the distribution around which the observations tend in some way to cluster. Since this clustering is a complex phenomenon, there are many sorts of averages, each with its own distinctive properties.

In common speech the word "average," if it is intended to be used exactly, is generally taken to signify only one of the many averages of technical usage, the arithmetic mean. It is, however, seldom used so exactly in the vernacular and generally implies that the average is a large group including all but a few strongly aberrant observations. Hence occasional outbursts of indignation that a third, or some other large fraction, of our human population lives in conditions "below the average," or has an income "below average," or is "below average" in intelligence. When one considers what averages really are, such statements are obviously ridiculous and tell nothing about the real distribution of living conditions, income, or intelligence. Obviously if the average in question is the median or (more approximately) if it is the mean, half the population must inevitably be below average in every respect. If the mean is high, a person may be far below average and yet be living luxuriously. This and analogous widespread fallacies in the use of words are both amusing and dangerous in the mouths of legislators. The reason for emphasizing them here is that zoologists sometimes tend to carry over this looseness of thought into their work and to confuse vernacular and technical usages of such words as "average."

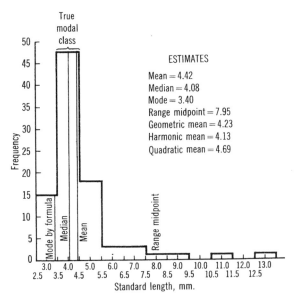

FIGURE 9. Diagram illustrating the relationship between various measures of central tendency in an asymmetric distribution. Standard lengths of specimens of the fish *Parexocoetus brachypterus hillianus*, April collection. Note the following points: The range midpoint gives no conception of the actual distribution; the mean is on the right of the median in the diagram, that being the side of the distribution with the longer tail; the mode is on the other side of the median; the mode, calculated from the rough formula Mode=3 median—2 mean, is poorly estimated, because of the extreme skew.

The word "typical" is also subject to such confusion, so much so that it is hard to give it an exact sense. It implies the existence of a standard of comparison, a type. In the vernacular the type is sometimes supposed to be the average, or a member of a large group around the mean, and sometimes is a sort of ideal by no means really average. The usual description of the "typical American" has more to do with what the speaker or writer wishes were the mean in our population than with what the mean really is. In somewhat better defined usage, one as common in scientific as in popular language, the "typical" condition is taken to be that most frequent. "Typical" then signifies in more technical language belonging to a modal group. This usage, proper but requiring definition, is in turn often confused with the strictly technical use of "types" in zoology. The "type" of a taxonomic group is simply a legalistic device under the rules of nomenclature. It need not necessarily be and very frequently is not in a modal class in the frequency distribution for the taxonomic division. It may be far removed from any average and is quite likely to be, since it is usually the first specimen that came to hand by chance. The "type" of a species is thus not "typical" in any of the more usual senses of the word, and it has no special biological significance.

The word "normal" in the vernacular is subject to a curious dual usage, in which two mutually exclusive ideas are confused and confounded. It is supposed, in the first place, that the normal is a sort of average and, in the second place, that it means the absence of some particular sort of variation regardless of the fact that such variations do occur and hence do in some degree characterize the average. Physicians are the worst technical offenders in this sense, and medical literature is full of equivocations resulting from this double usage. It is assumed that the "normal" condition is the mean condition and also that the "normal" condition is one without any pathological factors. "Normal" cannot mean both these things at once. If anyone in a population is ill in any way, as of course is true of all populations of any size, then the mean condition of the whole population is one of partial illness. The typical condition, in the sense of the modal condition, may or may not include pathological factors, but the mean condition always does. In practice the modal condition usually does also. Perfect health is relatively rare. It is in an extreme, not a middle, position in the frequency distribution of health, and normal health in this sense is an unusual and not an average condition.

It is more reasonable in such a distribution to think of all the observations that really fit into the distribution as "normal." It is as "normal" to be on the point of death as to be in perfect health. The smallest member of a species is as "normal" as the largest or as one of mean size. It is unfortunate that the word "normal" is used in a still more special and logically unrelated way in the name of the "normal distribution." (See page 133.) This is a highly special and technical use of the word, meaning not only conforming to a pattern but to one particular pattern of distribution. By no means all normal variations, in any but this one special sense, fall into a normal distribution.

CHAPTER SIX

Measures of Dispersion and Variability

The determination of any of the various averages gives a point or a small group around which observations are arranged in some way. In most cases the arrangement is such that they tend to cluster around this value, to be crowded toward it, or to pile up on it, with frequencies falling away from it in both directions. The determination of such a point, essential as it is, does not tell enough about the real nature of the distribution. It is necessary to know also about how far the observations extend on each side of this point and about how fast the frequencies fall away from it, or, expressing the same thing from a different point of view, to what extent they are piled up around it. It makes a great difference in the conclusions to be drawn from a series of measurements whether they run, say, from 2.8–7.6 or from 4.9–5.7, although in both cases the mean may be the same.

It also is essential to know whether the frequencies are rather evenly scattered or are strongly concentrated at some point, even though the range and the mean of the observations may be the same in either case, as shown by Example 28.

The adequate measurement of these important characteristics of frequency distributions is one of the greatest problems of zoology. There are good methods of making such measurements, called measures of dispersion, and this section is devoted to the most useful of these.

Range

The observed range is the difference between the highest and lowest observed values of a variate. It is usually and most usefully expressed by giving these extreme observed values although, strictly speaking, the observed range is not these values themselves but the difference between their limits. Thus in Example 29, which gives data that will be used throughout this chapter so that the different measures and means of calculation can be easily compared, the observed range is best recorded as 52–68 mm.

The difference between these, the actual value of the observed range, is 17 mm.[1]

EXAMPLE 28. Hypothetical distributions to show different dispersion with identical ranges and means.

A.		B.	
MEASUREMENTS	f	MEASUREMENTS	f
4.9	2	4.9	1
5.0	3	5.0	1
5.1	2	5.1	2
5.2	1	5.2	5
5.3	4	5.3	11
5.4	3	5.4	3
5.5	2	5.5	2
5.6	1	5.6	0
5.7	2	5.7	1

EXAMPLE 29. Tail length of specimens of the deer mouse *Peromyscus maniculatus bairdii*, both sexes, in the 1-year age class, taken near Alexander, Iowa. (Data from Dice, 1932)

TAIL LENGTH (X)	NUMBER OF OBSERVATIONS (f)
52 mm.	1
53	0
54	3
55	0
56	3
57	8
58	7
59	11
60	11
61	10
62	6
63	14
64	6
65	3
66	1
67	1
68	1

---

[1] Not 16 mm., as is usually stated in analogous cases. The range really lies between the implied limits of the extreme measurements, i.e., between 51.5 and 68.5 mm. in the example, not between the recorded measurements, which are midpoints and not limits.

The observed range, usually but with some danger of confusion simply called the range, is a useful datum and should be given whenever pertinent, but it has many drawbacks and is not a good measure of dispersion. Because of its simplicity, obvious meaning, and requirement of no calculation, it is frequently given in zoological publication, which is desirable; but unfortunately it is often given without any way to assess its value, and it may be assumed to be an adequate representation of a distribution and a significant measure of variability, which it is not.

In the first place, it is clear that the observed range is dependent on the number of observations made. If only one is made, the observed range is zero. Certainly this does not mean that the species, or other category, measured does not vary at all in nature. If two observations are made, the observed range may be large or small but will probably be small. In general the probability is that the more observations are made the larger will be the observed range. Unless the total frequency is also given, an observed range is thus meaningless. Even if the total frequency is given, the meaning of the observed range is uncertain, for its increase with increased number of observations depends in large measure on chance, and its value with any given N is a matter of probability, usually with a large element of uncertainty, rather than of any simple and easily calculable relationship.

Any variate does have a real range. In any given species, for instance, there really does exist in nature one individual that is the largest and one that is the smallest. The difference between these—the real as opposed to the observed range—is an important significant character of the species or, more generally, of any variate; but it is never surely available. The chances of actually observing the largest and smallest of all existing values of any variate are obviously very small, and in most cases it would be impossible to know that they were the extreme values even if they were observed. The population range is changing all the time for at any time an animal may be born or mature with a value of the variate in excess of the then existing limits. Even the existence of a precise or calculable theoretical limit for a variate is problematical. What is the greatest age to which an animal can live? To say that it is 100 years implies that an individual can live to this age but not an instant longer, clearly an absurdity. On the other hand, we can be quite sure that no mouse, for example, has ever reached the age of 100 years, and this does imply the existence of a theoretical limit at some unknown age less than 100 years.

In any event, as the distributions in Example 28 show, the range, especially the observed but even the real, does not give all the desired or necessary information about dispersion and variability. In terms of a frequency curve, it shows at best only where the curve ends and tells nothing of the equally or more important shape of the curve between the ends. For all these reasons, the observed range is the poorest of all the measures of dispersion.

Despite its drawbacks as a measure of dispersion, the range, or at least the highest and lowest observed values, provide information that no other measure of dispersion can.

Most of taxonomic procedure is dependent upon the existence of characters which fall into sharply divided classes without overlaps. Should one species of vertebrate have between 24 and 25 caudal vertebrae while another has between 20 and 23, then the number of caudal vertebrae would distinguish unambiguously between the species, and any individual specimen could be assigned to the correct species on this character. Should the ranges overlap, however, no such distinction could be made for a specimen with a vertebrae count within the range of overlap. While the other measures of dispersion to be discussed in this chapter can be used in conjunction with the mean as a basis for deciding whether two species differ *on the average* in the number of caudal vertebrae, they are often insufficient to determine whether there is in fact any overlap between species.

There is a real, practical, and biological difference between a character that sharply and unambiguously differentiates two groups and one which differs only on the average between them.

Example 30 shows the observed frequency distributions of numbers of caudal vertebrae in 5 species of the fresh-water sculpin *Cottus*. From these distributions it is clear that the group comprising *C. rotheus*, *C. gulosus*, and *C. beldingii* differs sharply from both *C. asper* and *C. aleuticus* in the number of caudal vertebrae. On the basis of these observed frequency distributions, "20–23 caudal vertebrae" and "24–28 caudal vertebrae" would be useful terms in a dichotomous key.

EXAMPLE 30. Variation in number of caudal vertebrae in certain western species of *Cottus*, the fresh-water sculpin. (Partial data from Hubbs and Schultz, 1932)

SPECIES	STATE	NUMBER OF CAUDAL VERTEBRAE									
		20	21	22	23	24	25	26	27	28	N
C. aleuticus	Washington					1	12	25	5		43
	Oregon						1				1
	California					1	5	5			11
C. asper	Washington						4	23	4	1	32
	Oregon							14			14
	California					4	36	16	3		59
C. rotheus	Washington	3	53	42	6						104
	Oregon	1	3	4							8
C. gulosus	Washington		1	42	31						74
	Oregon		16	27	18						61
	California	20	40								60
C. beldingii	Washington		6	17	9						32

There is, of course, no assurance that one specimen of *C. gulosus* may not be found with 24 caudal vertebrae and this is the great fault of observed ranges; the observed range will often be smaller than the real range. In Example 30, however, a fairly adequate sample—195 specimens—of *C. gulosus* has been examined, and despite the fact that 49 individuals, 25 per cent of the entire sample, had 23 caudal vertebrae, not a single individual had more.

In sum, then, the range, although poor as a measure of dispersion, does serve a useful purpose in systematics which cannot be served by other measures. The best procedure in publishing records, then, is to include the range not as a measure of dispersion *per se* but for its own peculiar value and to show in addition another measure of dispersion, like the standard deviation (see below).

Mean Deviation

The observed range is dependent on only two values, the most extreme of the sample distribution. Clearly a better measure of dispersion can be obtained if all the values are taken into consideration, and if their distribution enters into the measure. Of such measures, the simplest, although not a particularly useful one, is the *mean deviation*. As its name implies, it is the average distance that an observation will be from some fixed value, usually the mean of the distribution. The fact that some observations are above the mean (have larger values) and some below is represented usually by making the former positive and the latter negative. If this were done in defining the mean deviation, it follows from the definition of the mean that the mean deviation would always be zero. In fact, the concern here is with distances from the mean and not their direction so that all the deviations are taken to be positive, or as it is usually but less logically expressed, the signs are ignored. The sample mean deviation is defined as follows:

$$\text{M.D.} = \frac{\sum fd}{N}$$

in which M.D. is the sample mean deviation, d is any one deviation from the mean (in either direction), and the other symbols are as previously explained.

The mean deviation may also be taken from the median rather than the mean. If it is so defined, that fact should be specified, because mean deviation is otherwise understood as taken from the mean of the sample. The mean deviation around the median is always smaller than around the mean, a property already mentioned in connection with the median.

It is easy to understand what the mean deviation signifies, and this measure of dispersion is relatively easy to calculate, although the difference from the standard deviation (discussed below) in ease of calculation will not be found great. If there are large erratic deviations beyond the bulk of the distribution, they usually disturb the mean deviation less than the standard deviation, and if they are not considered significant for the problem, the mean deviation may in such cases be preferable. In general, however, the mean deviation is not the best measure of dispersion. Its use in algebraic calculations is inconvenient, and its relationships to the normal curve and the theory of errors and its use in comparing means or other constants are also relatively inconvenient and not so well worked out as for the standard deviation. In almost every case, the standard deviation is preferable. The mean deviation has been introduced at this point and explained at this length not so much because its use is recommended as because it provides a simple introduction and logical background for the problems of dispersion and the use of deviations in general.

Variance and Standard Deviation

By far the most widely used measure of dispersion or variability is the *variance* of a sample. It is calculated from the formula

$$s^2 = \frac{\sum f(X - \bar{X})^2}{N - 1}$$

where s^2 is the variance of the observations and the other symbols are as previously defined.

The difference between each value of the variate and the sample mean is taken, and then this deviation is squared. As a result all deviations are treated as if they were positive since the square of a negative number is positive. Each squared deviation is then multiplied by the frequency with which it occurs in the sample and the sum of the resultant weighted squares is divided by $N - 1$ to produce a sort of average squared deviation. There may be some confusion between our use of s^2 for sample variance and the usage of some other authors. Many books give the following as a formula for s^2

$$s^2 = \frac{\sum f(X - \bar{X})^2}{N}$$

and then give a corrected form:

$$\frac{N s^2}{N - 1}$$

as the quantity to be used in calculating the sample variance. Still others, especially older works, do not make this correction at all. In this book however, s^2 will always have the meaning of the sum of squared deviations divided by $N - 1$.

The reason for using s^2 as a measure of dispersion is certainly not obvious. Clearly it has the desirable characteristic of making all deviations effectively positive, but this virtue is shared by the mean deviation. In addition, s^2 puts more weight on large deviations than on small ones, a property that is of no clear advantage. Finally, the use of $N-1$ instead of N in the denominator would seem to be entirely without intuitive justification. In fact, any attempt to justify s^2 as a measure of dispersion on intuitive grounds is doomed to failure. The use of the sample variance is entirely dictated by considerations of the theory of probability and the testing of statistical hypothesis which will be discussed in the following chapters. A detailed explanation of its basis will therefore be put off until that discussion and it must be sufficient for the present simply to observe that s^2 obviously does measure dispersion.

In practical use, it is often more convenient to use the square root of the variance. This reduces the magnitude to one directly comparable to the deviations themselves and hence is particularly adapted to such uses as considering the significance of individual deviations and in general serves better the purposes for which a measure of dispersion is wanted. This quantity, the square root of the sample variance, is the standard deviation of the sample, symbolized by s.

The calculations of s^2 and s are shown in Example 31 where d is simply a short hand notation for the deviation, $X - \bar{X}$. An extremely important note on calculation is that squaring numbers and then summing these squares is *not* the same as summing the numbers first and then squaring the result. This is a tempting error since it considerably shortens the amount of calculation, but it is wrong. Even if all the deviations are taken to be positive these two procedures are not equivalent. When a variance is calculated by the deviations method, each deviation must be squared first and then the results summed as in Example 31.

While the method of calculating s^2 shown in Example 31 illustrates the basic operation involved, the squaring of the deviations, it is not the best method of calculation. The process of taking the deviations can be eliminated completely by use of a formula which is algebraically identical with the deviations formula. The particular method used will depend upon the availability of a mechanical aid to calculation.

When no calculator is available the following equivalent formula is the best:

$$s^2 = \frac{\sum f X^2 - \frac{(\sum fX)^2}{N}}{N-1}$$

where $f =$ frequency of a given class
$X =$ value of the variate for that class
$N =$ number of observations (total frequency)

EXAMPLE 31. Calculation of variance and standard deviation by means of squared deviations, from the data of Example 29.

X	f	fX	d	d^2	fd^2
52	1	52	−8.43	71.0649	71.0649
53	0	0	−7.43	55.2049	0
54	3	162	−6.43	41.3449	124.0347
55	0	0	−5.43	29.4849	0
56	3	168	−4.43	19.6249	58.8747
57	8	456	−3.43	11.7649	94.1192
58	7	406	−2.43	5.9049	41.3343
59	11	649	−1.43	2.0449	22.4939
60	11	660	− .43	.1849	2.0339
61	10	610	.57	.3249	3.2490
62	6	372	1.57	2.4649	14.7894
63	14	882	2.57	6.6049	92.4686
64	6	384	3.57	12.7449	76.4694
65	3	195	4.57	20.8849	62.6547
66	1	66	5.57	31.0249	31.0249
67	1	67	6.57	43.1649	43.1649
68	1	68	7.57	57.3049	57.3049
	$N = \Sigma f = 86$	$\Sigma fX = 5,197$			$\Sigma fd^2 = 795.0814$

$$\bar{X} = \frac{\Sigma fX}{N} = \frac{5,197}{86} = 60.43$$

d (for each class) $= X - \bar{X}$ $s^2 = \frac{\Sigma fd^2}{N-1} = \frac{795.0814}{85} = 9.3539$

$s = \sqrt{9.3539} = 3.0584$

It is the value of the variate itself, not the deviation from the mean, which is squared, multiplied by its frequency, and then summed. The deviations are taken care of by the second term. The advantages of this method are two. First, the entire process of finding deviations is avoided, saving time and decreasing the chance for error. In any calculation, every operation added represents an increase in the opportunity for error. Second, the numbers to be squared are the original observations themselves which are often whole numbers. Whether or not they are whole numbers, there are usually fewer digits in the observations than in deviations because these deviations must have as many significant figures as the mean. Since the observations have fewer digits, they are more easily squared, again with less chance of error. When the observations have three or fewer digits, regardless of the placement of the decimal point,

86 QUANTITATIVE ZOOLOGY

the squares are best found from a table of squares and square roots. Since the observations generally are one step apart in the last place, the squares can be copied down directly from the table in order. A book of tables of squares and square roots is invaluable and should be obtained by every person engaged in quantitative work.

Example 32 shows the calculation of s^2 for the same data that appeared in Example 31, using the hand calculating formula. The squares were read from a table of squares and square roots.

EXAMPLE 32. Hand calculation of the variance and standard deviation from the data of Example 29.

X	f	fX	X^2	fX^2
52	1	52	2704	2704
53	0	0	2809	0
54	3	162	2916	8748
55	0	0	3025	0
56	3	168	3136	9408
57	8	456	3249	25992
58	7	406	3364	23548
59	11	649	3481	38291
60	11	660	3600	39600
61	10	610	3721	37210
62	6	372	3844	23064
63	14	882	3969	55566
64	6	384	4096	24576
65	3	195	4225	12675
66	1	66	4356	4356
67	1	67	4489	4489
68	1	68	4624	4624
	$N = 86$	$\Sigma fX = 5{,}197$		$\Sigma fX^2 = 314{,}851$

$$\bar{X} = \frac{\Sigma fX}{N} = \frac{5{,}197}{86} = 60.43$$

$$s^2 = \frac{\Sigma fX^2 - \frac{(\Sigma fX)^2}{N}}{N-1} = \frac{314{,}851.000 - \frac{(5{,}197)^2}{86}}{N-1} = \frac{795.0820}{85}$$

$s^2 = 9.3539$
$s = 3.0584$

It might appear from Example 32 that more work rather than less is involved in this method, since it is necessary to find the total of the observations. It should be remembered, however, that the deviations method involves calculating the mean first since this value enters into the formula for s². As Example 32 shows, the mean is a by-product of the calculation of s² since $\sum fX$ is part of that calculation. Thus the two most important sample characteristics, $\bar{X}$ and s², are calculated in one operation. The example shows that the number of digits operated on in this method is consistently smaller than for the deviation method, a fact which makes hand calculation both rapid and less subject to error.

When a desk calculator capable of addition, subtraction, multiplication, and division is available, a slightly simpler form of the calculating formula is used which does not involve forming a frequency distribution. This formula

$$s^2 = \frac{\sum X^2 - \frac{(\sum X)^2}{N}}{N-1}$$

is based not on the class values but on each observation itself. Here, X stands for every observation so that in our example there will be 86 numbers to square. While there is considerable repetition in this method since many of the X values are equal (there are only 15 *different* values of X in the example), it is actually a more rapid method on a machine. Calculating machines can be made to accumulate simultaneously the sum of the observations and the sum of the squares of the observations without the necessity of transcribing the individual entries. The precise method varies for each machine. In any event, squaring each observation and accumulating these squares on the machine will yield directly the basic quantities which can be put in a tabular form as in Example 33.

EXAMPLE 33. Machine method of calculating the variance and standard deviation from the data of Example 29.

$\sum X^2$	$\sum X$	$\sum X/N = \bar{X}$	$(\sum X)^2$	$(\sum X)^2/N$	$\sum X^2 - (\sum X)^2/N$	s^2	s
314,851	5,197	60.43	27,008,809	314,055.9180	795.0820	9.3539	3.0584

The only drawback of this method is that if an error is made in squaring one of the numbers, the whole process of squaring and accumulating must be done over since the individual squares have not been recorded and so cannot be corrected individually. This method of calculating should always be repeated to be sure that no error has been made and is not recommended until a thorough familiarity and competence with a calculator is assured.

Since it is usually desirable to construct a frequency distribution of the observations in order to see the pattern into which they fall, the best general method for the calculation of the mean and standard deviation is the hand calculating formula. It strikes a balance between the unnecessary calculation of deviations in the first method and the greater chance of error in the third.

The standard deviation, like the mean deviation, is an absolute figure in the same units as those of the original measurements or counts. Although usually written as in Example 32, where it is recorded as simply 3.0584, it must be remembered that this is not an abstract number or a relative value but is itself a measurement, in this case 3.0584 mm.

Semi-Interquartile Range

Quartiles measure the values of a variate below which lie one-fourth, two-fourths, or three-fourths of the observations and are designated respectively as the first, second, and third quartiles. Obviously the second quartile, with two-fourths or one-half of the observations below it, is the median, and it is usually called by that name, only the first and third quartiles being explicitly called quartiles.

The more refined estimate of the first and third quartiles is the same as for the median except that a different value is given to n for each:

$$Q_1 = L_l + \frac{(n_1 - 1/2) i}{f}$$

$$Q_3 = L_l + \frac{(n_3 - 1/2) i}{f}$$

in which n_1 is found by subtracting the total frequency below the first quartile class from $(N + 1)/4$ and n_3 by subtracting that below the third quartile class from $3(N + 1)/4$.

The distance between Q_1 and Q_3 is a measure of dispersion. Half the observations are piled up within this distance, a quarter below it and a quarter above. The measure usually taken, however, is the semi-interquartile range, half of this distance, which is the average distance (regardless of direction) from the median to the first and third quartiles. It is also called the quartile deviation.

$$\text{Q.D.} = \frac{Q_3 - Q_1}{2}$$

in which Q.D. is the quartile deviation or semi-interquartile range. Using the data in Example 32,

$$\text{Q.D.} = \frac{Q_3 - Q_1}{2} = \frac{62.84 - 58.39}{2} = \frac{4.46}{2} = 2.23$$

The quartiles and the semi-interquartile range have much the same advantages and disadvantages as the median. They are relatively simple and obvious characteristics of a distribution and are not infrequently employed on that account; but they do not lend themselves to calculation and comparisons so well as do the mean and the standard deviations, and their general use is not recommended.

In place of quartiles, quintiles, (dividing the frequencies into 5 equal groups), deciles (dividing them into 10 groups), and centiles (dividing them into 100 groups) are sometimes useful in special cases, mostly outside the field of zoology. They are calculated like the median, getting an appropriate n for each by starting with multiples of $(N + 1)/5$ for quintiles, of $(N + 1)/10$ for deciles, and of $(N + 1)/100$ for centiles. If such figures are of any use, it is easier and usually sufficiently accurate to get them graphically, plotting the distribution as a cumulative distribution and drawing horizontal lines at heights corresponding to the desired fractions of the total frequency. The median can also be obtained in this way, as already mentioned, and so, of course, can the quartiles.

Relative Dispersion and Variability

All the measures of dispersion so far discussed are absolute. They are themselves measurements or counts of the variate, and their interest is that they are particular values of the variate lying at certain fixed and characteristic points in the given distribution. Valuable as such measurements are, they do not make possible a direct comparison of the dispersion and variability of variates of different absolute mean sizes and just such comparisons are what the zoologist usually has in mind when he talks of "variability." In most fields a measure of relative variability is useful, and in zoology it is indispensable.

The fact that elephants, for instance, may have a standard deviation of 50 mm. for some linear dimension and shrews a standard deviation of .5 mm. for the same dimension does not necessarily mean that the elephants are more variable, in the essential zoological sense, than the shrews. The elephants are a hundred times the size of the shrews in any case, and we should expect the absolute variation also to be about a hundred times as great without any essential difference in functional variability. The solution of this problem is very simple: it is necessary only to relate the measure of absolute variation to a measure of absolute size. The best measures to use for this purpose are the standard deviation and the mean, and since their quotient is always a very small number it is convenient to multiply it by 100. The resulting figure is a coefficient of variation, or of variability.[1] The formula is

$$V = \frac{100\,s}{\overline{X}}$$

V being the standard symbol for this coefficient of variation. The value of V, unlike that of s, is usually spoken of as a pure number, divorced from any unit of measurement. This is because both the numerator and denominator of the expression for V are dimensionally equivalent, so that the units of measurement cancel in the ratio.

The valid use of V depends on the assumption that variation as a biological function is relative to absolute size or, in terms of distributions, that absolute dispersion increases in direct proportion to the mean in variates of essentially the same variability from a biological point of view. It is logical in our understanding of what variability is that this should be so. It has also been empirically determined by many calculations on samples of all sorts of animals that this is at least approximately true of most of the variates used in zoology.

The comparison of values of V derived from different distributions is almost invariably valid and useful if the variates are homologous. If they are not, experience suggests that the comparison is still generally valid if the variates are analogous and belong to the same category—for instance, if they are all linear dimensions of anatomical elements, the usual case. It is also helpful that the units of measurement have no influence on the comparison as long as they are in one category: a V derived from measurements in millimeters is directly comparable to one for measurements in feet. As a rule, however, coefficients of variation for variates of essentially different categories cannot be usefully compared. V for a continuous and V for a discontinuous variate, V for a temperature and V for a mass, V for a linear dimension and V for an area, and the like are not to be considered comparable unless this is shown to be warranted by logic and by experience. It is also necessary to bear in mind that the implication that absolute dispersion relative to the mean measures biological variability is only a broad rule, mainly empirical in foundation, and that it is always open to exception. V is in every case a good measure of relative dispersion, but relative dispersion is not always a good measure of variability. It is, however, usually so, and V is by far the most useful measure of this characteristic that has yet been proposed. Its use involves only the

---

[1] Proposed by Karl Pearson (1857-1936), English leader in biometrics. As much as any one man, he was responsible for the extension and development of statistical and other advanced quantitative methods in the life sciences.

commonsense necessity of remembering what V is and being sure that this is really what one wishes to compare.

Discernment of the meaning of a value of V is largely a matter of experience. Its interpretation on functional zoological grounds depends on nonnumerical biological knowledge. We have compared hundreds of V's for linear dimensions of anatomical elements of mammals. As a matter of observation, the great majority of them lie between 4 and 10, and 5 and 6 are good average values. Much lower values usually indicate that the sample was not adequate to show the variability. Much higher values usually indicate that the sample was not pure, for instance, that it included animals of decidedly different ages or of different minor taxonomic divisions. If the sample is adequate and reasonably unified, then different values of V generally represent in a clear and useful way inherent differences in variability.

For the data of Example 32 (page 86), V is calculated as follows:

$$V = \frac{100\,s}{\bar{X}} = \frac{100 \times 3.0584}{60.43} = 5.06$$

This is a usual value for such a variate in a relatively unified sample. It shows this dimension to have about average variability.

The interesting series of values in Example 34 also shows about how values of V commonly run.

EXAMPLE 34. Coefficients of variability for skull measurements of males, adult unless otherwise noted, of the northern white-tailed deer, *Odocoileus virginianus borealis*, from New England (all but one or two from Maine). (Data from Phillips, 1920)

VARIATE	V
1. Palatal length	4.32
2. Audito-basal length	4.55
3. Length lower tooth row	4.97
4. Zygomatic width	5.11
5. Length upper tooth row	5.39
6. Orbital width	5.62
7. Mastoid width	7.10
8. Length of nasals	9.89
9. Antler length, adults	13.78
10. Antler length, 95 adults and 13 juveniles	18.66

The first six variates listed show about average variability, and such differences in V as exist show moderate increase in this characteristic in the order of listing. The seventh and eighth variates must be considered markedly variable for an adult sample of one sex of a single subspecies. In the case of length of nasals the variability is doubtless influenced by variability of suture form, always rather high. The figure for adult antler length is very high. Biologically there are doubtless two factors here: (1) variability proper, which is evidently high for this dimension; and (2) the fact that antler size in individual deer does not, like the other dimensions listed, reach a maximum and then remain nearly constant but declines markedly in adults past their prime. To get a measure strictly of variability for this dimension it would be necessary for all the animals observed to have been in the same year of their age and merely selecting adults is not adequate. The last figure given has no real usefulness; for to the two factors just mentioned is added marked heterogeneity by mixing in some juvenile specimens, an unwarranted procedure. This is stressed to point out once more a fact that cannot be overemphasized: that the mathematical procedures in this book are not mystic formulae and have in and for themselves no usefulness or validity except as they help to record and interpret useful and valid zoological facts. They are not to be viewed as producing numbers important *per se*, but only in the light of their zoological (and logical) meaning.

The means for the dimensions of Example 34 run from 4.27 to 49.8. The second dimension listed has a mean (26.65), about three times that of the third (8.31), but their V's (4.55 and 4.97) and variability are almost the same. This tends to corroborate the belief that this measure is a valuable one even for widely unequal and for nonhomologous but broadly analogous dimensions.

Example 35 shows V's for relatively small samples, such as are more common in zoology.

EXAMPLE 35. Frequencies and coefficients of variation for width of M^1 in samples of the African tree hyrax *Dendrohyrax dorsalis emini*. (Calculated from measurements by Hatt, 1936)

SAMPLE	N	V
1. Males only, one locality: Niapu	10	4.2
2. Combined males, four localities: Niapu, Akenge, Gamangui, and Ngayu	16	4.0
3. Females only, two localities: Niapu and Medje	8	3.1
4. All specimens, males and females: all five above localities	24	3.8

Sample 1 of this example is as homogeneous as a zoological sample could well be. The specimens are all of about the same age (adult), are of one sex, come from a single locality, and belong to a single subspecies.

V is relatively low, as would be expected because it is strictly a measure of variability, other causes of dispersion being excluded by the purity of the sample. Even within the pure sample this is one of the least variable dimensions, as other measurements given by the original author show. Sample 2 is less homogeneous as to locality, but all the specimens were taken within the range of a single race and the resulting V is not significantly different from that of the entirely homogeneous sample. The females (sample 3) are, in this dimension, even less variable than the males, probably a real sex difference although proof of this would involve analysis not pertinent in this chapter. The combined sample of males and females (sample 4) is slightly (not significantly) less variable than a sample of males alone. The reason for this is that the females are in this case less variable than the males without differing significantly in absolute size. If the females did differ in mean size in this dimension from the males, the V of the combined sample would be higher than for either sex alone and would not be exclusively a measure of variability but of this plus sex divergence.

As is discussed at greater length in Chapter 7, it is practically impossible to get a paleontological sample as homogeneous as a well-selected group of recent animals. The sexes usually cannot be separated, age groups may not be clear, and there will in many cases be unavoidable small differences in geological age and race. It may therefore be expected that V's based on fossil material will run higher than those based on well-recorded and selected recent material. They do so in general, but the differences are not great if the paleontological sample is also as well selected as the exigencies of collection permit. Example 36 is typical of many values of V based on fossils.

EXAMPLE 36. Coefficients of variation of linear dimensions of teeth of the extinct mammal *Litolestes notissimus*, sexes indistinguishable, derived from a single quarry but not exactly the same stratigraphic level. (Original data)

VARIATE	V
Length P_4	6.6
Width P_4	7.1
Length M_1	5.1
Width M_1	6.5
Length M_2	5.6
Width M_2	5.6
Length M_3	6.9
Width M_3	4.3

One of the essentials in good taxonomy is to select characters that are relatively little variable within a taxonomic group, for taxonomic comparisons are more easily and more reliably based on these than on highly variable characters. The coefficient of variation is very useful as a guide in the selection of such characters, too often merely guessed at or accepted with no real criterion. For instance, in dealing with fossil mammals of the family Ptilodontidae, order Multituberculata, there is a strong temptation to use the third upper premolar in taxonomic definition because it shows such strong and clear-cut differences. In a sample of *Ptilodus montanus*, a member of this group, the V's for other linear dimensions are for the most part around 5 or 6 with only one as high as 10; but for length of P^3 this figure is 18.5. Obviously the tooth is extremely variable and is not a good taxonomic guide or, expressed in other words, its variations reflect intraspecific variability and not reliable taxonomic differences.[2]

A measure of relative dispersion that is implied in some zoological procedures is the ratio of the extreme observed values. Such a statement as "the largest individual of a species is 15 per cent larger than the smallest" is a use of this sort of measurement of relative dispersion. The same sort of measurement is implied in comparing species when, for instance, it is given as a specific character that the type of one species is in some dimension 20 per cent larger than the type of another species. Stated in this usual way with no other data, the implication is that individuals within one species do not differ so much in size as do the types of the two species. Usually no evidence for this implied belief is given; and it is, in fact, usually wrong.

The best way to record this measure of relative dispersion would be $(100 \times$ highest observed value$)/($lowest observed value$)$, giving the highest as a percentage of the lowest value. Of all measures of relative dispersion, this is undoubtedly the worst; and it is discussed only to draw attention to the fact that it is so poor and yet has been implied in many earlier zoological works. It is also important to note that the figures thus obtained even for good samples of relatively low variability are usually higher than many zoologists have suspected. It used commonly to be stated or implied that 115 per cent, or 115, is a high value for a single species (or other natural group). Actually it is an unusually low value. Example 37 illustrates this, as well as the irregularity of such a measurement.

These samples are large for zoology (91 to 96 individuals), and so the observed limits probably average not far below the true limits. For the more usual small samples, such a measure bears little rational relationship to the real variability.

Several other measures of relative dispersion are in occasional use but

---

[2] The biological explanation of this extraordinarily high V seems to be that P^3 in this family is not functional and is being lost. Organs in this condition are usually extremely variable.

they have no advantages over V; and it is urged that the latter alone be employed, keeping all such data comparable.

EXAMPLE 37. The ratio of extreme observed values (100 × highest observed value/lowest observed value), using data of Example 34, with corresponding coefficients of variation.

	RATIO	V
1.	125	4.32
2.	127	4.55
3.	130	4.97
4.	129	5.11
5.	142	5.39
6.	129	5.62
7.	143	7.10
8.	179	9.89

CHAPTER SEVEN

Populations and Samples

The first requirement for intelligent work in quantitative zoology—indeed, in any quantitative science—is an understanding of the meaning of numbers and of the methodology of observation. It is therefore with these matters that we have begun this book. Having properly made and recorded his observations, the zoologist is now faced with the task of relating these observations to some generalized universe of facts with which he is trying to deal. Any particular set of observations usually has little or no interest unless it reveals characteristics of broader scope and wider application than those actually observed. Even observations that are truly unique, such as those of abnormalities not repeated, have no value unless they cast light on more normal and widespread processes like heredity and embryology. Undue preoccupation with what is actually observed and failure to relate it to broader issues and conclusions are a constant danger. The logical transition from the particular to the general is the most difficult part of research, and it is the point where the uninitiated is most likely to go astray. Observation, in itself, is not a science and has no value except as a basis for interpretation and some degree of generalization. *For this reason, the remainder of this book will be devoted to the methodology of making general statements from particular observations.* The most fundamental notion in this process of inference is that of the relationship between a population and a sample.

Zoology is, or should be, a study of populations. The word "population" in this sense is not only literal; it applies to all existing phenomena of which a few are observed. When specific characters of a population are determined from a sample, the population is literal—the assemblage of all animals of the species. When an individual's behavior is studied, the population is figurative and twofold: it is (1) the whole of the individual's behavior in this respect, before, during, and after actual observation; and (2) the behavior of all animals in which that behavior follows recognizably similar patterns. A whole population cannot be brought into the laboratory or examined in the field, so that the only practical approach is by the method of samples. In previous chapters we have discussed methods of securing observations, of properly recording them, and of arranging them in some semblance of order. Having performed these operations we no longer have any interest in the sample, *per se*, except as it may provide information on the population from which it has been taken. In the early history of statistics no distinction was made between the characteristics of a sample and those of the population from which it was taken. R. A. Fisher clearly differentiated these two concepts and it is to this insight of Fisher's that the modern science of statistics owes its success.

Estimation and Calculation

A parameter is a constant which describes some quantitative aspect of a population. For example, the population mean, median, mode, range, and variance are parameters whose values vary from population to population but which are fixed for a given population and which describe central tendencies and dispersion peculiar to each assemblage of animals. Sample characteristics, on the other hand, like sample means and sample variances, will vary from sample to sample even when these samples are drawn every time from the same population.

Because an entire population can never be observed (except in special cases of complete censusing), the characteristics of the population can never be known exactly. It is impossible, for example, to *calculate* the mean size of animals in a population. What can be calculated is the mean size in some sample but the result of this calculation will be a number which is the *sample* mean, not the population mean. This sample characteristic may be considered as an *estimate* of the population parameter however. Such an estimate will never be equal *exactly* to the true value, but methods that will be developed in later chapters can be used to determine the range within which the true value probably lies.

To avoid confusion between parameters of populations and the sample values which estimate them, a separate notation will be used for parameter and sample values. *Parameters* will always be represented by Greek letters like μ, σ, ρ, etc., while *sample values* will be denoted by Latin letters like $\bar{X}$, s, r, etc. Usually these letters will be the Latin equivalents of the Greek parameter symbols.

To illustrate the differences between populations and samples drawn from them Example 38 was constructed from the data of Example 29. The individuals of *Peromyscus maniculatus bairdii* taken by Dice form a small sample of all the deer mice in the natural population. For purposes of illustration, however, we have regarded these individuals as representing an entire population and have then taken a number of samples from this artificial "population." This was done by writing the observed tail length of each mouse on a separate card and mixing these cards in a box. A card was drawn, the value on it noted, and the card returned to the box where it was well mixed with the others in preparation for the next drawing. In this way four samples were chosen from the population, two of 43 and two of 86. The results of this sampling procedure are shown in Example 38. The second and third columns are the two samples of 43, the fourth and fifth show the two samples of 86, the sixth column represents a sample of 258 made by treating all the samples as a single large one, and the last column shows the "population" for comparison. The means and variances are shown for each sample at the bottom of the appropriate column.

EXAMPLE 38. Results of artificial sampling from a "population" constructed from the data of Example 29.

X	SAMPLES					"POPULATION"
	A	B	C	D	E	
52	0	0	2	1	3	1
53	0	0	0	0	0	0
54	7	1	1	4	13	3
55	0	0	0	0	0	0
56	1	3	3	3	10	3
57	2	8	6	9	25	8
58	2	3	8	5	18	7
59	6	4	17	10	37	11
60	2	4	8	7	21	11
61	5	4	8	8	25	10
62	3	2	3	4	12	6
63	12	9	15	19	55	14
64	2	4	7	8	21	6
65	0	0	3	1	4	3
66	0	1	2	3	6	1
67	1	0	3	2	6	1
68	0	0	0	2	2	1
ΣX	2,580	2,585	5,210	5,227	15,602	5,197
N	43	43	86	86	258	86
$\bar{X}$	60.00	60.12	60.58	60.78	60.47	$60.43 = "\mu"$
$\bar{X} - "\mu"$	−.43	−.31	.15	.35	.04	
s^2	12.34	8.69	9.96	11.73	10.65	$9.35 = "\sigma^2"$
$s^2 - "\sigma^2"$	2.99	−.66	.51	2.38	1.30	

Each sample mean and variance is an estimate of the corresponding parameter of the population and each differs from the parameter by a greater or lesser amount. Sample variances differ more from this parameter than do sample means, and larger samples tend to approximate the true population more closely than do small ones. These are points to which we will return in the discussion of confidence intervals and statistical testing, but the essential point to be made here is that two samples from the same population will differ from each other and from the parent population in mean, variance, or whatever quantity may be calculated.

Properties of Estimates

If the characteristics of a sample, like the sample mean or sample variance, are useful not in themselves but only as estimates of population parameters, then the sample quantities to be calculated should be those which in some sense are the "best estimators" of the population values. For example, the average of the largest and smallest observation in a sample (range midpoint) is also an estimate of the mean of the population, and there is no *a priori* reason for preferring the sample mean.

There are a number of criteria that may be applied to sample quantities, and although these criteria have some intuitive appeal as defining a "good" estimator of a population parameter, none of them can be said to be a rigorous requirement.

Consistency. It is theoretically, and in some cases even practically, possible to take larger and larger samples from a population to a point where the sample is, in fact, the entire population. If a sample does include the whole population, then any estimate calculated from that sample ought to be exactly equal to the population value. An estimate which exactly equals the population parameter when the sample includes the entire population is said to be a *consistent estimate*. A corollary of this requirement is that as the sample size gets larger and larger the estimate should fall within smaller and smaller limits on either side of the true value, in the limiting case being equal to that true value. In other words, the distribution of the estimate should be concentrated within narrower and narrower limits as the sample size increases, until for an infinite sample all the probability is concentrated at one point, that point being the true value of the population parameter.

The range midpoint is an example of an estimate of the population mean which is not consistent. If the distribution of the population is skewed, then no matter how large a sample is observed, the average of the largest and smallest values will not tend toward the population mean.

The artificial example, Example 38, in which samples of various sizes were chosen from the "population" of *Peromyscus* is a demonstration of the way in which consistent estimators behave. For the two samples of 43, the sample means deviated from the "true value" by .43 and .31 centimeters. In the samples of 86, these deviations were somewhat smaller, .35

and .15, while in the large sample of 258 individuals, the deviation was only .04 centimeters.

Some inconsistent estimators do not approach any fixed value at all as the sample size increases but simply wander about and may even have wider and wider limits as the sample size grows. Inconsistent estimators should be avoided like the plague, for consistency is the easiest requirement to satisfy in estimation.

Unbiasedness. If many samples are chosen from a population and an estimate of some parameter is calculated from each sample, these estimates will vary as we have seen, but it is reasonable to expect that the *average* of an indefinitely large number of such estimates will be equal to the parameter that is being estimated. If, in the long run, the deviations of the estimates from the true value cancel out in this way, the estimate is said to be *unbiased*. Sometimes the average amount of bias of an estimator is known, so that an unbiased estimator can be constructed by a correction. This is the case for the sample variance. The population variance, σ^2, is defined as:

$$\sigma^2 = \frac{\sum f(X - \mu)^2}{N}$$

where μ is the *population* mean. The analogous quantity for a sample would be

$$\frac{\sum f(X - \bar{X})^2}{N}$$

where $\bar{X}$ is the *sample* mean.

In Chapter 6, however, it was stated that the usual sample quantity calculated to estimate the variance is

$$s^2 = \frac{\sum f(X - \bar{X})^2}{N - 1}$$

which differs from the first expression only in having $N - 1$ instead of N as its denominator. The reason for the choice of s^2 is that it is an unbiased estimator of σ^2, the true value, while the first expression tends on the average to be a little too small. This bias is only of the order of $(N - 1)/N$, so that for very large samples the two expressions are nearly equivalent. For samples of 10, however, the biased estimate would be 10 per cent too small on the average, a considerable difference.

Efficiency. It must be emphasized that unbiasedness, like consistency, is only a statement about the average behavior of an estimator. An unbiased estimate will in any particular case be somewhat larger or smaller than the parameter, but in the long run these deviations tend to cancel out. It is clearly desirable, however, that in addition to consistency and lack of bias, an estimate should have the property of deviating as little as possible from the parameter in any given sample. Because estimates like $\bar{X}$ and s^2 are variates, they will have frequency distributions. Just as the frequency distributions of observations have means and variances, so the frequency distribution of estimators like $\bar{X}$ or s^2 will itself have a mean and a variance. If the estimator is unbiased, its mean will be equal to the parameter estimated. If the estimator is *efficient* it will have a small dispersion as measured by its variance. If the variance of an estimator is small, then in all samples the estimate will have about the same value, while if the variance is great, the estimate will be large in some samples and small in others. Since the zoologist usually takes only a single sample from a population, it is very desirable that the estimate that he calculates from the sample be an efficient one. It is of very little use to him to know that his estimate is unbiased, if the particular value that he obtains may lie at a very great distance from the population value. While bias and consistency are absolute properties which an estimator either possesses or does not, efficiency is relative. Some estimators are more efficient than others, and whenever possible the most efficient should be used.

Particular Estimators

In Chapters 5 and 6, a number of sample quantities were discussed. Those which were recommended are all consistent and usually unbiased estimates. In addition they are the most efficient estimators where this requirement is consistent with the other requirements. $\bar{X}$, the sample mean, is the best estimate in every sense of μ, the population mean; s^2, the sample variance, is both a consistent and unbiased estimate of σ^2, the population variance, but it is not quite the most efficient estimator. The most efficient, in this case, would be the biased quantity discussed above, but it is better in this case to correct the bias, as it introduces an element of conservatism in statistical testing.

The sample median is the best estimate of the population median and in a perfectly symmetrical distribution it is also a consistent and unbiased estimate of μ although it is not as efficient as $\bar{X}$ and so should not be used as a substitute for the sample mean, even in symmetrical populations.

We have already discussed at some length the fact that the sample range is a biased estimate of the true range as it tends to be too small. In the general case, however, there is no adequate substitute for it, although in the discussion of the normal frequency distributions (Chapter 8) another method of estimating the true range will be discussed.

The sample value of V, the coefficient of variation, is not a perfectly unbiased estimate of the population coefficient of variation. Haldane (1956) has shown that V calculated in the usual way is somewhat too small and that most of the bias is removed if the usual estimate is multiplied by

$(4N + 1)/4N$. This correction is obviously quite small for samples larger than 10.

In later chapters new parameters like regression and correlation coefficients will be discussed. In each case the sample quantities described are as far as possible the best estimators of their respective population parameters.

Characteristics of Good Samples

The ideal representative of a population is a sample that is homogeneous, adequate, and unbiased, three requirements which mean that:
1. All individual observations in the sample belong to a single, defined population.
2. These observations include all the essential variations within the population.
3. These variations occur in the sample with about the same relative frequencies as in the population.

The more nearly a sample meets these three requirements, the better it is and the more reliable are conclusions based on it.

The requirement of homogeneity is in practice subject to at least one qualification and one exception. Often it is impossible to say that a given population is itself homogeneous, and the very purpose of sampling is to learn from the sample whether the population is pure or mixed with respect to the problem in hand. When, for instance, all the generally similar fossils from one horizon and locality are laid out, the first concern is to see whether they are a homogeneous taxonomic group or really represent two or more groups and hence populations. Samples have in this respect one very important limitation. They can frequently prove beyond reasonable doubt that the population is heterogeneous, but they can never strictly prove that it is homogeneous. That is why assurance of homogeneity depends primarily on the specifications of the population and not on observations on the sample.

If a distribution of a sample is definitely bimodal and can be shown to combine two distributions with different parameters, then the sample and the populations are surely heterogeneous—some tests for this are discussed in Chapter 10. If many different distributions can be made for different variates of a single sample and if none indicate heterogeneity, it becomes probable that the sample and population are homogeneous. This, however, is frequently impossible because only one variate is pertinent to the problem, or for various other reasons; and in any case the negative cannot be really proved: it cannot be established that the sample is not heterogeneous.

In taxonomic work, unless population homogeneity is determined, two closely related subspecies may be mingled in a sample and their combined

variation may be such that it is impossible to separate them or even to establish the fact that two subspecies are present. This exemplifies the importance of population specification. For if samples were from populations each known beforehand to be completely homogeneous in origin, this confusion could not arise. Knowledge of homogeneity is derived from collection records, and in taxonomic studies it involves homogeneity
1. of place (derivation from a single locality or small area)
2. of environment (ecological unity)
3. of time (contemporaneity of the animals studied)
4. of age (animals all in about the same stage of their lives)
5. of sex.

Sometimes all of these specifications cannot be met or clearly determined, and this may introduce uncertainty into the results. Other specifications may also be necessary for the problem in hand, such as homogeneity of physiological condition (all animals gravid females, all undiseased, all with some specified disease, and the like). Good sampling starts with a decision as to what homogeneity is for the problem to be attacked—in other words, with specification of the population. Good collecting involves as complete a record of all specifications as possible, so that samples of a population meeting all necessary requirements for homogeneity can be drawn from the collection.

Another common purpose and result of sampling is to prove that difference in specifications does not produce evident heterogeneity. For instance, it may be found that samples from different localities (otherwise homogeneous) are not different. In taxonomy, the conclusion will be that the same species (or smaller unit) does occur at both localities. If the populations compared are well specified, this may amount in practice to proof of homogeneity, although without good specification such proof is impossible.

Exceptions to the requirement of homogeneity are provided by problems in which heterogeneity of population is itself an element. Obviously this is not an exception to the requirement for specification of population, for such problems cannot be attacked at all without such specifications. In the study of body temperature, if the problem is individual variability, the population should be specified as homogeneous—in race, sex, age, perhaps weight or other growth characteristics; in physiological conditions except body temperature; and in environment, particularly environmental temperature. But if the purpose is to measure homeothermy, the samples should deliberately be selected so as not to be homogeneous in environmental temperature but, on the contrary, to be as heterogeneous as possible in this respect, the environmental temperature being, however, exactly recorded for each body temperature observation.

Even in such problems, in which heterogeneity is deliberately and necessarily sought, homogeneity is also of prime importance Every effort should be made to keep all the population specifications except the one under consideration as homogeneous as possible. For instance, in determining homeothermy, heterogeneity in any respect except environmental temperature may obscure or falsify the result. If some animals are old and some young, some male and some female, some hibernating and some not, some resting and some exercising, some well and some ill, or some fat and some thin, the results will not be valid. The influence of each of these or of any other factors on homeothermy could be determined by keeping it as the variable and making the sample as homogeneous as possible in every other respect. The element of individual variability must also be eliminated or measured and allowed for. In fact, this problem would be best attacked by multiple samples, each derived from a population comprising the body temperatures of a single animal over a short period and in one physiological state with varying environmental temperatures. Body temperature is taken here only as an example; the same sort of considerations applies to any sampling to determine relationships between two variable factors (see Chapters 11 and 13).

It should hardly be necessary to add that observations for which an essential specification is lacking should never be included in a sample. If, for instance, there is any reason to believe that fossil animals from two successive geological horizons differ or if it is desirable to test whether they differ or not, specimens of unknown or inexactly known horizon must be omitted even if the student thinks that he can make a subjective separation of them. If recorded population specifications are inadequate, the problem simply cannot be studied in this form, and either the sample itself must prove heterogeneity (which it often will not do, even though heterogeneity be a fact), or the problem must be abandoned as insoluble from the data available.

This is why such a large amount of material in collections is entirely unfit to provide samples that will really solve urgent and legitimate problems to which the collections are related. It is why poor collecting or collection by inadequately informed amateurs or by careless, venal, or dishonest persons is more likely to make problems insoluble than to help to solve them. It is why so many species and subspecies, especially those of a century or more ago but also a painfully large number up to the present day, are completely and permanently unrecognizable on the basis of the type specimens and why the achievement of a valid and stable nomenclature under the International Rules of Zoological Nomenclature requires frequent suspension of the rules provided in them and promulgation of "official lists" of names by fiat.

The requirement that a sample should include all the essential variations of the population does not mean that it must include observations exactly at the range limits and at every possible point in between. This would, it is true, be ideal; but it is impossible either to obtain such observations in every case or to know them to be such if they are at hand. For quantitative observations, it is enough if the observations are well distributed within the range so that they permit reasonable inference as to the population range. Even from single specimens or inadequately small samples, some idea of the population range can be obtained (see Chapter 10) although of course the information is less exact and less reliable than for larger samples.

As regards qualitative characters, the basic sampling requirement is that each important variation should occur at least once in the sample. The adequacy of the sample in this respect can be determined from the theory of sampling limits, which is discussed on page 199.

In zoology the sample size is usually fixed in practice by what can be obtained. Instances in which a sample can be made of any desired size are rare, and it is a good general principle to use all the available observations. The question of adequacy is not so much that of deciding how large a sample is desirable, but of how adequate the sample actually in hand is and whether it suffices to solve a given problem. There is very seldom any excuse or reason for not using all the observations that can be obtained with proper specification. Availability means observations possible on all the collections accessible to the student or the observations on living wild or laboratory animals that can be made in the time available and before conditions change so as to involve a change in specifications. Even one observation can throw important light on a problem, and hundreds of observations are seldom too many to handle or so many that equally good results can be obtained from fewer. In the rare cases where the available sample is really too large or unnecessarily large, subsampling can be carried out on the same principles as sampling in general, considering the unduly large sample as if it were a population and sampling it to reach conclusions regarding its characteristics.

Bias may enter into qualitative sampling, but it is a still greater danger in quantitative sampling. If an essential variation of the population cannot be inferred from the sample or if the sample is such that inferences based on it as to the frequencies of variation in the population are incorrect, then the sample is biased, and conclusions based on it are unreliable or wrong. Sampling bias may be very difficult or even impossible to detect. If bias is suspected, it is sometimes possible to obtain a new sample from exactly the same population, preferably by a different sampling technique. The bias may then appear from significant differences between the samples, which should give essentially the same results if the sampling were unbiased in both cases.

The most practical way to avoid bias is to give careful consideration to the specifications and sampling technique so that they cannot conceivably bias the result. A few examples of specifications and technique will suggest the innumerable ways in which bias can arise and the importance of considering all possibilities before selecting a sample. It was pointed out above that specimens of unrecorded horizon should be rejected in studying the relationships between fossils of successive strata. It is unlikely that there is any relationship between having or not having a record and any morphological characters of the specimens. It is therefore usually safe to assume that this selection does not bias the result. But suppose, for instance, that specification of essential homogeneity of individual age were omitted and that, as often happens, the animals of one stratum were mostly juvenile and those of the other mostly mature. Many morphological characters are influenced by age, and the samples are therefore biased with respect to these characters by faulty specification. Or suppose that growth of a mammal limb bone is being studied and that it is specified that the limb bones be complete. A common reason for incompleteness is loss of epiphyses. This occurs only in young animals and so is directly related to the problem of growth, and the specification strongly biases the sample for the study of that problem. Again, some detailed experiments were made on heritability of acquired skill in animals, and the sampling was done by opening a cage and taking the first animal that came to hand. Now the coming of the animal is evidently likely to be affected by its past experience, intelligence, or activity; and these are elements in its acquisition of skill. The sampling technique therefore very probably introduced bias into the results. Similarly field collecting sometimes tends to get an unduly large proportion of the more bulky or more active animals or to be otherwise biased. Often there is little that can be done about this, but it must be kept in mind in interpreting the results.

Some investigations are biased in unavoidable yet subtle ways, and there is probably a good deal more bias than zoologists realize in their work. For example, any attempt to estimate the proportions of rare forms in a polymorphic species from museum collections is bound to result in an overestimate of these proportions, for collectors, even highly competent ones, will usually collect such forms because they *are* rare and museum collections sometimes tend to be disproportionately weighted with unusual specimens. In like manner, estimation of the relative frequencies of different species in nature by baited traps will be biased by the nature of the bait. Closely related species of insects, for instance, show markedly different preferences for food baits or light traps so that a single technique cannot possibly produce an unbiased faunal sample.

The theory of probability on which inference from sample to population depends assumes that the sample is taken at random. Conscious selection is involved only in specifying the population. The sample should be selected only by chance, and any element not chance and random may introduce bias. The sampling of laboratory animals, as in the learning-inheritance experiment, would be unbiased if each animal were numbered and the numbers written on identical balls or cards, thoroughly mixed, and the required sample taken at random from these. In collecting wild animals, the method used should be as little selective as possible so that the collection will be random, or as nearly so as possible, within each specified population. Once the specimens are collected, it is necessary to assume that they are a random sample or to allow for this if it is known that the collecting was biased. Then the whole collection is used as far as it meets population specifications or, in rare cases, subsampled by chance methods, such as drawing numbered cards or balls.

Sampling is random and meets the requirements for being unbiased when within a population any one individual observation is as likely to be drawn in a sample as any other. It follows that a population from which a biased sample has already been drawn and not returned cannot give another absolutely unbiased sample, because there is no chance that the second sample will include any of the observations already withdrawn but nevertheless pertinent to the population. For instance, it is impossible to obtain an unbiased sample of male deer in this country, because the game laws assure the withdrawal from the population of a biased sample by hunters at each open season.

Sampling Without Replacement

All of the populations in the real world are finite. It is impossible to conceive of a population of physical entities which contains an unlimited number. Populations of observations, however, can be infinite if after an animal is drawn from the population, and measured, it is replaced and allowed to mix again with original population. In this way the process of sampling and measuring could go on forever without exhausting the individuals. In practice, it is hardly ever possible or convenient to replace each individual after it is drawn, consequently sampling is nearly always *without replacement* and the result of such a sampling scheme is to give biased estimates of some parameters while not affecting others. If a certain estimate is said to be unbiased, it must be understood that this may be *strictly* true only if based upon sampling with replacement.

In general, no matter what the distribution of the observations is in the population, an unbiased estimate of the *mean* is not affected by sampling without replacement. An estimate of variability like s^2 is very slightly affected, so that for a population of size 100 this parameter will be overestimated by about 0.5 per cent if the usual estimate for sampling with replacement is employed. For a population of size 1000 this bias is only 0.05 per cent, so that for most practical purposes it can be ignored.

Other estimates may be biased in proportion to the relative size of sample and population. If, then, the sample is small in relation to the population, sampling without replacement is practically the same as with replacement.

Although these biasing effects of the sampling scheme are recognized, they will be ignored, first, because samples are usually small compared to populations and, second, because one is mostly concerned with comparisons between samples so that a roughly equal bias in two samples being compared will not affect the comparison.

Faunal Sampling

Most field collecting is a sampling operation in which the population is specified as all the animals of some zoological group (such as all insects, all fishes, all birds, all mammals, all carnivores, all cats) living in a given area. Data are recorded with the collections whereby smaller samples with more limiting specifications can be drawn from the collection. The requirements of different sorts of animals and of different types of problems are so various that no general schedule of data for all vertebrates, for example, is useful, but the following include the most commonly useful specifications:

Date of collection and, if possible, time of day
Weather (temperature, light, humidity, and precipitation)
Place (geographic region)
Height or depth from sea level and from local surface of land or water
Station (local habitat)
Field identification (for help in sorting; not an adequate population specification)
Individual age, as closely as possible
Sex
Field measurements and counts, color notes, and other observations of variates best taken on recently dead material
Method of taking, including such details for trapping, for instance, as type of trap, number of trap, interval since trap last visited, number of days for trap in same location, exact character of trap set, bait used, etc.
Physiological condition (oestrous condition, sleeping, feeding, hibernating, shedding, etc.)

For fossils, the exact locality and stratigraphic horizon are universally pertinent field-sample data, but others may also be required, such as nature of matrix, attitude of the fossil in the rock, or immediate association with other fossils.

With good field data of this sort, a collection can become a source of samples with almost any pertinent specifications. The collection is, however, a sample in itself—properly, a faunal sample. Almost all older collections and many of those still made are intended only as qualitative faunal samples. The purpose is to include at least one representative of each taxonomic unit pertinent to the collecting, with no concern for relative frequencies. The collector simply moves into an area and employs as many different collecting methods in as many different places and ways as hold any promise of getting different species or small groups of the animals that interest him. The complexity of such qualitative sampling for large zoological divisions is suggested by methods used to get a qualitative sample of the fishes of Panama (Meek and Hildebrand, 1923)—drag net, set net, tide net, dip net, dynamite, two kinds of poison, hook and line, and dredge—and even this extensive list does not include all the methods of taking faunal samples of fishes. Each method takes some species collected by no other.

Collection of this sort is called "skimming," the intention being, so to speak, to skim off a few representatives of each kind of animal in an area. If the samples could be of the same size for each kind, so much the better. The collection is not really considered as a sample itself but only as a series of samples specified by taxonomic considerations, to be sorted and studied separately in the laboratory.

In recent years increasing attention has been paid to faunal sampling as such, or to quantitative faunal sampling in which the purpose is not only to learn what kinds of animals live in a given area but also to observe their relative frequencies in that area and in the various local habitats, or stations, within it. This is not necessarily effected by killing and bringing back a faunal sample with the same relative composition as the fauna; in fact, when possible it is preferable to collect a qualitative sample and to make the quantitative observations on the living animals. Good quantitative sampling, however, often does involve making a quantitative collection.

Quantitative sampling, whether on living animals or by collecting, involves counting individuals of each species under such conditions that (1) the observed counts for the various species have the same relative sizes as do numbers of those species living in the area, or (2) the observed count for each species has an approximately known ratio to the whole number of individuals of that species in the area. Fulfilling either of these conditions always involves considerable difficulty, and except with very localized and special groups neither one can ever be achieved with complete accuracy.

Collections made in the ordinary sort of qualitative sampling often give approximate quantitative data. For instance, with fishes that are all easily taken on hooks of the same size with the same bait, the catch will be a good quantitative sample, or a drag net with large enough mouth and small enough mesh to catch both larger and smaller fishes will generally give a fair quantitative sample of the water it traverses. But the drag-net sample will not be comparable with one taken by dredging or by spearing, and the hook-and-line sample will not be comparable with any of these.

The principal methods specially devised for quantitative sampling are the quadrat, station, and traverse methods and various adaptations and combinations of these. In the quadrat method, the area to be sampled is subdivided into squares, or quadrats, and an effort is made to count or collect all the animals living on each of a selected number of these quadrats. If the quadrats are well chosen and the observation or collection is made rapidly and well, the total number of animals in the area can be closely estimated by multiplying the quadrat observations by the ratio of the total area to the area of the selected quadrats.

The desirable size and number of such sampling quadrats necessarily vary greatly, depending on the animals sought, nature of the country, and other factors. The size of the quadrat must depend largely on the normal range of individuals of the species sought and should be large enough so that several individuals will occur on each quadrat. Collecting or observation should be as rapid and yet thorough as possible in order to include most of all of the animals in the quadrat when the work is begun but not to give time for others to wander in. The quadrats should themselves be an adequate random sample of the area, including its various local habitats in the right proportions and chosen by chance, if possible by laying out on a map and taking every fifth, tenth, etc., quadrat mechanically. No fewer than 10 quadrats is advisable, and a higher number is better if possible.

The distance between the chosen quadrats should also be great enough, relative to the size of the quadrats, so that the operations do not seriously disturb the animal population and, if collecting is the method, so that the local fauna will not be depleted or seriously unbalanced—this applies to any type of collecting, for the collector as such is not a proper agent of animal control. If sampling quadrats are well selected, even the complete extermination of the animals in them does no harm, for the relative numbers of animals of various species are not changed and the empty quadrats quickly fill up by normal increase around them. Skimming sampling is much more likely to upset a fauna.

It is not necessary or even desirable that each quadrat include a representative assemblage of all the species of the area. It is much better both as a sampling operation and as a conservation measure that the sampling quadrats be small and widely scattered, each with only a random and not complete representation of the fauna. If the quadrats are a good sample of the area, the animals of all combined will be a good sample of the fauna even though those of any one quadrat are not.

Perfect sampling is an impossible ideal. With relatively sedentary and easily captured or observed animals, the quadrat method seems to approach the ideal most closely. In other circumstances, it may be poor or even wholly impractical.

The principle of the traverse method is the same as that of the quadrat method, but the observations are made along lines instead of in squares. The lines should be parallel, numerous enough to sample the area properly and far enough apart for a single individual rarely to be recorded on different lines or, in collecting, for one line not to draw animals from the same area as another line. It is preferable to run the lines simultaneously if possible and, in any event, as rapidly as may be. When the area has settled down again, another set of traverses may be run as a check, at right angles to the first. The system of moving trap lines covering a band is a combination of the quadrat and traverse methods, usually difficult to evaluate but sometimes giving better absolute results.

The great drawbacks of simple traversing are that the size of the area of observation relative to the whole area of the region sampled is very difficult or impossible to evaluate with any accuracy and that different species will be drawn from areas of different sizes. More wandering species will appear relatively more abundant (as they are likely to by any method, but especially by this one); and larger, less timid, or more distinctive animals will be observed at greater distances. This cannot wholly be avoided by restricting observations to the path in which small and obscure species can be identified, for many animals tend to move out away from the line. Distance from the line is therefore an essential datum, and the interpretation of results may become very complex and uncertain. In favorable circumstances, however, traversing gives reasonably good data on relative abundance. It seldom gives a very adequate idea of absolute abundance, although it may set a minimum.

Adapted forms of both quadrat and traverse methods are used in the station method. The principle of this is to establish a number of stations of observation or collection and to record for each species the number of different stations at which it was taken (not, as in the quadrat and traverse methods, the number taken at each station). The stations may be quadrats, traps or trap lines, unit linear distances along traverse lines, scattered points or circles of observation, or time units. In using time stations, the period of observation or collection for any one region or place is divided into equal intervals, and the number of separate intervals in which each species was seen is the record.

This apparently simple method is more complex numerically than may appear at first sight. The relation of the data to either relative or absolute abundance is not direct and may be indeterminate; and it seems likely that in most cases the data of ordinary qualitative sampling are more satisfactory, even quantitatively, than are those of the relatively elaborate station method. Surely well-conducted quadrat or traverse studies are preferable. Unlike these, the station method requires a high number of units, preferably 100 or more; and it logically requires units so small that the occurrence of any one species in them is relatively uncommon. If this condition is met, then most really rare species will be missed altogether; and if it is not met, then all the species except those really rare will be given high and about equal frequencies. Communal animals which are very abundant but only at limited spots will appear rare by the station method, and rare but solitary and widely dispersed species may appear abundant. In fact the station method really measures not relative frequency, but relative dispersion, which is quite a different thing in faunas as it is in variate distributions. As a measure of faunal dispersion it may prove to have considerable value. It is also fairly well adapted to studies of faunal fluctuation from hour to hour, season to season, or year to year, where frequency is a function of dispersion or where dispersion is the important factor.

The Number of Animals in a Population

An important aspect of faunal sampling is the estimation of the total size of a population. Apart from a total census of the population, a technique often used for large animals in open terrain, there have been many suggested schemes that estimate in one way or another the total size of a population. Among these, perhaps the simplest and most efficient is that of *capture-mark-recapture* which has been extensively applied in insect, fish, bird, and small mammal investigations.

The technique is exceedingly simple. A group of specimens is collected from the population in such a way as to avoid any injury to them, and these are then marked in some distinctive way, taking care that the marking procedure does not alter the mobility or life expectancy of the animals in nature. The marked individuals are then released into the population. Following a period sufficient to insure adequate mixing of the marked animals with the remainder of the population, a new sample is drawn and the proportion of marked individuals is noted. The estimated population may then be calculated as follows:

Let n = number of individuals marked at first

p = proportion of marked individuals in the second, or recapture, sample

N = total size of population

Then the proportion of marked individuals in the population as a whole is n/N, while that in the sample is p. Setting the sample proportion equal to the population proportion,

$$p = \frac{n}{N}$$

we get an estimate of N:

$$N = \frac{n}{p}$$

An instructive example of the use of this estimate, especially as regards the technique of sampling, is the work of Lamotte on the land snail *Cepaea nemoralis*. He estimated the population size of a number of colonies of these snails in France by the capture-mark-recapture technique. Live snails were collected in a locality and the shells marked (presumably with a spot of colored paint—the author does not say). These marked snails were then replaced in the population to await the second sampling. As the animals move at a proverbially slow rate, the problem of adequate mixing of marked and unmarked individuals becomes acute. Lamotte's solution was to replace the individuals more or less randomly over the entire area which the colony inhabited. The second sample (recapture sample) was then taken by a different observer so that there would be no tendency either to revisit or to avoid the spots at which the marked snails were released. Such precautions are not in vain. One might assume, for example, that this technique would be ideal for flies whose apparent high rate of activity would insure adequate dispersal. It has been found, however, that some species of *Drosophila*, a small dipteran fly, move quite slowly, 88 per cent of released marked individuals remaining within 60 meters of the release point after 24 hours. Even after 10 days, 95 per cent of the released animals are within about 300 meters of the release point. *A priori* assumptions about the mobility of animals are often misleading.

Example 39 shows some of the results obtained by Lamotte in sampling snail populations and the estimated sizes of these populations.

Paleontological Sampling

Sampling in paleontology is very different from sampling among recent animals and constitutes a special problem, although the treatment of the samples, once obtained, is often the same in both fields. The most marked sampling difference is that in paleontology field collecting cannot meet any of the usual specifications except those of place and geological age. To these may be added a few very special data such as type of matrix and manner of occurrence, to some extent permitting specifications analogous to those of habitat or method of taking for recent animals. But many specifications often well filled from recent data, like those for sex and individual age and those requiring that values of variates shall be from those of living (or recently killed) animals, cannot be met at all by paleontological collecting; and heterogeneity in these respects can be determined, if at all, only from operations with the sample itself. Paleontological samples are thus always somewhat heterogeneous in fact, as they come to the laboratory, and it is seldom possible to make them completely homogeneous by any amount of study and selection. They usually demand broader treatment, but within these broader limits work on them may be and should be just as accurate and just as useful as on recent materials. Another peculiarity of paleontological samples is that they are always biased in some respects. They may not be biased for a particular problem, and it is often possible to determine the bias roughly and to allow for it; but its existence demands recognition.

EXAMPLE 39. The number of marked snails (n) released in populations of *Cepaea nemoralis*, the proportion (p) of marked individuals in the recapture sample as calculated from the size of the recapture sample (r), and the number of marked individuals (m) in the recapture sample. N is calculated to the nearest 10 individuals. (Data from Lamotte, 1951)

COLONY	n	r	m	$p = \dfrac{m}{r}$		$N = \dfrac{n}{p} = \dfrac{n\,r}{m}$	
Samoëns	110	271	38	.140	Samoëns:	$N = \dfrac{110}{.140} =$	780
Orsay	40	62	6	.097	Orsay:	$N = \dfrac{40}{.097} =$	410
Fontaine	131	80	11	.138	Fontaine:	$N = \dfrac{131}{.138} =$	950
Coquerel	300	430	10	.026	Coquerel:	$N = \dfrac{300}{.025} =$	12,900

The agencies by which paleontological faunal samples are biased are: (1) biotic areas or facies, by which the animals were to some extent sorted out while still alive; (2) agencies of burial and fossilization, by which the animals whose remains are actually preserved were selected; (3) agencies of exposure, by which some, but in no case all, of the preserved animals are made available to the collector; and (4) the collector, who finds and collects some but never all of the available fossils. This multiple sieve through which the animals are, so to speak, sifted before they become a sample for investigation exerts a pronounced influence on the nature of the sample. Nothing can be done about this (except, to some extent, as regards factor 4); but bearing these various factors in mind permits some judgment as to the degree and nature of the imperfection of a paleontological sample and may lead to considerable modification of the inferences based on it.

Factor 1—biotic areas or facies—means that every paleontological sample is biased at the start. Seldom or never does any geological formation yield all the types of animals living at any given time, even in the general area in question. It includes only animals that lived in some one type or a few types of surroundings, perhaps marshy, or desert, or (in practice, extremely rarely) mountainous. Study of the fauna itself, its entombing sediments, analogous recent faunas, etc., permits some judgment on this point. Another sort of facies bias exemplified in several collections is due to the fact that some species frequented the area in question only at certain periods of the year or of their lives.

Factor 2—agencies of burial and fossilization—is analogous to facies or tends to emphasize or modify facies differences, in that some types of animals are much less likely to be fossilized than others. The usual relative scarcity of bird, bat, or higher primate remains and the virtual absence of worms in the paleontological record doubtless result principally from this cause. Other related factors also may have a strong and fairly obvious selective influence; for instance, small animals necessarily predominate in deposits in narrow fissures, and large animals are likely to predominate in coarse gravels. Also to be considered is that any assemblage of fossils was formed after the animals were dead; it is a thanatocoenose (death community) and not a biocoenose (life community). It may include diverse species that were not associated in life, and it may have been formed in places and under conditions quite different from those of the living animals. Forces of decay, transport, and burial have usually sorted out species and individuals, so that the representation of the fauna as a whole and of particular species in it may have been strongly biased. For example, in some instances, at least, the so-called "dwarf faunas" of invertebrates reflect only a post-mortem sorting out of smaller individuals and not a real dwarfing of the living animals. These factors greatly complicate, although they are far from precluding, faunal and ecological studies of extinct animals.

As far as it is natural, factor 3—exposure—is generally, but not invariably, an unbiased sampling agency. In a well-exposed formation, fossils are likely to be exposed in about the proportions in which they actually occur in the rock. Some incidental factor may, however, cause bias. For instance, a stratum laid down at some particular time may have been formed under conditions that gave it a distinctly different fossil content from adjacent strata and also gave it a physical nature promoting better or worse exposure, thus definitely biasing the collection from the formation as a whole. Weathering of the fossils after exposure may also be differential, those of some particular zoological or size group breaking down more rapidly under weathering and thus being abnormally few in the collection.

Collecting bias, factor 4, depends on ability and on attitude and purpose. It is well known that two collectors may consistently differ in their results, one finding, perhaps, a much higher percentage of small animals than the other. Also, a collector who is instructed or who naturally tends to collect only for exhibition or for any specified and limiting aim will inevitably make a biased collection of the fauna as a whole and often even of any particular species in the fauna (for instance, may tend to collect chiefly large variants). Methods of collecting also bias the result. For example, collecting individual specimens in the field often gives results radically different from collecting masses of matrix with subsequent washing or acid treatment.

Two other factors may modify a paleontological sample for practical purposes, both affecting chiefly the apparent relative number of individuals of various groups in the collection. The first is that some animals actually have a much greater number of hard parts suitable for fossilization than have others. Thus an armadillo, thanks to its armor, has hundreds of bones more than has a rodent; therefore, a much greater number of pieces in a collection may not represent a correspondingly large number of individual animals in the population. The other factor might be called "bias of identifiability." Although two species might be actually equally abundant in a collection, one would be recorded as more abundant if it were more readily identified from poorly preserved material. Or the apparent size of a species might be considerably greater than the real mean size if, for instance, its larger or adult specimens were more easily distinguished from some related form than its smaller or juvenile specimens.

CHAPTER EIGHT

Probability and Probability Distributions

Definition of Probability

If a coin is tossed, two events or outcomes are possible: it may come up "heads" or "tails." (We will assume that the coin, a very worn and thin one, can never land balanced on its edge.) If the coin were tossed 100 times, the result might be 47 heads and 53 tails, say. If tossed 1000 times, 493 heads and 507 tails is a likely outcome. As the coin is tossed more and more times the number of heads will tend to stabilize at some fraction of the total trials, that is, the relative frequency of heads will approach some fixed value. We will define the *probability of the event "heads"* as the relative frequency with which the event occurs in an indefinitely large number of trials. In this sense, to say that the probability that an event will occur is .9 is the same as saying that it will occur "9 times out of 10." This latter expression is one in common usage, and it is really an expression of empirical probability. It does not mean that in exactly 10 trials there will be exactly 9 occurrences, although it is often mistakenly thought to do so. If the probability of an event is .9, it will occur 9 times out of 10 trials only as an average over an infinitely large number of sets of 10 trials. This is the empirical definition of probability and it is the one to which we will adhere.

With such a definition, it is clear that no probability can ever be exactly known, because no action can be performed an infinite number of times. In addition many statements using the word "probability" are actually not valid because by their nature they cannot be subjected to an empirical check even in theory. For example, the phrase "the probability that this specimen belongs in the genus *Peromyscus* is .73" is not a valid probability statement. Either the specimen does belong in the genus or it does not. It is absurd to say that 73 per cent of the time it will be in the genus and 27 per cent of the time it will not. Statements such as "he is probably correct," "there are probably 30 birds in that tree," or "it will probably rain tomorrow," although perfectly acceptable in general usage, cannot be allowed as exact statements. It is curious that although the notion of probability is fundamental in making inductive inferences, probability statements themselves are *always deductive*. Thus, it is incorrect to say that "judging from the observations, the probability is .82 that my hypothesis is correct." The correct statement is "if my hypothesis is correct, the probability that I will obtain such observations is .82." Obviously, this last statement is capable of empirical justification, for observations may be taken again and again and the frequency with which a certain set occurs may be observed. The hypothesis, on the other hand, if made again and again will either always be correct or always wrong. It cannot be correct 82 per cent of the time.

Events in Probability

In discussing the definition of probability, the turning up of "heads" in coin tossing was termed an "event." If two coins are tossed, the result of each may be considered an event but the combined results may equally well be so considered. In the tossing of two coins "one head and one tail" or "two heads" are descriptions of events, because they are two possible outcomes of the experiment. A probability can be assigned to the event "two heads" by tossing a pair of coins a great many times and noting the results in terms of both coins simultaneously. A length of "2.53 cm." and a count of "12 scales" are also events because they are possible outcomes of the action of counting or measuring. Every value of a variable which is or can be observed is then an event and to each a probability can be assigned. From this broad concept of events in probability, two rules for the defining of events arise. First, events to which a probability is to be assigned must be *mutually exclusive*. A coin cannot turn up both heads and tails, a measurement cannot be both 2.53 cm. and 2.54 cm. (assuming that the accuracy of measurement is as great as has been implied in the numbers), nor can an animal have both 12 and 13 scales of a given sort. Second, the events must be *exhaustive* of the possibilities. In defining the events in a given situation, all possible outcomes must be included. In tossing a die, the events are clearly "1," "2," "3," "4," "5," and "6" and not simply "1," "2," "3," "4," and "5." If events are properly defined in this way it immediately follows that the sum of all the probabilities in a given situation is unity. In coin tossing, if the probability of "heads" is one-half, then the probability of "tails" must also be one-half, for if "heads" comes up half the time, the other half the time the event must be "tails." We have not allowed any other possibilities.

Probability Distribution

The review of "events" and the probabilities assigned to them suggests a very close parallel between these two concepts and those of "class" and "frequency" discussed previously. In fact, a class in a frequency distribution is simply an "event." Moreover, an empirical probability was defined as the relative frequency of an event when an infinite number of observations are made. One is led, then, to the concept of a *probability distribution* which is nothing but the relative frequency distribution of an infinitely large sample of observations. A probability distribution is the limit to which an observed relative frequency distribution tends, as more and more observations are made. In a heuristic sense, an observed relative frequency distribution is a probability distribution observed "through a glass, darkly." Whereas this relation between probability and frequency distribution is fairly obvious for relative frequencies, it is not so clear for absolute frequencies, dependent as they are upon the total number of observations.

Any sort of variable, whether it is a directly observed phenomenon or a quantity calculated from the observations, has a probability distribution. For an observed variable like tail length or scale number or color, this probability distribution is simply the frequency distribution of the entire population of animals from which the observations have been taken. In Example 38 of Chapter 7, an artificial "population" of *Peromyscus maniculatus* was set up, and from this population samples of various sizes were chosen in order to show that samples differ from populations and from each other. If the results of this artificial sampling scheme are tabulated in a somewhat different way, as in Example 40, they serve to illustrate what is meant by the empirical definition of probability and probability distributions. Each column in the table represents one of the samples, but the figures shown are the *relative* frequencies in each class in order to facilitate a comparison with the frequencies in the "population" shown in the last column.

In general, the relative frequencies in each case are closest to the "probabilities" in the large sample of 258 and differ most radically from the population values in the smallest samples. Presumably, if a very large sample of, say, 1000, were chosen, the frequencies in this sample would come even closer to those in the population. In this artificial example, we know the probabilities because the "population" has been artificially constructed, but in nature they are unknown. Nevertheless, from this artificial situation, it is possible to see that the probabilities of each class and thus the probability distribution will be better and better approximated by larger and larger samples from the population.

The probability distribution of an observed variate is represented by the frequency distribution of that variate in nature, but the probability distributions of calculated values like $\bar{X}$, s^2, V, and so on, do not have such direct natural parallels. For a given measurement, every natural population of animals has a mean μ, and a variance σ^2, but these are fixed constants for that population. Every sample chosen from such a population will have a sample mean $\bar{X}$ and a sample variance s^2; these will differ from sample to sample, and if many samples are chosen, the observed values will form a frequency distribution. If a very large number of samples are chosen, infinitely many, in fact, a probability distribution will result, but there is no population of real objects to which it corresponds. Such *sampling distributions* have forms that are predictable on the basis of the mathematics of probability theory. They differ essentially from distributions of observed variates in that the latter are populations of real objects that are examined by a sampling procedure, while the former are products of the sampling procedure itself. Different sampling methods will produce different probability distributions of $\bar{X}$, but the probability distribution of an observed variate is, at least before the zoologist disturbs the probability, independent of the method of observation.

EXAMPLE 40. Relative frequencies resulting from the artificial sampling procedure shown in Example 38.

X	A	B	C	D	E	"POPULATION"
52	0	0	.023	.012	.012	.012
53	0	0	0	0	0	0
54	.163	.023	.012	.047	.050	.035
55	0	0	0	0	0	0
56	.023	.070	.035	.035	.039	.035
57	.047	.186	.070	.105	.097	.093
58	.047	.070	.093	.058	.070	.081
59	.140	.093	.198	.116	.143	.128
60	.047	.093	.093	.081	.081	.128
61	.116	.093	.093	.093	.097	.116
62	.069	.047	.035	.047	.047	.070
63	.279	.209	.174	.221	.213	.163
64	.047	.093	.081	.093	.081	.070
65	0	0	.035	.012	.016	.035
66	0	.023	.023	.035	.023	.012
67	.023	0	.035	.023	.023	.012
68	0	0	0	.023	.008	.012
TOTAL	1.001	1.000	1.000	1.001	1.000	1.002
N	43	43	86	86	258	

Continuous Probability Distributions

An observed frequency distribution is always constructed in terms of discrete classes even when the basic variate is continuous. This, as we have seen, is due to the impossibility of making an indefinitely refined measurement. This problem does not exist for a probability distribution because it is, after all, an idealized situation. Just as the probability of a given class was arrived at by allowing the number of observations to grow infinitely large, so, hypothetically at least, the width of the class interval may be made smaller and smaller by increasingly refined measurements. If both of these processes are assumed to go on simultaneously—the number of observations growing larger and larger, the refinement of observation growing greater and greater—the end result will be a *continuous probability distribution* or, more properly, a *probability density function*.

In a discrete distribution, the probability of falling in a given class can be read directly from the histogram (see p. 51) representing it. Since the classes are all of equal width it follows that the area of a given column is also proportional to the probability of falling into that class. In a continuous probability distribution, it is impossible to speak of the area of the column representing a single class. The classes have an infinitesimally small width, since each class is only one point in a continuum. The result is that the probability of falling in a given class, let us say 3.4999 . . . cm., is itself infinitesimally small.

What can be done, however, is to specify the probability of falling between two points, let us say, between 3.4999 . . . and 4.5000 . . . cm. This probability is, as in a discrete distribution, proportional to the area under the curve between the points. Figure 10 shows a continuous and a discrete distribution with the relationship of area under the curve to probability. The cross hatched area A is proportional to the probability that the variate X falls between 3.000 . . . and 5.999 . . . . The stippled area B is proportional to the probability that the variate takes a value of 6.000 or greater. The area C is proportional to the probability that the variate X takes a value less than 3.000 . . . .

In the discrete case, the area can be measured exactly by multiplying the height of each column by its width. For the continuous curve, this is impossible since the height is continuously changing within each interval. It must therefore be calculated from the mathematical function which represents this curve.

Cumulative Distributions

For those continuous distributions of interest in the quantitative treatment of data, areas have been calculated and tabulated by statisticians and these tabulations are reproduced in Appendix Tables I-V. It would be inconvenient and unnecessary to tabulate the areas between all possible combinations of values of the variate. Instead, it is the cumulative distributions of these variates that are given, and these in turn can be used to find the probability that the variate in question falls between two preassigned limits.

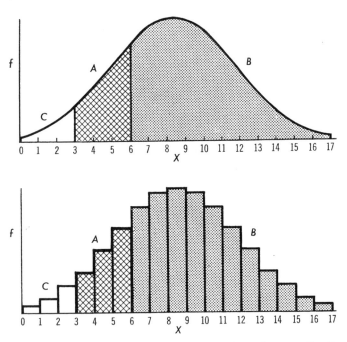

FIGURE 10. Comparison of a continuous and a discrete distribution, illustrating the relationship between area under the distribution and probability. Area A is proportional to the probability that X lies between 3.000... and 5.999.... Area B is proportional to the probability that X is 6.000... or greater. Area C is proportional to the probability that X is less than 3.000....

The construction of a cumulative probability distribution is exactly the same as for any frequency distribution. In the discrete case, the value on the ordinate is the sum of all the probabilities up to and including the class in question. For the continuous case, the value is the area under the curve from one end of the distribution up to the point in question. Such a cumulative distribution begins at 0 at one end of the range of the variate and rises finally to 1 at the other end. This simply means that the total probability from one end of the range to the other is unity. Notice that the value of the *ordinate* of a cumulative distribution gives the value of the *area* under the original distribution. The area under a cumulative distribution has no special meaning. The relation between these two kinds of distribution is shown in Fig. 11 for the continuous case.

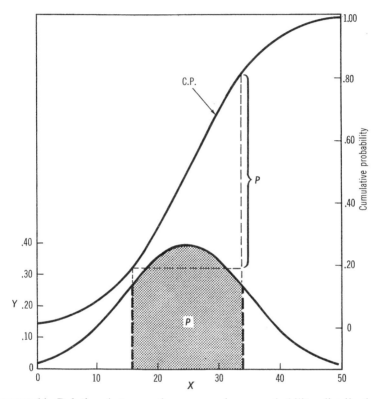

FIGURE 11. Relation between the area under a probability distribution and ordinates of a cumulative distribution. The difference P in ordinates of the cumulative distribution is the probability corresponding to the area P under the frequency distribution.

To find the probability that a variate falls between two limits, it is only necessary to subtract the cumulative probability of the smaller value of the variate from the cumulative probability of the larger value of the variate. By reference to Fig. 11, it can easily be seen that the result is proportional to the area between the given values of the variate and is, therefore, the probability that the variate will fall within the specified range.

There are a number of probability distributions that are of special importance for the quantitative treatment of zoological data. The use of some of these will only appear later on and a discussion of them will be deferred until then. There are three closely related distributions, however, of immediate interest which will illustrate many of the properties of probability functions: the *binomial*, *Poisson*, and *normal* distributions.

Binomial Distribution

If a single coin is tossed, the outcomes may be recorded as 0 or 1 head; if two coins are tossed, 0, 1, or 2 heads are possible; and so on. For the purposes of illustration, suppose that the coin and the manner of tossing are such that the probability of a head coming up for any particular coin is .5 (probability will generally be recorded as decimal fractions rather than proper fractions). If a single coin is tossed a very large number of times, the relative frequency of "heads" then will be very close to .5, as will the frequency of "tails." Simply recording the number of heads, we will find the value 1 to have a frequency of .5, as will the value 0. Now if two coins are tossed together a very large number of times, there will be a certain frequency of 0 heads, 1 head, and 2 heads, which frequency will be the probability of 0, 1, and 2 heads. If n coins are tossed an indefinitely large number of times, there will be some probability associated with events 0, 1, 2 . . . up to n heads. Example 41 contains the probabilities associated with various numbers of heads using from 1 to 7 coins. Figure 12 represents this sort of result as a histogram.

EXAMPLE 41. Theoretical binomial probabilities of obtaining various numbers of heads when tossing from 1 to 7 coins.

NUMBER OF HEADS	WITH 1 COIN	WITH 2 COINS	WITH 3 COINS	WITH 4 COINS	WITH 5 COINS	WITH 6 COINS	WITH 7 COINS
0	.500	.250	.125	.063	.031	.016	.008
1	.500	.500	.375	.250	.156	.094	.055
2	—	.250	.375	.375	.313	.234	.164
3	—	—	.125	.250	.313	.312	.273
4	—	—	—	.063	.156	.234	.273
5	—	—	—	—	.031	.094	.164
6	—	—	—	—	—	.016	.055
7	—	—	—	—	—	—	.008

The probabilities in Example 41 were not arrived at by an actual coin-tossing procedure but are the mathematical result of the assumption that for any one coin the chance of "heads" or "tails" is one-half. If the probability of one event (heads, for example) is p and the number of chances for its occurrence is n, then the probability that it will happen exactly X times is given by

$$\frac{n!}{X!(n-X)!} p^X (1-p)^{n-X}$$

! is called the "factorial" operator and is a symbol which will be used again. When used in conjunction with a variate like X, it is shorthand notation for

$$X! = (X)(X-1)(X-2)(X-3)\cdots(1)$$

Noting that we define 0! as equal to 1, not 0, the reader may easily check the entries in the table in Example 41 with this formula.

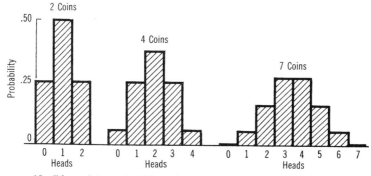

FIGURE 12. Binomial probability histograms showing the probabilities of obtaining various numbers of heads when tossing 2, 4, and 7 coins.

The binomial distribution is so named because the probabilities that the event will occur $n, (n-1), (n-2), \ldots, 0$ times are given by the successive terms in the binomial expansion

$$(p+q)^n = p^n + np^{n-1}q + \frac{n(n-1)}{2} p^{n-2} q^2 + \cdots q^n$$

where q, the probability of nonoccurrence, equals $1-p$. A more detailed explanation of how the formula for this distribution is derived will be found in a simple and quite readable form in Feller (1950).

The coin-tossing case is special, in that p is 1/2, giving rise to a symmetrical probability distribution. In the more general case the distribution will be skewed over to one side. An example of such a nonsymmetrical case is that of tossing dice. Consider the number of times that a 1 will appear in throwing one die. Here there are six possible events, as a die has six faces but only one satisfies the condition. If it is assumed that the construction of the die and the method of throwing is such that all faces are equally likely to come up, the probability of a 1 is 1/6, while the probability that the specified event will not occur is 5/6. Probability distributions for 1 to 5 dice are given in Example 42 and represented graphically in Figure 13.

EXAMPLE 42. Theoretical binomial probabilities of obtaining a one in throwing from 1 to 5 dice.

NUMBER OF ONES APPEARING	WITH 1 DIE	WITH 2 DICE	WITH 3 DICE	WITH 4 DICE	WITH 5 DICE
0	.883	.694	.579	.482	.402
1	.167	.278	.347	.386	.402
2	—	.028	.069	.116	.161
3	—	—	.005	.015	.032
4	—	—	—	.001	.003
5	—	—	—	—	.000+

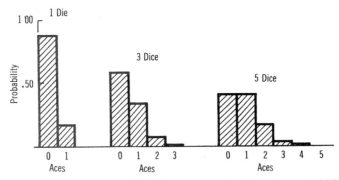

FIGURE 13. Binomial probability histograms showing the probability of obtaining various numbers of aces when tossing 1, 3, and 5 dice. Note the marked asymmetry of these distributions in contrast to Fig. 12.

Notice that although there are six possible primary events in the throw of a die, this problem is still described by the binomial distribution. This is because the number of primary events has been reduced to two—the occurrence and nonoccurrence of 1. The appearance of 2, 3, 4, 5, or 6 are all lumped under "nonoccurrence." It would be possible by lumping the events 1, 2, and 3 into one class and 4, 5, and 6 into another to return to the symmetrical case identical to coin tossing. Tables of both the ordinary and cumulative binomial distributions can be found in Romig

(1953) and the Harvard Computing Laboratory Tables (1955) for many values of p and n. Detailed instructions for their use are included with the tables. The direct applications of the binomial distribution to zoological problems are not as numerous as for some other distributions, but there are some zoological processes that fit the binomial model quite well. One obvious example is the number of males and females in a family. In terms of the binomial model, each birth in the family is a trial. At each trial there is a probability p that the offspring will be a male and a probability $q = 1 - p$ that it will be a female. In a family of size n (excluding parents), there are n trials so that the probability that a family will be made up of exactly X males and $n - X$ females is the binomial probability.

$$\frac{n!}{X!(n-X)!} p^X (1-p)^{n-X}$$

If one examined a very large number of families each of size n, the relative frequency of families with exactly X male offspring would then be given by this expression, since probability as we have defined it is simply the relative frequency of an event in a very large number of observations. Example 43 shows just such an observed distribution as compared with a theoretical binomial distribution. The observed distribution is that of the number of boys in German families of 8. The theoretical distribution is constructed in the following way. The total number of families, N, is 53,680. Since each family contains 8 children, there are a total of 8 × 53,680, or 429,440 children in the entire sample. The total number of boys in the sample is clearly

$$(0 \times 215) + (1 \times 1{,}485) + (2 \times 5{,}331) + \cdots + (8 \times 342) = 221{,}023$$

EXAMPLE 43. Distribution of the number of boys in German families of size 8 (excluding parents) as compared with the theoretical binomial distribution. (Data from Geissler, quoted by Fisher, 1950)

NUMBER OF BOYS IN FAMILY	NUMBER OF FAMILIES OBSERVED	THEORETICAL BINOMIAL FREQUENCIES
0	215	165.22
1	1,485	1,401.69
2	5,331	5,202.65
3	10,649	11,034.65
4	14,959	14,627.60
5	11,929	12,409.87
6	6,678	6,580.24
7	2,092	1,993.78
8	342	264.30

Thus the proportion of boys in the sample is

$$\frac{221{,}023}{429{,}440} = .5147$$

which is then the value of p used in construction of the theoretical binomial distribution. This distribution is then

$$\frac{8!}{X!\,(8-X)!}\,(.5147)^X\,(.4853)^{8-X}$$

These probabilities or relative frequencies are made directly comparable to the observed distribution by converting them to absolute frequencies. This is simply done by multiplying each probability or relative frequency by N, the total sample size.

While there is a fair agreement between the observed distribution and the theoretical binomial, the example shows that there is an excess of families with few boys and with many boys, while those families containing close to the average number of boys are less frequent than they should be.

The example also illustrates the calculation of the mean and variance of a binomial distribution. As for any frequency distribution the mean $\bar{X}$ is

$$\frac{\sum fX}{N}$$

If the numerator and denominator are multiplied by n, the family size, we can write the mean as

$$\bar{X} = n\,\frac{\sum fX}{nN}$$

but

$$\frac{\sum fX}{nN}$$

is the total number of boys divided by the total number of children, that is, the observed proportion of boys. Symbolizing this by $p_{observed}$,

$$\bar{X} = np_{obs.}$$

For the theoretical binomial distribution,

$$\mu = np$$

where p is the true proportion of successes.

Since the theoretical binomial distribution was constructed by using $p_{obs.}$ in place of p, the true mean, μ, is equal to $\bar{X}$ in this case, both being equal to 4.118.

The variance of a *theoretical* binomial distribution is

$$\sigma^2 = np\,(1-p)$$

but this will not be equal to the variance of the observed distribution unless that distribution fits the binomial probabilities exactly. In Example 43 the theoretical variance, σ^2, is 1.998 while the variance of the observed distribution, s^2, is 2.067. This somewhat larger variance is a reflection of the excess of extreme classes in the observed distribution over what is predicted on the binomial model.

Some care must be taken not to confuse n and N. The quantity N is, as usual, the total frequency or sample size. In our example, N is 53,680, the total number of families or units observed. On the other hand, n is the number of primary events in each unit, 8 children in each family, for example, or 3 eggs in each clutch. It is this n that is a parameter of the binomial distribution and which is used in determining the mean and variance of the binomial distribution.

The Poisson Distribution

The Poisson distribution is an approximation to an extremely asymmetrical binomial distribution. Suppose that p, the probability of success on any one trial, is very small but that a very large number of trials or chances for success, n, occurs.

Then the binomial probability of exactly X successes in n trials,

$$\frac{n!}{X!(n-X)!} p^X (1-p)^{n-X}$$

becomes a very tedious quantity to evaluate. If n were 1000 and p .001, for example, the calculation of the binomial probabilities would be prohibitive. However, in such a case the binomial probability is very closely approximated by the Poisson probability

$$\frac{e^{-np}(np)^X}{X!}$$

where e is the number 2.71828 (base of natural logarithms), and the other symbols have the same meaning as in the binomial distribution. Since n and p do not enter the expression separately but only as the product np, they do not have to be separately known. The product np, the mean of the Poisson distribution, is identical with the mean of the binomial distribution and as we have already noted, it is estimated by

$$\overline{X} = np_{\text{obs.}}$$

in any particular case. Thus, the theoretical Poisson distribution corresponding to an observed distribution can be constructed from a knowledge of the sample mean alone, while for a binomial distribution both n and p must be known separately. Example 44 shows a distribution that illustrates this point. In terms of the binomial model, p is the probability that an individual of *Litolestes notissimus* will be entombed and fossilized in a particular square meter of ground. We do not have the slightest notion of what this probability is, but certainly it is not great since the number of square meters available for the fossilization of *Litolestes* was immense. On the other hand, n, the number of *Litolestes* fossilized in some square meter, although unknown again, must have been very large. Since neither n nor p can be estimated, it is not possible to fit a binomial distribution to the observations but because n and p are undoubtedly quite large and small respectively, and because np can be estimated from the mean of the observed distribution, the Poisson approximation can be used.

EXAMPLE **44**. A Poisson series. Distribution of the number of specimens of the extinct mammal *Litolestes notissimus* found in each of thirty 1 m. × 1 m. squares of horizontal quarry surface. (Original data)

NUMBER OF SPECIMENS PER SQUARE (X)	NUMBER OF SQUARES
0	16
1	9
2	3
3	1
4	1
5 and over	0

Example 45 shows the theoretical frequencies for the Poisson series as compared with the observations, a comparison made also in Figure 14.

The steps in setting up the theoretical Poisson distribution are quite similar to those for the binomial. First, the sample mean

$$\bar{X} = \frac{\sum fX}{N}$$

is calculated and used as an estimate of np. This is then substituted into the general expression for the Poisson probabilities. The square, cube, fourth power, etc. of $\bar{X}$ are easily found, and since X is never very great unless an immense sample is taken, the number of such terms to be evaluated is small. In the case of Example 45, only the fourth power of $\bar{X}$ needs to be calculated. The term $e^{-\bar{X}}$ is easily found from the table of exponentials in any book of mathematical tables. The probability of 5 or more successes is found by summing up the previous probabilities and subtracting that sum from unity. Finally, the probabilities are made directly comparable with the observed distribution by multiplying them by the total frequency, N, converting them thereby into absolute frequencies.

EXAMPLE 45. Comparison of observed and theoretical frequencies from Example 44.

NUMBER OF SPECIMENS PER SQUARE (X)	OBSERVED FREQUENCIES (f)	POISSON PROBABILITY	THEORETICAL FREQUENCIES
0	16	.48	14.4
1	9	.35	10.5
2	3	.13	3.9
3	1	.03	.9
4	1	.01	.3
5 and over	0	.00	.0
	$\Sigma f = N = 30$		30

$$\bar{X} = \frac{\Sigma fX}{N} = \frac{0 \times 16 + 1 \times 9 + \cdots + 4 \times 1}{30} = .73 \text{ specimens per square}$$

$$s^2 = \frac{\Sigma f X^2 - N \bar{X}^2}{N-1} = \frac{46 - 15.987}{29} = 1.034$$

$$e^{-\bar{X}} = e^{-.73} = .482$$

Then the Poisson probabilities are calculated as

$X = 0 : e^{-\bar{X}} = .482$

$X = 1 : \bar{X}e^{-\bar{X}} = .482 \times .73 = .35$

$X = 2 : \dfrac{\bar{X}^2}{2} e^{-\bar{X}} = .482 \times .269 = .13$

$X = 3 : \dfrac{\bar{X}^3}{6} e^{-\bar{X}} = .482 \times .066 = .03$

$X = 4 : \dfrac{\bar{X}^4}{24} e^{-\bar{X}} = .482 \times .014 = .01$

$X \geq 5 : 1 - (\text{sum of preceding terms}) = 1 - 1.00 = .00 +$

The value of the remainder of the series is always greater than zero, but the record .00 + indicates that it is less than .005 and for the purposes of the present calculations is negligible.

As the example shows, the observed distribution is quite similar to the theoretical Poisson distribution, especially considering the small size of the sample. As in the case of the binomial distribution discussed in the last section, there is an excess of extreme classes in the observed distribution. A reflection of this fact is that the sample variance of the observed distribution is 1.034. In a theoretical Poisson distribution, the variance is equal to the mean np. The observed variance is then about 1.42 times what it would be if the observation fit the Poisson distribution exactly.

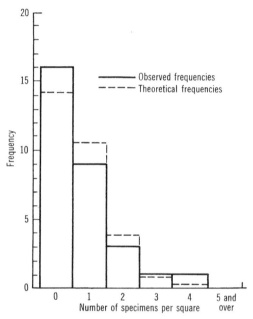

FIGURE 14. Histograms of a Poisson series and of an observed distribution approximating it in form. The solid lines represent numbers of specimens of the extinct mammal *Litolestes notissimus* found in each of 30 square meters of quarry surface (data of Example 44). The broken lines show a Poisson distribution fitted to the observations.

While the Poisson distribution is an approximation to the binomial distribution for very large n and very small p, it is quite adequate even for n of 100 and p as large as .01, the error in each class being only in the third decimal place. Example 46 shows the exact binomial probabilities and Poisson approximation for such a case.

In general, the Poisson approximation is often an adequate description of a markedly skewed distribution of a discrete variate. The most frequent application of this distribution in zoology is in the result of faunal sampling operations where the variate in question is the number of animals or species per unit of observation as in Example 44.

At any event, whether the Poisson distribution is an adequate description of any particular sample is best told by statistical comparison of the sample with the theoretical frequencies, a method discussed in Chapter 13.

EXAMPLE 46. Comparison of binomial and Poisson probabilities for $p = .01$ and $n = 100$. (Abridged from Feller, 1950)

X	BINOMIAL PROBABILITIES	POISSON APPROXIMATION
0	.36603	.36788
1	.36973	.36788
2	.18487	.18394
3	.06100	.06131
4	.01494	.01533
5	.00290	.00307
6	.00046	.00051
7	.00006	.00007
8	.00001	.00001

The Normal Distribution

By far the most important probability distribution in the quantitative treatment of observations is the *normal distribution*, also called the Gaussian curve,[1] or Laplace's normal curve.[2] This distribution has a two-fold importance summed up in the biometrician's common saying that the normal distribution is used by biologists in the belief that it is a mathematical necessity and by the mathematicians because they believe it to be a biological reality. It is a curious fact that this curve does adequately represent the way in which many variates, especially continuous ones, are distributed in nature; and at the same time, it closely approximates the distributions of many sample quantities like $\bar{X}$ and s^2 from large samples. Specifically, the sample mean, $\bar{X}$, is normally distributed irrespective of the distribution of the observations themselves, while quantities like the sample variance, median, and a number of others to be discussed

[1] K. F. Gauss (1777-1855), German mathematician and geodesist, who published on numerical series, including that from which the normal curve is derived.

[2] P. S. de Laplace (1749-1827), French astronomer and mathematician, who studied the theory of probabilities and laid the foundation on which many statistical procedures are based.

in later chapters are normally distributed when they are based on very large samples and when the distribution of the original observations is not far from normality itself.

The formula for the normal curve is

$$Y = \frac{1}{\sqrt{2\pi}\,\sigma} e^{\frac{-(X-\mu)^2}{2\sigma^2}}$$

where

π = the familiar constant 3.1416
μ = the mean
σ = the standard deviation
e = the base of natural logarithms
X = the value of the variate
Y = the height of the ordinate for a given value of X.

The particular normal curve will vary depending upon the values of μ and σ, the two parameters. Figure 15 shows three normal curves with the following parameters

A: $\mu = 0$ $\sigma = 1$
B: $\mu = 0$ $\sigma = 2$
C: $\mu = 1$ $\sigma = 1$

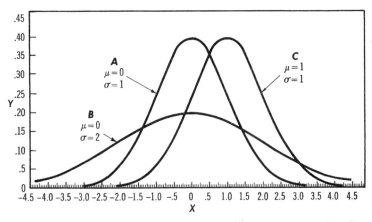

FIGURE 15. Comparison of three normal distributions to show the effect of changes in the mean μ and standard deviation σ. A: $\mu=0$, $\sigma=1$; B: $\mu=0$, $\sigma=2$; C: $\mu=1$, $\sigma=1$.

The value of μ, the mean, simply locates the curve along the abscissa but has no effect upon its shape. The standard deviation, σ, on the other hand, determines how disperse or concentrated the curve is but does not affect its location. These relations are completely in accord with the notion that the mean is a measure of central tendency only, irrespective of variability, while the variance or standard deviation measures dispersion independent of the mean.

At first glance there would seem to be two considerable limitations to the usefulness of the normal distribution. As we have discussed on pages 120-21, the distribution of a continuous variate differs from that of a discrete variate in one very important respect. While it is possible to find the probability of, say, 10 successes in a binomial distribution by simple substitution in the binomial probability formula, it is not possible to find the probability that an observation drawn from a normally distributed population will be equal to 10.000 . . . cm. The probability of an observation being exactly 10 centimeters is infinitesimally small. What can be done, however, is to find the probability that an observation will be between, say, 9.500 . . . and 10.499 . . . cm. by finding what portion of the total area under the normal curve lies between these two limits. The process of calculating this area is extremely complex mathematically but it can be done and the results tabulated for future use. This raises the second difficulty. Like the binomial distribution, the normal curve depends upon two parameters so that there is not one but an infinity of normal curves, each corresponding to a pair of values for μ and σ. Clearly it would be impossible to tabulate them all or any considerable fraction of them. This difficulty is overcome by use of the *standardized normal deviate*.

Suppose the mean, μ, and standard deviation, σ, of a normal curve are known, Then it is possible to form a new quantity

$$\tau = \frac{X - \mu}{\sigma}$$

where τ, the standardized normal deviate, is the difference between the value of the original variate, X, and the mean of the distribution, divided by the standard deviation. This quantity is distributed with the following formula

$$Y = \frac{1}{\sqrt{2\pi}} e^{-\tau^2/2}$$

That is, τ is normally distributed with $\mu = 0$ and $\sigma = 1$. Thus, all normal distributions can be reduced to a single distribution that does not contain μ or σ by the simple expedient of subtracting each value of the original variate from the mean and dividing this difference by σ.

The quantity τ is measured in units of standard deviations. To say that τ is equal to 3 is identical with saying that the original variate X lies at a distance 3 standard deviations from the mean. Then the probability that a variate X lies within 3 standard deviations of the mean is identical with the probability that τ takes a value between $+3$ and -3.

136 QUANTITATIVE ZOOLOGY

The areas under the standardized normal distribution are tabulated in Appendix Table I in the following way. The first column lists the values of τ, that is, of deviations from the mean in terms of the number of standard deviations. The second column contains the areas between the tabulated value of τ and the mean of the distribution ($\tau = 0$). Since the normal curve is exactly symmetrical around the mean, the area between the mean and $+\tau$ is identical with the area between the mean and $-\tau$. The last column gives the value of the *ordinate* of the standardized normal curve at the corresponding value of τ.

With this table, then, it is possible to find the probability that τ will be between any two values. There are two cases to consider. First, suppose that the limits are on opposite sides of the mean. For concreteness, we may find the area between $+.70$ and $-.30$. From the table, the probability that τ falls between 0 and $+.70$ is .2580 while the probability that τ lies between 0 and $-.30$ is .1179, so that the total area between these points is $.2580 + .1179 = .3759$. The second possibility is that the two limits are on the same side of the mean (both positive or both negative) The area between the mean and 1.30, for example, is .4032 while that between the mean and 1.00 is .3413, so that the probability of falling between 1.00 and 1.30 is .4032 *minus* .3413, or .0619. There are other methods of tabulating the standardized normal distribution, but the method we have adopted is the most flexible. A similar but much more complete tabulation giving τ to two decimal places is contained in the *C.R.C. Standard Mathematical Tables*.

In summary, to find the probability that a normally distributed variate X falls between the limits X_1 and X_2: first, convert X_1 and X_2 to standardized variates by subtracting the mean and then dividing by the standard deviation; second, use the tables of the standardized normal deviate in the manner just outlined.

If it is desired to construct that normal distribution which most closely resembles an observed distribution, the most obvious choices for values of μ and σ are the observed sample values $\bar{X}$ and s. It is these quantities which are then substituted into the expression for τ.

The construction of such a normal distribution is shown in Example 47 for the data of Example 29. The first column contains the true limits of the classes symbolized by the measurements. The second column expresses these limits as deviations from the mean. The third column contains τ, the ratio of deviation to standard deviation, and the fourth shows the probability associated with these intervals as determined from the table of the standardized normal distribution. These probabilities are multiplied by N, the total sample size, to convert them to absolute frequencies for direct comparison with the observed distribution. The agreement between observed and expected is very good indeed and confirms the fact that observed distributions of continuous variates may be very close to normal.

EXAMPLE 47. Fitting a theoretical normal distribution to the observed distribution of Example 29, page 79.

TAIL LENGTH (MM.)	CLASS LIMITS AS DEVIATIONS	$\dfrac{d}{s}$		NORMAL PROBABIL- ITIES	THEORET- ICAL FRE- QUENCIES	OBSERVED FRE- QUENCIES
51.50–53.49...	−8.93 – −6.93	−2.92 –	−2.26	.0102	.88	1
53.50–55.49...	−6.93 – −4.93	−2.26 –	−1.61	.0418	3.59	3
55.50–57.49...	−4.93 – −2.93	−1.61 –	−.96	.1148	9.87	11
57.50–59.49...	−2.93 – .93	− .96 –	− .30	.2136	18.37	18
59.50–61.49...	− .93 – 1.07	− .30 –	.35	.2547	21.90	21
61.50–63.49...	1.07 – 3.07	.35 –	1.00	.2045	17.59	20
63.50–65.49...	3.07 – 5.07	1.00 –	1.66	.1102	9.48	9
65.50–67.49...	5.07 – 7.07	1.66 –	2.31	.0381	3.28	2
67.50–69.49...	7.07 – 9.07	2.31 –	2.96	.0089	.77	1
			TOTALS	.9968	85.73	86

$\bar{X} = 60.43$
$N = 86$
$s = 3.06$

In the example, the probabilities do not quite add to unity nor do the theoretical frequencies quite add to N. This is because the ordinates of the normal curve are not zero except at positive and negative infinity. Theoretically then, if a variate is really normally distributed there is some small but positive probability that it will take a very large positive or negative value. However, the proportion of the area under the curve lying between $+4$ and -4 standard deviations is .99994. That is, only .00006 of the total probability lies outside these limits. Even the limits $+3$ and -3 standard deviations include .997 of the total probability, so that for all ordinary purposes it can be assumed that a measurement lying outside the limit of ± 3 and certainly of ± 4 standard deviations from the mean has no probability; that is, it does not exist.

If animal size were exactly normally distributed, this would mean that we had only to collect enough specimens to find a mouse with ears a mile long, or, more absurd, one whose ears had a great negative length! The statement that a natural variate is "normally distributed" must be understood to mean that it is so for all practical purposes.

The Normal Approximation to the Binomial Distribution

Not only is the normal distribution adequate to describe many continuous variates, but it is, like the Poisson distribution, an approximation to the binomial. Whereas the Poisson approximation depended upon n

growing quite large while p grew correspondingly small, the normal distribution approximates the binomial as n grows larger, irrespective of the size of p. It is sometimes stated that the normal distribution approximates only a symmetrical binomial distribution ($p = .5$) but, as a matter of fact, for sufficiently large n, the value of p is unimportant. Even for moderate n, the approximation is quite good, as shown in Example 48, which is the binomial distribution for $p = 1/6$, $n = 12$ compared to the normal approximation. To find the normal distribution corresponding to a particular binomial, it is only necessary to set

$$\mu = np$$

and

$$\sigma = \sqrt{npq}$$

the mean and standard deviation, respectively, of a binomial distribution. The number of successes, X, can then be converted to standardized normal deviates by the relationship

$$\tau = \frac{X - np}{\sqrt{npq}}$$

and the two distributions directly compared, always remembering that X must be considered the midpoint of range of values for calculation of the normal probabilities.

EXAMPLE **48**. Comparison of a binomial distribution with its normal approximation for $n = 12$, $p = 1/6$. X is the number of successes.

X	CLASS LIMITS	CLASS LIMITS AS DEVIATIONS	STANDARDIZED DEVIATIONS	BINOMIAL PROBABILITIES	NORMAL PROBABILITIES
0	−.500 − .499	−2.499 − −1.500	−1.94 − −1.16	.1122	.0968
1	.500 − 1.499	−1.499 − −.500	−1.16 − −.39	.2692	.2253
2	1.500 − 2.499	−.500 − .499	−.39 − .39	.2961	.3034
3	2.500 − 3.499	.500 − 1.499	.39 − 1.16	.1974	.2253
4	3.500 − 4.499	1.500 − 2.499	1.16 − 1.94	.0888	.0968
5	4.500 − 5.499	2.500 − 3.499	1.94 − 2.71	.0284	.0228
6	5.500 − 6.499	3.500 − 4.499	2.71 − 3.49	.0066	.0032
7	6.500 − 7.499	4.500 − 5.499	3.49 − 4.26	.0011	.0005
8	7.500 − 8.499	5.500 − 6.499	4.26 − 5.03	.0002	.0000
9	8.500 − 9.499	6.500 − 7.499	5.03 − 5.81	.0000	.0000
10	9.500 − 10.499	7.500 − 8.499	5.81 − 6.58	.0000	.0000
11	10.500 − 11.499	8.500 − 9.499	6.58 − 7.36	.0000	.0000
12	11.500 − 12.499	9.500 − 10.499	7.36 − 8.13	.0000	.0000
			TOTALS	1.0000	.9741

$\mu = np = 2.000$
$\sigma = \sqrt{npq} = \sqrt{1.667} = 1.291$

As the example shows, the agreement between binomial and normal probabilities is reasonably good despite the very small value of n and the high degree of asymmetry. It is this asymmetry which causes the normal probabilities to add to .9741 rather than 1.000, since the larger negative deviations are missing. Had n been 10 times as large so that the mean number of successes was 20, then 0 successes would lie about 5 standard deviations from the mean rather than 1.5 standard deviations as it does in this example. Such a large deviation would correspond to a zero probability for this extreme class, and the normal probabilities would add to unity for all practical purposes.

Special Properties of the Normal Distribution

If, as Quételet discovered, many continuous variates in nature are normally distributed for all practical purposes, then many of the particular properties of the normal distribution are also properties of natural distributions, and some use may be made of them.

Range and standard deviation. In our discussion of the range, it was pointed out that the sample range is a downwardly biased estimate of the population range. For a normally distributed variable, however, there is an adequate substitute for the sample range as an estimate. A glance at the table of the standardized normal distribution shows that about 95 per cent of the population falls between $\mu + 2\sigma$ and $\mu - 2\sigma$, and that 99.7 per cent of the area falls between $\mu + 3\sigma$ and $\mu - 3\sigma$.

In other words, if the actual distribution of the natural population is nearly normal, about 3 individuals in every 1000 will have a measurement of a given variable more than 3 standard deviations above or below the mean. Since natural populations usually number many thousands and may run into the millions, there may thus be a large absolute number of individuals outside the range $\bar{X} \pm 3s$ even though their proportion in the population is small.

The theoretical normal curve has an infinite range, but in a natural population there is, at any given moment, a finite largest and a finite smallest truly existing value for every variate, and hence a real and finite range. As a rule the absolute maximum and minimum values in the natural population cannot be determined, but every sample drawn from the population has an observed highest and lowest value of each measured variate, and hence a known *observed range* (often symbolized O.R. in tabular publication). The probability that any sample will contain both the highest and the lowest value from the natural population is usually exceedingly small, virtually nil. Hence the observed range is practically always smaller than the population range; that is the basis for the previous statement that this estimate of range is biased downward.

The extent of that bias depends on the size of the sample. A single specimen as a sample has O.R. = 0, obviously the greatest possible underestimate. Two specimens have O.R. > 0, but still, in practice, far below the natural population range. Ten specimens will, as a rule, have a considerably larger O.R., but still well below the population range. As the sample size increases the O.R. tends more and more closely to approach the population range, which is the maximum possible O.R. If many samples of the same size are taken from the same population, their O.R.'s will of course vary but the average of those O.R.'s will tend toward a fixed value, which depends on the sample size, N, and the standard deviation, estimated by s. For $N = 10$, for instance, the mean value of O.R. $= 3.08\ \sigma$ and is of course estimated by 3.08s. For $N = 100$ the estimated mean O.R. is 5.02s. (Note that an observed range is almost completely meaningless unless the sample size is known; never publish O.R. without the corresponding value of N.)

Table 1 shows the relationships among some values of N, σ, and O.R., on the assumption that the population from which the samples are taken is really normally distributed. In addition to the average value of O.R./σ, the table gives the limits between which sample values of O.R./s will fall 99 per cent of the time.

Such a table of relationships can be of some use in estimating true population ranges or at least in comparing the observed ranges from different populations. A real, finite population of animals can be regarded as a sample from an infinite population of possible animals, so that the real population range could be estimated from Table 1. Thus, if the real population contained 1000 individuals, and if this population were regarded as a sample of 1000 from an infinitely large, normally distributed population, then the real population range could be estimated as $\bar{X} \pm 3.24$s. Even this method of estimating population range has serious drawbacks. First, the size of the population is rarely if ever known, nor can it be closely estimated in most cases. This is not too serious an objection, since for very large populations the ratio of range to standard deviation is fairly insensitive to changes in population size. For example, a doubling of population size from 500 to 1000 results in a change in the mean ratio of only 6.67 per cent. Although it is not shown in the table, a population of 10,000 has an expected ratio O.R./σ of only about 7.75. A second and more serious objection to the estimation of population range by this method is that the real population is almost certainly not limited in its range because of accidents of sampling alone. There are biological restrictions on the range of any variate in nature quite apart from the statistical restrictions, so that no population, no matter how many individuals it contains, can ever have a range larger than a relatively few standard deviations.

TABLE 1. Sample frequency (N), mean observed range ($\overline{O.R.}$), and standard deviation (σ). (Data from Tippett, 1925, and Pearson, 1932)

A. For $N \leq 100$, with 99 per cent confidence intervals

N	$\overline{O.R.}/\sigma$	99% INTERVAL FOR O.R./s	N	$\overline{O.R.}/\sigma$	99% INTERVAL FOR O.R./s
2	1.28	.02–3.64	22	3.82	2.36–5.71
3	1.69	.22–4.10	24	3.90	2.45–5.76
4	2.06	.47–4.38	26	3.96	2.53–5.82
5	2.33	.70–4.59	28	4.03	2.61–5.87
			30	4.09	2.69–5.91
6	2.53	.89–4.74	35	4.21	2.84–6.01
7	2.70	1.07–4.87	40	4.32	2.97–6.09
8	2.85	1.22–4.98	45	4.42	3.09–6.16
9	2.97	1.36–5.07	50	4.50	3.19–6.23
10	3.08	1.48–5.15			
11	3.17	1.59–5.22	55	4.57	3.28–6.29
12	3.26	1.69–5.28	60	4.64	3.36–6.34
13	3.34	1.78–5.34	65	4.70	3.43–6.38
14	3.41	1.86–5.39	70	4.76	3.50–6.43
15	3.47	1.94–5.44	75	4.81	3.56–6.47
16	3.53	2.01–5.49	80	4.85	3.62–6.50
17	3.59	2.08–5.53	85	4.90	3.67–6.54
18	3.64	2.14–5.57	90	4.94	3.72–6.57
19	3.69	2.20–5.61	95	4.98	3.77–6.60
20	3.74	2.25–5.64	100	5.02	3.81–6.63

B. For selected larger values of N

N	$\overline{O.R.}/\sigma$	N	$\overline{O.R.}/\sigma$	N	$\overline{O.R.}/\sigma$
150	5.30	400	5.94	600	6.18
200	5.49	442	6.00	700	6.28
250	5.64	450	6.01	800	6.35
300	5.76	500	6.07	900	6.42
350	5.85	550	6.13	1000	6.483

In practical use, then, estimates of range are not estimates of *real population range* but of mean range for samples of a given size or for a so-called *standard population* to which a conventional (and generally fictitious) size is assigned. Thus $\bar{X} \pm 3s$ is an estimate for a "standard population" of 442 individuals. Another estimate sometimes used is $\bar{X} \pm 3.24s$, called *standard range* (S.R.), which assumes a "standard population" of 1000 individuals. Even though there is no practical unbiased estimate of real population range, *comparisons* of ranges among different populations are valid and unbiased if they are based on the same convention as to "standard population" size.

Some idea as to how these estimates work out can be obtained from the data of Example 29, on tail length in *Peromyscus*, where $\bar{X} = 60.43$, $s = 3.06$, and $N = 86$. The mean observed range for samples with $N = 86$ is $(\bar{X} + 2.45s)$ to $(\bar{X} - 2.45s) = 67.9$ to 52.9. The standardized estimate from $\bar{X} \pm 3\sigma$ is $(\bar{X} + 3s)$ to $(\bar{X} - 3s) = 69.6$ to 51.2. Standard range (based on $N = 1000$) is S.R. $= (\bar{X} + 3.24s)$ to $(\bar{X} - 3.24s) = 70.3$ to 50.5. The range in the actual sample is O.R. $= 68$ to 52. That is very close to the mean O.R. for samples of this size. It is, as would be expected, a somewhat smaller range than the estimates based on "standard populations" larger than this sample.

The relationships given in Table 1 also make it possible to estimate σ, the population standard deviation, without calculating the sample standard deviation, s. The observed range is not as efficient as is the sample standard deviation, s, but for samples of 10 or less its efficiency is reasonably high.

For the example just cited of tail length in *Peromyscus*, the observed range was from 51.5 to 68.5, or 17. For a sample of 86 the expected ratio of σ to range is approximately .204. The estimate of σ is then
$$(.204)(17) = 3.468$$
which is larger than the estimate of 3.06 based upon s.

The observed range is not really an adequate substitute for s in most cases, but it can be used for a rough estimation of σ where more accuracy is not required and as a check on gross mistakes in calculation of the sample standard deviation.

Special Parameters

Most zoological variates are distributed in accordance with Quételet's rule, that is, they have one class which is most frequent, the frequency of classes falling away on each side of this class. In this sense, the distributions roughly resemble a normal distribution, and it is usually assumed (in fact, *must* be assumed for many statistical operations) that they are normally distributed. Since it is unlikely that any real distribution will be exactly normal, it is occasionally useful to describe the deviation of the distribution from normality. In actual fact, any measure of such deviation is not particularly useful except as a rough description, since no really adequate measure of these deviant tendencies has been devised.

Moreover, it is doubtful that anything of real interest can be learned from the degree of nonnormality of a distribution except in very special cases. Nevertheless, there are two measures, *skewness* and *kurtosis*, which have been used in the past, so that a short discussion of them is necessary.

Skewness. Skewness is a deviation from the normal curve by asymmetry of the distribution, which falls off more rapidly on one side of the high point than on the other. Its high point, the mode, no longer coincides with the mean, as in the normal distribution, for the mean is strongly affected by extreme values and the mode is not affected at all by them. Advantage is taken of these facts to devise a measure of skewness. The distance between the mean and the mode is such a measure, but it is not satisfactory. In the first place, it depends not only on skewness but also on the degree of dispersion, and in the second place, it is an absolute figure and hence is not readily used in comparison of variates of different magnitudes. Both these objections are met by dividing the difference between mean and mode by the standard deviation of the sample, giving $S_k = $ (mean $-$ mode)/s, S_k being the usual symbol for skewness.

It was shown in Chapter 5, however, that exact estimation of the mode is impractical and that it can most easily be approximated by the relationship

$$\text{Mode} = 3 \text{ median} - 2 \text{ mean}$$

Substituting this in the equation for skewness gives

$$S_k = \frac{\text{mean} - (3 \text{ median} - 2 \text{ mean})}{s} = \frac{3 (\text{mean} - \text{median})}{s}$$

This is the best form for calculating this measure. In a symmetrical distribution, mean and median are equal; so S_k is 0. In a right-skewed distribution there are extreme values to the right of the averages; hence the mean is larger than the median and the coefficient of skewness is positive. Similarly, in a left-skewed distribution the coefficient is negative. There is no theoretical limit to the magnitude of the coefficient, but in fact it seldom is less than -1 or more than $+1$ in the distributions even distantly approaching the normal.

Example 49 shows the calculation of this coefficient for a distribution that is clearly skewed.

Anthropological data are generally not used in the examples, but it is desired to show the calculation for a large sample and to exemplify an obviously skewed curve of a type common in zoology. In any case, Cambridge undergraduates are zoological materials. This distribution is nearly normal, certainly near enough to use the parameters of a normal curve in dealing with it, but it does have a moderate positive skew. The approximate mode is 6.02, falling in the 6.0 class and there are 8 classes above and only 5 below this, so that skewness is visible on inspection.

EXAMPLE 49. Calculation of the coefficient of skewness for the distribution of head breadth, in inches, of Cambridge students. (Data from Macdonell, 1902)

X	f	fX	X^2	fX^2
5.5	3	16.5	30.25	90.75
5.6	12	67.2	31.36	376.32
5.7	43	245.1	32.49	1,397.07
5.8	80	464.0	33.64	2,691.20
5.9	131	772.9	34.81	4,560.11
6.0	236	1,416.0	36.00	8,496.00
6.1	185	1,128.5	37.21	6,883.85
6.2	142	880.4	38.44	5,458.48
6.3	99	623.7	39.69	3,929.31
6.4	37	236.8	40.96	1,515.52
6.5	15	97.5	42.25	633.75
6.6	12	79.2	43.56	522.72
6.7	3	20.1	44.89	134.67
6.8	2	13.6	46.24	92.48
TOTALS	1000	6,061.5		36,782.23

$$\overline{X} = \frac{\Sigma fX}{N} = 6.062$$

$$s^2 = \frac{\Sigma fX^2 - \frac{(\Sigma fX)^2}{N}}{N-1} = \frac{36,782.23 - 36,741.78}{999} = \frac{40.45}{999} = .0405$$

$$s = .201$$

$$\text{Median} = 5.95 + \frac{23.1}{236} = 6.048$$

$$S_k = \frac{3(\overline{X} - \text{median})}{s} = \frac{3(.014)}{.201} = \frac{+.042}{.201} = +.21$$

Moderately right-skewed distributions of this type are common in zoology, and although an adequate census has not been made, they appear to be more common than are left-skewed distributions. In biological terms, a right-skewed distribution indicates that large variants are more common than small variants within the sample representing the population. Although there are abundant exceptions, this may be a general tendency of morphological characters in zoology.

With the smaller samples usual in zoology, the values of S_k are inevitably somewhat erratic, and a general tendency can only be detected by comparing a considerable series of such values. Example 50 shows typical values for such a series with the usual small samples of zoology and paleontology.

EXAMPLE 50. Coefficients of skewness for a series of small samples of teeth of the extinct mammal *Litolestes notissimus*. (Original data)

VARIATE	N	S_k
Length P_4	10	+.19
Width P_4	11	−.52
Length M_1	19	−.21
Width M_1	19	+.52
Length M_2	28	−.17
Width M_2	29	0
Length M_3	24	+.45
Width M_3	24	+.19
Length P^4	7	0
Width P^4	7	−.51
Length M^1	10	+.29
Width M^1	10	+.83
Length M^2	13	−.18
Width M^2	13	−.54
Length M^3	8	+.60
Width M^3	9	+.43
Mean S_k		+.09

It may be noted that the assumption on which the use of V and the study of variability in general are based involves a constant tendency for such characters to show a small positive skew. This assumption is that the dispersion is proportionate to the absolute value of the variate. If this is true as between different samples and different variates, it should be true also within a single distribution. The absolute dispersion should tend to be, or for an average of many different homologous distributions should be, greater for higher values of the variate than for lower and should increase steadily from the left-hand end of the graphic distribution through to the right-hand end. Since the values thus tend to be spread farther on the absolute scale above the mode than below it, a positive skew is involved. A given value below the mode will bear the same ratio to the mode that the mode will bear to a corresponding value (one of theoretically the same frequency) above the mode, instead of being simply equidistant from the mode as in the normal curve. From this relationship of ratios as opposed to simple linear distances, it follows that the geometric mean and the mode will tend to coincide, whereas in the normal distribution the mode and arithmetic mean coincide (the geometric mean being smaller than the arithmetic for any distribution).

It is probable that the slightly skewed form of curve thus determined is a better theoretical description of most zoological variates than is the normal curve. Basing calculations on the skewed curve would, however, be exceedingly difficult, and it would rarely make the results significantly better in practice, however preferable in theory. The skewed curve thus determined differs little from the normal—so little as to be wholly obscured by chance fluctuations in most zoological samples. Even with the large sample of Example 49, skewness caused by this relationship is too small to have a really appreciable effect. The geometric mean for this distribution is 6.058, the arithmetic mean 6.062. Rounded to the number of places really accurate and significant, these two are exactly the same—6.1 or 6.06—and there is no demonstrable effect of this sort of skewness in the data. The observed significant skewness of this distribution is thus due not to this phenomenon but to some other biological factor.

Kurtosis. Kurtosis is the property of being more pointed or flatter than a normal curve with the same parameters. It does not involve symmetry and usually cannot be detected by inspection unless it is very great. Unfortunately, the measurement of this characteristic is somewhat laborious, although not unduly complicated. The best measure of kurtosis, K_s, is the value of the expression

$$\frac{\sum d^4}{N s^4} - 3$$

It will be remembered that

$$s = \sqrt{\frac{\sum d^2}{N - 1}}$$

so that

$$s^4 = \left(\frac{\sum d^2}{N - 1}\right)^2$$

and this expression may sometimes be more advantageous in calculation.

When the coefficient of kurtosis is 0, the distribution is neither more peaked nor flatter than the normal curve and is sometimes spoken of as *mesokurtic*.

When this value is positive, the distribution is more peaked or sharper than the normal curve and is called *leptokurtic*.

When it is negative, the distribution is flatter than the normal curve and is called *platykurtic*.

Platykurtic distributions generally tend to have relatively large s and V. The flattening usually reflects either high variability or some heterogeneity in the sample. Conversely, leptokurtic curves usually have lower s and V; the peakedness generally reflects low variability. The calculation of a coefficient of kurtosis is shown in Example 51.

EXAMPLE 51. Calculation of the coefficient of kurtosis from the data of Example 29, page 79.

TAIL LENGTH X	f	fX	d $(X-\bar{X})$	d^4	fd^4
52	1	52	−8.43	5,050.22	5,050.22
53	0	0	−7.43	3,047.58	0
54	3	162	−6.43	1,709.40	5,128.20
55	0	0	−5.43	869.36	0
56	3	168	−4.43	385.14	1,155.42
57	8	456	−3.43	138.41	1,107.28
58	7	406	−2.43	34.87	244.09
59	11	649	−1.43	4.18	45.98
60	11	660	− .43	.03	.33
61	10	610	.57	.11	1.10
62	6	372	1.57	6.08	36.48
63	14	882	2.57	43.62	610.68
64	6	384	3.57	162.43	974.58
65	3	195	4.57	436.18	1,308.54
66	1	66	5.57	962.54	962.54
67	1	67	6.57	1,863.21	1,863.21
68	1	68	7.57	3,283.85	3,283.85
TOTALS	86	5,197			21,860.69

$$N = 86 \qquad \bar{X} = \frac{5,197}{86} = 60.43 \qquad s = 3.0584$$

$$K_S = \frac{\Sigma fd^4}{Ns^4} - 3$$

$$K_S = \frac{21,860.96}{(86)(3.0584)^4} - 3 = 2.91 - 3 = -.09$$

CHAPTER NINE

Confidence Intervals

The Principles of Inference

Not only statistical inference but all scientific inference is of a nature not generally understood by the layman, or indeed by many scientists themselves. It is often said that science has "conclusively proved (or disproved)" something. It is unfortunately true that very few general theories or laws can be proved or disproved in the realm of natural science (as opposed to mathematics where the opposite is true), and that those which are of a provable nature are seldom worth the proof. What is true of theories or laws is that greater or less confidence can be placed in them by virtue of repeated observation. If very many observations are made and they all support a theory, the confidence in the truth of the theory may be very great, but it is not proved, for there is no way to exclude absolutely and for all time the possibility that a contrary observation may be made.

The process of theory formation is that a number of observations on some phenomenon suggest a common relationship among them. This first step between the initial observations and the general law is not a rigorous one for there exists no methodology for proceeding from the particular to the general in an exact way. To reason that because a coin has come up heads ten times in a row it will always do so and in fact has two heads is an intuitive, not an exact, result.

From the general law, various specific results which ought to follow can be deduced or predicted. This process of deduction is logically rigorous in contrast to the inductive reasoning which went into making the hypothesis.

General statements—those which are formed implicitly or explicitly in terms of "always," "sometimes," or "never"—vary in the degree to which observation can substantiate them. As an example, we may take the following four statements about the plague grasshopper *Austroicetes cruciata:*

    1. In 1960 there will be an outbreak of the plague grasshopper *A. cruciata* in the vicinity of Hawker, Australia.

2. There are always six years between outbreaks of the plague grasshopper.
3. There may be as many as 1000 adults of *A. cruciata* in a square meter during an outbreak.
4. The average number of adults in a swarm of *A. cruciata* in a plague year is 10,343.

Statement 1 is highly specific and it can be either proved or disproved. Either there will be an outbreak in 1960 or there will not, so that a single observation will suffice to determine the truth of the statement.

Statement 2 is more general and, in fact, specifically contains the term "always." Such a generalization can be disproved only. A single occurrence of a plague five or seven years after the previous one will show the theory to be false, but many observations of a six-year cycle can never prove that the cycles will always be six years long, although one might become more and more confident of the theory's validity as the number of observed six-year cycles grew greater and greater.

In a similar way, statement 3, which contains an implicit "sometimes" can be proved but not disproved. If a single square meter is found to contain 1000 individuals, proof is at hand, but the failure to find such a concentration does not show conclusively that it can never exist.

Finally, statement 4, which is by far the commonest sort in quantitative work, is not susceptible of proof or disproof at all. It is really a statement about a population which includes all the swarms of *A. cruciata* that have ever been or will be. Any statement about a population parameter is of this sort.

Statements like 2, 3, and 4 are termed *hypotheses* in statistical usage, and this is the term that we will use in referring to any assertion about a population when only samples are actually observable.

As hypotheses become more general, less can be said about them in a definite way. Hypothesis 2 makes a very general statement ("always") about the periodicity of outbreaks, but this generality occurs at the expense of the degree of definiteness with which the hypothesis can be established. The same contrast holds between statements 3 and 4. While 3 can be proved, it only states that some swarms will have 1000 individuals in a square meter, not a particularly useful piece of information. Hypothesis 4 about the mean is far more informative, but its truth can never be established.

This choice between knowing a small amount exactly and a large amount poorly is one with which the zoologist is continually faced. There is no formula, no equation, no rule which can make this decision for the zoologist. The rules and formulae can provide an idea of how accurately any value is known and how much confidence may be placed in a given

150 QUANTITATIVE ZOOLOGY

hypothesis, but the choice of hypothesis rests entirely with the scientist.

While it is true that most hypotheses cannot be proved, there must certainly be a point at which a hypothesis is either supported or not to such an extent that it is "proved beyond a reasonable doubt." That is, one must recognize that although it may be unprovable, it is necessary to act as if a hypothesis were true or false in any particular case. Although one cannot know exactly the value of a population parameter like the mean but can only estimate it, at least some limits for the value must be specifiable along with the degree of confidence that the value really is contained within those limits. For if some statement cannot be made about the truth of hypotheses or the true value of parameters, there would be very little sense indeed in measuring or testing anything.

The science of statistical inference is designed to make somewhat more rigorous this concept of "reasonable doubt." At the same time, it cannot tell the zoologist how reasonable the doubt must be before he should accept or reject a hypothesis. Rather it is a method for making quantitative the degree of confidence in a given hypothesis or in a given range of possible values for a parameter. Whether that confidence is sufficiently high to convince the zoologist is another matter indeed and one that involves intuition.

The Basic Form of Statistical Inference

A parameter is a fixed value so that no valid probability statement can be made about it. Either the mean of a population is equal to some number X or it is not. One simply cannot speak of the "probability" that it takes the value X, under our empirical definition of probability. An estimate, on the other hand, is a variate with a probability distribution, and this probability distribution will depend upon the true value of the parameter. This dependence of the probability distribution of the estimate on the true value of the parameter is the only rigorous link between estimate and parameter, and it is this relationship which must be used in the process of inference.

Suppose that a fair coin is tossed 100 times. If the probability of heads is really one-half, then the probability that all 100 tosses will come up heads is vanishingly small (to express the probability as a decimal fraction would require 30 zeros between the decimal point and the first digit). To turn the example around, let us suppose that when a coin is tossed 100 times it actually *did* come up heads every time. Then, on the basis of the previous probability statement, it would be considered extremely unlikely that the coin was a fair one. The judgment that the coin is not fair (in fact, has two heads!) is based upon the contrast between what would be expected if the coin *were* fair and what really happened. The process of reasoning is as follows: An hypothesis is set up, in this case, that the coin is a fair one with the probability of heads equal to one-half. As a consequence of this hypothesis, a distribution can be set up giving the probability of observing any particular sample. A sample is then taken and the result observed. If the result is one with a very small probability under the hypothesis, then the hypothesis itself is considered doubtful, or, to put it in another way, there is very little confidence in the correctness of the original assumption. The *measure* of the confidence in the original hypothesis is the probability of observing what was actually observed. It is important to notice that the probability is not assigned to the hypothesis but that a special term, "confidence," is used. We have greater or less confidence in any hypothesis, depending upon whether the observed results have greater or less probability under the hypothesis. *This is the basic form of all statistical inference.* It is not simply quibbling about the words "confidence" and "probability," because an inexact understanding of the process of inference can lead to erroneous results and in the early history of statistics did lead to such results at the hands of very competent mathematicians, who failed to understand the logical difference between a sample and a population.

Again the problem arises as to how small the probability of the observation must be before the hypothesis is rejected, and again there is no answer to this problem except by reference to the zoologist's own judgment. In some cases, it may be important to be quite sure about the hypothesis, so that even a moderately large probability will be considered too small. In other cases one may wish to have very strong reasons for rejecting the hypothesis, so that a very small probability will be required for rejection. In any event, no matter what decision is made, it must be paid for in the chance of making an error. If a very small probability is demanded before a hypothesis will be rejected, this increases the chance that an incorrect hypothesis will still be considered acceptable. For instance, if it is demanded that a coin turn up heads 100 times before the hypothesis of $p = \frac{1}{2}$ is rejected, there is a very strong probability that a highly erroneous hypothesis will be maintained. Conversely, if the hypothesis of fairness of the coin were rejected when 45 heads and 55 tails appeared, there is a strong chance that the rejection is unwarranted. In general, the more stringent the requirements for rejection the greater the chance of keeping a false theory; the more easily a hypothesis is rejected the greater the chance for throwing away a true one.

In summary, the methods to be discussed under the general heading of statistical inference provide, first, a quantitative estimate of the degree of confidence which may be placed in any hypothesis by consideration of the observations, and, second, an estimate of the chance that a given decision is erroneous in one way or another. This is all that statistical methods can do.

Confidence Intervals

Methods have been discussed for estimating various parameters of distributions from characteristics of the samples. A distinction was drawn between the sample values which are estimates of the true values and the true values themselves. We do not know the true values; we can never know them except in those few cases where the entire population is fixed over time and could be exhaustively examined. Is it possible, then, to say anything at all about the true values of the parameters, when only estimates are available? The answer, fortunately, is that a good deal can be *inferred* about the population parameters from the evidence of the samples, along the lines suggested in the last chapter. Although it is impossible to know the exact value of a parameter, it is possible to describe a range of values or an interval in which the parameter is likely to lie. This range or interval is called a *confidence interval*, and the specification of such an interval is quite as important in quantitative work as the estimation of the parameter. To say that the estimate of the mean of a certain population is 2.73 cm. is of not much interest unless some idea is also given about how far away from this estimate the true mean may really be. The construction of a confidence interval can best be understood by reference to Fig. 16.

Suppose that the true mean of a certain population is μ and that all of the other information needed to specify the population is also known. Different samples drawn from this population will have different values of $\bar{X}$, the sample mean, and the probability distribution of $\bar{X}$ can be constructed from the knowledge of the characteristics of the original population. Fig. 16 shows the probability distribution of $\bar{X}$ in a schematic way. The mean of this distribution is the mean of the original population, μ, since $\bar{X}$ is an unbiased estimate of μ. It should be noted that Fig. 16 is *not* the probability distribution of the original population but of $\bar{X}$ itself. This distribution is dependent upon the form of the original distribution and on the parameter μ, but it is not identical with the former. The abscissa denotes the values which $\bar{X}$ may take and the points $\bar{X}_a$ and $\bar{X}_b$, which mark off the shaded area, include between them 95 per cent of the probability. This means that 95 per cent of the time the value of $\bar{X}$ in samples taken from the original population will lie between $\bar{X}_a$ and $\bar{X}_b$.

Another way to describe this is that the interval $\bar{X} - b$ to $\bar{X} + a$ will cover the point μ 95 per cent of the time. Under the distribution are drawn four possible cases which make this relation clear. The horizontal line is of length $a + b$ with $\bar{X}$ near the center of the interval. When $\bar{X}$ lies slightly to the right of $\bar{X}_b$ the interval does not quite include μ. This is also the case when $\bar{X}$ lies to the left of $\bar{X}_a$. On the other hand, when $\bar{X}$ falls between $\bar{X}_a$ and $\bar{X}_b$, the interval does include μ. This interval between $\bar{X} - b$ and

$\overline{X} + a$ is the 95 per cent confidence interval for the parameter μ. Had a and b been made larger distances, the probability that the interval would cover the mean would be even greater and one may construct a 99 per cent confidence interval or a 99.9 per cent confidence interval or a smaller one if desired. The larger the interval, of course, the greater the confidence that the true value of the parameter will be included in it, but the less useful it is. Whereas one can be sure that a confidence interval from negative to positive infinity will include the true value, this is not a useful bit of information. The smaller the range of possibilities for the value of the parameter, the better. Unfortunately, as the confidence interval is made smaller and smaller, the lower becomes the confidence that μ really lies in this range. In each case a decision must be made as to whether it is better to be nearly sure that the parameter lies in some large range, or to be not so

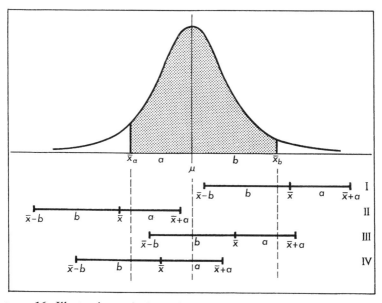

FIGURE 16. Illustration of the principle of confidence intervals. The probability distribution is that of the sample mean, $\overline{X}$. The cross-hatched area between $\overline{X}_a$ and $\overline{X}_b$ includes 95 per cent of the area under the curve. The four horizontal lines below the distribution represent four possible sample values of $\overline{X}$ with associated confidence interval $\overline{X}-b$ to $\overline{X}+a$. In cases I and II, $\overline{X}$ falls outside the 95 per cent range and the associated confidence intervals do not include μ, the true mean. In cases III and IV, $\overline{X}$ falls within the 95 per cent range and the associated confidence intervals do cover μ.

sure that it lies in a smaller one. In some cases, it is better to know somewhat less about the parameter but to be quite confident about what you do know. No statistical technique can be devised which will make this decision.

Once more the choice between "knowing a great deal poorly or a little bit well" is up to the zoologist. There has grown up in most quantitative work a tradition that 95 per cent is the confidence limit which is best. There is no particular reason for this either biologically or mathematically, but it is so firmly entrenched that it is now standard procedure. Almost all publications which give confidence intervals use 95 per cent as the confidence level, but this ought not to be a deterrent to the use of any value which seems reasonable in a given case. For many, if not most cases, it is not necessary to publish confidence intervals at all, because they bear a constant relation to the magnitude of the standard deviation. When this is the case, the reader may construct his own confidence intervals from the published values of $\bar{X}$, s, and N, the sample size. The cautious worker, assuming no statistical knowledge on the part of his reader, might better publish the 90 per cent, 95 per cent, and 99 per cent limits, as well as the basic data, allowing others to make the choice.

It might be supposed that the general method of confidence intervals discussed above assumes a knowledge of μ, the very quantity which is unknown in practice. In fact, the methods of finding confidence intervals in specific cases discussed below are such that they are totally independent of the true value of the parameter and the intervals constructed by these methods will cover the true parameter 95 or 99 per cent of the time, no matter what the true value is.

Binomial Distribution

In Example 52 are some results obtained in a study of the relation between clutch size and survival of eggs in the American eider (*Somateria mollissima dresseri*). For each clutch size, a number of nests was observed and each nest was examined to see whether any of the eggs had been destroyed by predators. Any nest in which one or more eggs had been attacked was scored as negative, while nests in which all eggs were intact were scored as positive.

The samples vary greatly in size, from 3 to 45, and the sample values of f/n, the proportion of undisturbed nests, vary from .13 to .33. Each sample value is an estimate of a true proportion, p, of nests which have not been preyed upon, and the author was interested in knowing whether this proportion is the same irrespective of clutch size. While that problem, strictly speaking, is one of statistical testing, a topic which is discussed in Chapter 13, we may use these data as an exemplification of confidence interval estimation for a binomial proportion.

EXAMPLE 52. Clutch size as related to predation pressure on eggs of *Somateria mollissima dresseri*. (Partial data from Paynter, 1951)

CLUTCH SIZE	NUMBER OF NESTS (n)	NUMBER OF NESTS WITH ALL EGGS INTACT (f)	RELATIVE FREQUENCY OF UNAFFECTED NESTS (f/n)	95% CONFIDENCE INTERVAL
1	3	1	.33	.01–.91
2	24	6	.25	.10–.42
3	37	8	.22	.10–.35
4	45	6	.13	.05–.26
5	17	5	.30	.10–.50
6	6	2	.33	.04–.78

For concreteness, we may fix 95 per cent as the confidence level. The lower confidence limit for p, denoted by p_l, is the value of p so low that the observed number of positive nests *or more* would occur only 2.5 per cent of the time. Similarly, p_u, the upper confidence limit for p, is the value so high that the observed number of positive nests *or less* will occur 2.5 per cent of the time. Then only 5 per cent of the time will the interval between p_l and p_u fail to cover the true p, while it will cover p 95 per cent of the time. For example, for a clutch size of 2, there were 6 unaffected nests out of 24 nests observed. Then p_l is the value of p in a binomial distribution with $n = 24$ such that the probability of 6-24 occurrences is .025. Conversely, p_u is the value of p which makes the probability of 0-6 occurrences equal to .025. To find these values, it is necessary to use tables of the cumulative binomial distribution such as the *Harvard Computation Laboratory Tables* for n less than 50 and Romig's *Tables* for n between 50 and 100.

Binomial tables are extremely useful and ought to be employed for confidence interval estimation far more often than they are. One reason for this neglect is that the tables are a little difficult to use. To save space, the publishers tabulate values for binomial distributions with $p = .50$ or less only. It then requires some mental agility to make completely general use of the tables. Fortunately, in the introduction to Romig's *Tables* (pp. xx-xxiii) there is a simple explanation of the most general use of binomial tables with examples of 8 different cases, covering all possible situations. A familiarity with this discussion will enable the reader to use not only Romig's *Tables* but other binomial tables which are differently arranged.

For the case of clutch size of 2, the Harvard Tables are used and it is found that when $n = 24$, the probability of 0-6 occurrences is closest to .025 for $p = .42$. From the table, it also appears that when $p = .10$, the probability of 6-24 occurrences is about .025. Then

$$p_u = .42 \text{ and } p_l = .10$$

Performing the same operation for the rest of the data results in the confidence intervals shown in the last column of Example 52.

The confidence intervals thus obtained are not exact 95 per cent intervals. It is not possible to find a p in the table for which the required probability for p_l and p_u is exactly .025. For $p = .10$, the tables actually show that the probability of 6 or more successes is .02766. When $p = .42$, the probability of 6 or less successes is .02587. The confidence level is then more precisely $1 - .02587 - .02766$ or .94647. This is, however, sufficiently close for our purposes. Moreover, there is nothing particularly sacred about the 95 per cent level of confidence.

The example shows that the confidence limits for the true value of p are fairly wide even for the large sample size of 45. For clutch size of 1, only three nests were observed and the confidence interval is so large as to make the observation virtually useless. Confidence intervals are generally large for a binomial proportion unless the number of observations is quite large. For example, if 100 nests with a clutch size of 4 had been examined rather than 45 and the same proportion of successes observed (13 successes out of 100), the confidence interval for p would have been .07-.20 which is some improvement over the interval .10-.26 but not an immense one.

The Binomial Probability in Large Samples

In Chapter 8 the normal distribution was derived from the binomial distribution as the number of observations (n) grew larger and larger. The *number* of successes in a sample of size n is then approximately normally distributed when n is large, and this normal distribution has a mean of np and a standard deviation of $\sqrt{npq}$. The *proportion* of successes will be normally distributed with mean p and standard deviation $\sqrt{pq/n}$. Since it is impractical to tabulate the binomial distribution for very large n, and since the normal approximation holds quite well for even moderately large n, the confidence limits for p in a binomial distribution may be found by using the normal approximation.

Recalling the previous discussion of the standardized normal distribution, it was shown that 95 per cent of the area under a normal curve lies between $+1.9600$ and -1.9600; 99 per cent of the area lies between $+2.5760$ and -2.5760; and so on. These relationships can be checked by referring to Appendix Table I of the cumulative normal distribution. If f/n denotes the proportion of successes or observed occurrences in a sample size of n, the 95 per cent confidence limits for p, the population probability, is given approximately by

$$\frac{f}{n} + 1.960 \sqrt{\frac{\left(\frac{f}{n}\right)\left(\frac{1-f}{n}\right)}{n}} \quad \text{to} \quad \frac{f}{n} - 1.960 \sqrt{\frac{\left(\frac{f}{n}\right)\left(\frac{1-f}{n}\right)}{n}}$$

That is, it is the observed proportion of successes plus or minus 1.960 times the standard deviation. The 99 per cent confidence limits would be

$$\frac{f}{n} + 2.576 \sqrt{\frac{\left(\frac{f}{n}\right)\left(\frac{1-f}{n}\right)}{n}} \quad \text{to} \quad \frac{f}{n} - 2.576 \sqrt{\frac{\left(\frac{f}{n}\right)\left(\frac{1-f}{n}\right)}{n}}$$

In general if K per cent of the area under the normal curve is included between the limits $\pm \tau$, then the K per cent confidence interval for p is given by

$$\frac{f}{n} \pm \tau \sqrt{\frac{\left(\frac{f}{n}\right)\left(\frac{1-f}{n}\right)}{n}}$$

To exemplify the use of this formula, consider Example 52. If the clutch sizes are ignored (as in this case they may be, since there is no difference in proportion of nests affected among the various clutch sizes), the whole set of observations can be treated as a single sample of 132 nests in which 28 nests are not affected. The sample proportion f/n, is then .21 and the 95 per cent confidence interval for p is given by

$$.21 \pm 1.96 \sqrt{\frac{(.21)(.79)}{132}}$$

which equals

$$.21 \pm .069$$

or from .14 to .28

This normal approximation is best for n greater than 100, since tables of the binomial distribution are incomplete for large values of n. Even for n between 50 and 100, the approximation holds fairly well. If binomial tables are available, it is far better to use them especially when the observed f/n is close to 0 or 1. In that case the true binomial confidence interval tends to be very asymmetrical, while the normal approximation is always symmetrical around f/n as is evident from its form. When n is large this asymmetry disappears even for f/n close to 0 or 1, so that the normal approximation is again applicable. It is interesting to compare the normal approximation with the binomial tables for the example just given. The closest sample size listed in the Harvard Tables is 130 and using this value the binomial confidence interval for p in the example turns out to be approximately .14-.28 which is identical with the result of the normal approximation calculated to two decimal places.

The Mean of a Normal Distribution

The fact that most quantitative measurements are roughly normally distributed has been discussed on page 133.

Most methods of statistical inference are dependent upon a reasonable approach to normality of the distribution of the variate, but there is one fact which makes this requirement not quite as restrictive as it may seem. Recent empirical work by statisticians has shown that most common statistical methods are not particularly sensitive to deviation from normality, and although the normal distribution is a basic assumption of these methods, they apply nearly as well to variates of many kinds. This is a particularly fortunate discovery since any great sensitivity of statistical methods to the distribution of the variate would make even the simplest statistical work a very complex procedure requiring the services of a professional mathematical statistician.

For most determinations of confidence intervals, it is then necessary, and not entirely unreasonable, to assume that any continuous variate of the sort which is directly observed is normally distributed.

If there is some very strong reason for supposing that a variate under consideration is very far from being normally distributed—for example, if it should be strongly bimodal—the standard methods of determining confidence intervals are not applicable, and certain special techniques must be invoked. There are two such techniques available. The first involves fitting the observations to one of a series of standardized curves known as *Pearsonian distributions* from which inference may be made. The second, or *distribution-free*, technique is predicated on the characteristics of cumulative probability distributions which are free of any assumption about the nature of the original distribution of the variate. This latter technique is quite weak in the precision with which inferences can be made, but it has the advantage of being independent of any assumptions about the variate. Both of these techniques are outside the scope of this book, but any zoologist who feels strongly that his observations are of such a nature that they require special treatment ought to consult with a statistician familiar with these methods. An excellent and readable discussion of distribution-free methods can be found in Mood (1950, Chapter 16), and many zoologists may find this discussion sufficiently comprehensible to be useful to them. Any use of Pearsonian distributions had best be left in the hands of the professionals.

Assuming, as always, that continuous variates are roughly normally distributed, it is a simple matter to find a confidence interval for the mean of a population.

If the standard deviation of the population were known, then the 95 per cent confidence interval for μ would be given by $\bar{X} \pm 1.96\sigma/\sqrt{N}$ where N is the sample size. Unfortunately, σ is not known but can only be estimated from s. Knowing neither μ nor σ would seem to make the problem of confidence intervals for μ insoluble since the distribution of $\bar{X}$ cannot be specified. This problem was for a long time unsolved until W. S. Gosset, a statistician for Guinness Brewery, writing under the pen name of "Student" published what is now known as the "Student's t" distribution. This distribution is given in a special cumulative form in Appendix Table II. The body of the table contains various values of Student's t. Across the top are given the probabilities of falling between $+ t$ and $- t$. Finally at the left are values of n, the *degrees of freedom*, a parameter which will be considered again for other distributions. In sample distributions of the kind here discussed, the degrees of freedom are equal to one less than the sample size N, so that if a sample of 31 is taken, n will be 30. A more general discussion of degrees of freedom will be found in the chapter on χ^2 tests (page 309).

The form of the confidence interval for the mean of a continuous variate (always assuming normality, of course) is simply

$$\bar{X} \pm t_n \left(\frac{s}{\sqrt{N}}\right)$$

where t_n is the value of t for n degrees of freedom which corresponds to the confidence level desired. As an example of the calculation of this confidence interval we may use Dice's measurements of the tail length in *Peromyscus maniculatus* (Example 32).

For this case
$$N = 86$$
$$\bar{X} = 60.43$$
$$s = 3.0589$$

Then the confidence interval will be of the form

$$60.43 \pm t_{85} \left(\frac{3.0589}{\sqrt{86}}\right)$$

or

$$60.43 \pm t_{85} \, .3298$$

To find the suitable value of t, choose a confidence level—let us say 95 per cent—and using the .95 probability column in the table, find the value of t corresponding to 85 degrees of freedom. This turns out to be approximately 1.99, so that the 95 per cent confidence interval for this example is

$$60.43 \pm (1.99)(.3298) = 60.43 \pm .66$$

With 95 per cent confidence, then the true value of μ lies between 59.77 and 61.09. Had the 99 per cent confidence interval been desired, the .99 probability column would be consulted, which for 85 degrees of freedom gives t equal to approximately 2.64. The resultant confidence interval is from 59.56 to 61.34.

The t distribution is symmetrical, like the normal, and as the number of degrees of freedom grows larger, the t distribution gradually becomes a normal distribution. For a very large sample (indicated by infinite degrees of freedom in the table), the value of t corresponding to 95 per cent probability is 1.96, which, it may be remembered, is exactly the value of the 95 per cent limits in the normal distribution. For a small number of degrees of freedom, however, the value of t for a given probability level is larger than the corresponding value of the normal variate. That is, it is necessary to go out a greater distance on either side of the mean in the t distribution to include a given area. In older works on biometry, the normal distribution was used as a basis for confidence interval estimation (although it was not called by this name) but such a method will provide confidence intervals that are consistently too small for a given confidence level. Of course, for very large samples the normal and the t distributions are virtually identical, but as long as the sample size is less than 120, the largest number of degrees of freedom given in the t-table, there is no excuse for basing confidence limits on the normal curve.

The confidence limits for a mean are of the general form

$$\bar{X} \pm t \frac{s}{\sqrt{N}}$$

In this expression, $s/\sqrt{N}$, the standard deviation of the sample divided by the square root of the number of observations, is known as the *standard error of estimate*, or, simply, the *standard error*. This standard error, which will be abbreviated by $s_{\bar{X}}$, is an estimate of the true standard deviation of the distribution of the sample mean, $\bar{X}$. The sample mean, $\bar{X}$, is normally distributed with a mean μ, and a standard deviation $\sigma_{\bar{X}} = \sigma/\sqrt{N}$. The original variate is distributed with mean μ and standard deviation σ. Thus the distribution of the sample mean, $\bar{X}$, has the same mean as the original population but has a standard deviation only $1/\sqrt{N}$ as large. This is entirely reasonable since we ought to expect the mean of a number of observations to fall closer to the true value than any single observation. If this were not so, it would be futile to take more than one observation since a single measurement would give just as good an estimate of μ.

It is a common practice to publish a sample mean, $\bar{X}$, along with the standard error of the mean, the standard deviation of the observations often not being given. Since $s_{\bar{X}} = s/\sqrt{N}$, they are equally useful, providing N is also given. It is also common practice to note this information in the form

$$7.32 \pm .43$$

for example. *This is not meant to be a confidence interval,* for the number .43 is the standard error and not t times the standard error. In general, unless an author specifically states that an expression like $7.32 \pm .43$ *is* a confidence interval, he means it only as a shorthand notation for "the mean is 7.32 and its standard error is .43." From such information, it is a simple matter to construct a confidence interval by an appropriate choice of t.

Many authors use, instead of the standard error, a figure that they call the "probable error" (P.E.). This is obtained by multiplying the standard error by .6745. The rationale of this procedure is that for the normal curve the area between $+ .67450$ and $- .67450$ is exactly .50. That is, the probability of falling between plus and minus one P.E. is .50. Unfortunately, even this doubtful value of the P.E. is incorrect, since it is Student's t and not the normal distribution which gives the correct probabilities for the sample mean. For example, for 15 degrees of freedom it is .691 times the standard error that marks off the 50 per cent limits, not .6745. Again we see that the use of the normal curve underestimates the true confidence limits. The probable error is useless and ought never to be calculated. Because of this difference of usage in the older literature, if an author fails to state whether he has used the standard or probable error, it is necessary either to ignore his figures, since the two are so different as not to be in any degree interchangeable, or else to calculate the standard error from his data if sufficient information is provided.

The Variance and Standard Deviation

In order to establish confidence intervals for a variance, it is necessary to know the distribution of s^2, assuming once more that the observations themselves are from a normal distribution. This distribution is known as the "chi-square distribution," symbolized by χ^2. Like Student's t, the χ^2 probabilities depend upon the number of degrees of freedom, which again is one less than the sample size. If N is the sample size and s^2 is the sample variance, the 95 per cent confidence limits for σ^2 are given by

$$\frac{(N-1)s^2}{{}_n\chi^2_{.975}} \quad \text{and} \quad \frac{(N-1)s^2}{{}_n\chi^2_{.025}}$$

where ${}_n\chi^2_{.025}$ signifies the value of χ^2 with n degrees of freedom, below which .025 of the area of the χ^2 distribution falls, and ${}_n\chi^2_{.975}$ is the value of χ^2 below which .975 of the area falls. The reason for choosing the 2.5 per cent and 97.5 per cent points is that 5 per cent of the area falls outside of these limits and 95 per cent falls within them. If the 90 per cent confidence interval is desired, the χ^2 values to be used are $_n\chi^2_{.05}$ and $_n\chi^2_{.95}$, while for a 99 per cent confidence interval they are $_n\chi^2_{.005}$ and $_n\chi^2_{.995}$. Appendix Table III is the cumulative χ^2 distribution for various values of n, the degrees of freedom. In this cumulative distribution, the values of P across the top are the areas of the distribution below the corresponding value of χ^2 in the body of the table. Once again the degrees of freedom must be specified. As an example, 5 per cent of the area of the distribution falls below $\chi^2 = 7.26$ for 15 degrees of freedom, while for 25 degrees of freedom this same area is found below 14.6.

As an illustration of confidence limits for σ^2, the data in Example 24 on the length of the third upper premolar of *Ptilodus montanus* can be used. The sample size N is 21 so that there are 20 degrees of freedom. The sample estimate of the variance, s^2, is .037 and the mean, $\bar{X}$, is 2.97. Assume that a 95 per cent confidence limit for σ^2 is wanted. This confidence interval is of the form

$$\frac{20(.037)}{_{20}\chi^2_{.975}} \text{ to } \frac{20(.037)}{_{20}\chi^2_{.025}}$$

From Appendix Table III $_{20}\chi^2_{.025}$ is found to be 9.59 and $_{20}\chi^2_{.975}$ is 34.2. Then the confidence interval in question is

$$\frac{.73}{34.2} \text{ to } \frac{.73}{9.59}$$

or

$$.021 \text{ to } .076$$

Since the smallest value that s^2 can take is 0, the χ^2 distribution does not extend below zero. This distribution is very different from a normal distribution which extends symmetrically in both positive and negative directions. However, as the number of degrees of freedom becomes very large, even this highly asymmetrical distribution approaches a normal curve, although very slowly. Since the χ^2 table is not given for degrees of freedom greater than 30, a normal approximation must be invoked for large samples. For this purpose the quantity $\sqrt{2\chi^2} - \sqrt{2n-1}$ may be regarded as having the standardized normal distribution. If, say, $\chi^2_{.025}$ is needed for a number of degrees of freedom greater than 30, it will be found by setting this expression equal to the value of the normal variate which cuts off .025 of the area under a normal curve.

Suppose $\chi^2_{.025}$ and $\chi^2_{.975}$ for a sample size of 86 as in Example 32 are needed. The table of standardized normal curve shows that .025 of the area falls below -1.96 and that .975 of the area falls below $+1.96$. To

find the two values of χ^2 required, it is only necessary to solve the equations
$$\sqrt{2\chi^2_{.025}} - \sqrt{2n-1} = -1.96$$
and
$$\sqrt{2\chi^2_{.975}} - \sqrt{2n-1} = +1.96$$
Now
$$N = 86$$
$$n = 85$$
$$\sqrt{2n-1} = 13$$
so that the equations become
$$\sqrt{2\chi^2_{.025}} - 13 = -1.96$$
and
$$\sqrt{2\chi^2_{.975}} - 13 = +1.96$$
which gives
$$\chi^2_{.025} = 60.94 \text{ and } \chi^2_{.975} = 111.95$$
It is these values, which are very close to the true values of χ^2, that are then used to find the confidence interval as in the previous example. For the sample size given in Example 32 of 86, the value of s^2 was 9.35. Then the 95 per cent confidence interval lies between
$$\frac{85\,(9.35)}{111.95} \text{ and } \frac{85\,(9.35)}{60.94}$$
which corresponds to an interval of 7.10 to 13.04.

This interval is very nearly symmetrical around the estimate of $s^2 = 9.35$, while for the smaller sample size of 21 the interval was highly asymmetrical around the observed value of s^2.

Having found a confidence interval for the variance σ^2, it is only necessary to take the square root of the limits in getting the confidence interval for σ, the standard deviation. In the first example, the confidence limits for σ^2 were .021 and .076. The 95 per cent limits for σ are the square roots of these numbers, or .144 and .276, the observed estimate, s, being the square root of $s^2 = .037$, or .192. In the second example, s is 3.06 and the 95 per cent limits are 2.66 and 3.61.

Ratios

The ratio between two measurements is a fairly common variate in zoology and some of the peculiarities of ratios have already been discussed. Another serious drawback in working with ratios is the difficulty in making any exact confidence statements about them. If the variates in the numerator and denominator of the ratio are independent—that is, if the way in which one varies does not depend upon variation in the other—there is a fairly straightforward method for finding confidence intervals. The confidence limits for the numerator and denominator are found separately. If the upper and lower limits for the numerator are written as A_u and A_l and the limits for the denominator symbolized as B_u and B_l, confidence limits for the ratio are given by

$$\frac{A_l}{B_u} \text{ and } \frac{A_u}{B_l}$$

The confidence value for this interval is the product of the individual confidence levels. If the confidence limits for numerator and denominator are those for a 95 per cent interval, the confidence interval for the ratio has a confidence level of 90.25 per cent ($.95^2$). If a 95 per cent level is wanted for the ratio, the confidence limits for numerator and denominator must represent the 97.5 per cent limits ($\sqrt{.95} = .975$). Suppose that the 97.5 per cent limits for the numerator and denominator were both 10.0 to 30.0. Then the 95 per cent interval for the ratio would be

$$\frac{10.0}{30.0} \text{ to } \frac{30.0}{10.0} \text{ or } .33 \text{ to } 3.00$$

The upper limit of the ratio is then nine times the lower limit, while for the numerator and denominator separately, the upper limits are only three times the lower limit. This bears out the statement in Chapter 1 that ratios are more variable than individual measurements. It also points out that this greater variability of ratios is a greater *relative* variability rather than a greater *absolute* variability. Obviously the interval from 10.0 to 30.0 is larger in *absolute* terms than the interval from .33 to 3.00.

What has been said about confidence limits for ratios only applies when the numerator and denominator are independent of each other. In such a case, however, it is usually foolish to use a ratio since it generally provides no more information than the individual measurements. Ratios are most widely useful precisely when the numerator and denominator are dependent upon each other, in which case confidence intervals derived as above are incorrect. The situation is not quite hopeless, however, because ratios are useful when the numerator and denominator vary together in the same direction, the increase in one being accompanied by an increase in the other. When this is the case, the ratio tends to be more stable than the measurements (the reason for using ratios), so that the confidence interval obtained in the manner we have suggested is too large. That is, with 95 per cent confidence, the true value of the ratio really lies in a smaller interval than indicated. The statement of confidence is then conservative, a highly desirable characteristic.

If the numerator and denominator should vary in the opposite direction from each other, the indicated confidence limit is deceptively small, but there is no reason to use a ratio in such a case as it will obscure differences which can be detected from the original measurements.

It is often possible to treat ratios as if they were direct observations in no way different from weights, lengths, and other continuous variates. Whether this can be done in any specific case depends upon how nearly normal the distribution of the ratio may be. If the numerator and denominator of a ratio are each normally distributed, the distribution of the ratio will tend to be considerably more platykurtic than a normal distribution, while if there is a positive association between numerator and denominator, this platykurtosis will be counteracted by a tendency toward piling up of values in the central classes. The result of these two opposing tendencies will often be to produce a distribution of ratios not very different from normal, in which case confidence intervals can be found in the same manner as for any other continuous, normally distributed variable. By forming the frequency distribution of the ratio and fitting it to a normal curve, the zoologist can check on this possibility in any particular instance.

Other Parameters

In constructing confidence limits for a binomial proportion, a mean, and a variance, the exact probability distributions of the estimates of these parameters were used. The entire logic of confidence interval construction is dependent upon a knowledge of how a given estimate is distributed. This must be so because the establishment of, say, a 95 per cent confidence interval for a parameter depends upon knowing between what limits an estimate will fall 95 per cent of the time. While exact distributions have been tabulated for s^2 and $\bar{X}$, there are a number of estimates discussed in Chapters 5 and 6 for which such distributions are not available so that it is not possible to construct exact confidence limits for the parameters which they estimate. Fortunately, it is possible to find approximate confidence intervals for the parameters because it can be shown that their estimates are roughly normally distributed in large samples. What is meant by *large samples* changes with the estimate, since some have distributions which approach normality quite rapidly with increasing sample size, while others do so more slowly. Assuming for the moment that the sample size is large enough to assure a reasonable approach to normality, then rough confidence limits for any parameter are given by

$$\text{estimate} \pm \tau \text{ (standard error of estimate)}$$

where τ is the value of the standardized normal variate which cuts off a per cent of the area under the normal curve equal to the desired confidence level. For example,

$$\text{estimate} \pm 1.96 \text{ (standard error of estimate)}$$

gives a rough 95 per cent confidence interval since 95 per cent of the area

166 QUANTITATIVE ZOOLOGY

under the normal curve is included between ± 1.96. A 99 per cent confidence interval would be given by a τ of 2.576 and a 90 per cent interval by τ equal to 1.645.

The following are the standard errors of estimate of the more important parameters so far discussed. When other measures are introduced in subsequent chapters, their standard errors, or else methods for finding exact confidence intervals, will be discussed with them. The derivations of these are somewhat complicated mathematical operations based on probability theory and will not be explained here.

Standard error of estimate of the arithmetic mean:

$$s_{\bar{X}} = \frac{s}{\sqrt{N}}$$

We have already discussed a method for finding exact confidence limits for the mean and in that discussion pointed out that when the sample size exceeded 120, the expression

$$\bar{X} \pm 1.96 \frac{s}{\sqrt{N}}$$

gives a very close approximation of the 95 per cent confidence interval. For smaller samples Student's t distribution with the appropriate number of degrees of freedom should be used, so that in place of the multiplier 1.96 the proper value of Student's t is inserted in this expression. A glance at the table of Student's t shows that the proper multiplier approaches 1.96 as the degrees of freedom grow large for a 95 per cent confidence interval, illustrating that the use of the standard error approach discussed in this section is an approximation to true confidence limits for large samples.

Standard error of the median:

$$s_{\text{median}} = 1.2533 \frac{s}{\sqrt{N}}$$

Standard error of first or third quartiles:

$$s_Q = s_{Q_3} = 1.3636 \frac{s}{\sqrt{N}}$$

Standard error of mean deviation:

$$s_{\text{M.D.}} = .6028 \frac{s}{\sqrt{N}}$$

Standard error of the coefficient of variation:

$$s_V = \frac{V}{\sqrt{2N}}$$

Standard error of a standard deviation:

$$s_s = \frac{s}{\sqrt{2N}}$$

While this formula is in older literature, there is no excuse for using it since an exact procedure is available, as explained in the last section.

The standard error of V is the least useful because it assumes that the mean and standard deviation of the sample are independent of each other. However, the only reason for using a coefficient of variation is because it is assumed that the mean and standard deviation are not independent. Were they really unrelated, there would be no advantage to this measure at all. There is one saving feature however. The kind of dependence of variability on mean which is assumed when the coefficient of variation is used is a positive one. That is, the larger the mean the larger the standard deviation, the smaller the mean the smaller the measure of variability. When this is true, the expression

$$V \pm 1.96 \frac{V}{\sqrt{2N}}$$

will give a confidence interval which is somewhat too large. This means that the true confidence in this interval is greater than 95 per cent which is precisely the direction in which the error should be if there must be one. The stated 95 per cent confidence interval is then conservative, in the sense that the chances are better than 95 in 100 that such an interval covers the parameter.

An Example of Adequate Records

It is seldom necessary to estimate all of the parameters discussed for a given problem, by far the most important being the mean and variance. If the mean and variance or standard deviation are given together with the sample size, this will generally be sufficient. From these three numbers, the confidence intervals for the mean and the variance or standard deviation can be calculated. The most complete picture of the observations, aside from publication of the observations themselves, would involve estimates of all the parameters mentioned, together with their standard errors, or confidence limits. All of these calculations are illustrated in Example 53 using the data of Example 29. The estimates have already been calculated in Chapters 5 and 6 and are repeated here for reference. The results of this example can be put in the form of a table suitable for publication, ás shown in Example 54.

EXAMPLE 53. Standard errors and confidence intervals for various estimates calculated from the data of Example 29.

$$N = 86$$
$$\bar{X} = 60.43$$
$$s^2 = 9.35$$
$$s = 3.06$$
$$M.D. = 2.46$$

$$V = 5.06$$
$$\text{Median} = 60.41$$
$$Q_1 = 58.19$$
$$Q_3 = 62.58$$

Standard Errors

$$s_{\bar{X}} = \frac{s}{\sqrt{N}} = \frac{3.06}{9.27} = .33$$

$$s_{\text{median}} = 1.25 \frac{s}{\sqrt{N}} = (1.25)(.33) = .41$$

$$s_{Q_1} = s_{Q_3} = 1.36 \frac{s}{\sqrt{N}} = (1.36)(.33) = .45$$

$$s_{M.D.} = .60 \frac{s}{\sqrt{2N}} = (.60)(.23) = .14$$

$$s_V = \frac{V}{\sqrt{2N}} = \frac{5.06}{13.11} = .39$$

95 Per Cent Confidence Intervals

MEAN

$$\bar{X} \pm \frac{1.99s}{\sqrt{N}} = 60.43 \pm .66$$

d.f. = 85

$t_{.95} = 1.99$

95% interval: 59.77 ——— 61.09

VARIANCE

$$\frac{(N-1)s^2}{\chi^2_{.975}} \quad\text{———}\quad \frac{(N-1)s^2}{\chi^2_{.025}}$$

$$= \frac{85(9.35)}{\chi^2_{.975}} \quad\text{———}\quad \frac{85(9.35)}{\chi^2_{.025}}$$

$$= \frac{794.75}{\chi^2_{.975}} \quad\text{———}\quad \frac{794.75}{\chi^2_{.025}}$$

EXAMPLE **53**, *continued*

To find $\chi^2_{.975}$ and $\chi^2_{.025}$ for 85 degrees of freedom

$\sqrt{2\chi^2_{.975}} - \sqrt{2n-1} = 1.96$ and

$\sqrt{2\chi^2_{.025}} - \sqrt{2n-1} = -1.96$

$\sqrt{2\chi^2_{.975}} - 13 \quad\quad = 1.96$ and

$\sqrt{2\chi^2_{.025}} - 13 \quad\quad = -1.96$

$\chi^2_{.975} = 111.95 \quad\quad \chi^2_{.025} = 60.94$

95% interval: $\dfrac{794.75}{111.90}$ ———— $\dfrac{794.75}{60.94} = 7.10$ ———— 13.04

STANDARD DEVIATION

95% interval: $\sqrt{7.10}$ ———— $\sqrt{13.04}$
$= 2.66$ ———— 3.16

MEDIAN

Median $\pm (1.96)(1.25)\dfrac{s}{\sqrt{N}} = 60.41 \pm (1.96)(.41) = 60.41 \pm .80$

95% interval: 59.61 ———— 61.21

FIRST AND THIRD QUARTILES

$Q_1 \% (1.96)(1.36)\dfrac{s}{\sqrt{N}} = 58.19 \pm .88$

95% interval: 57.31 ———— 59.07 for Q_1

$Q_3 \pm (1.96)(1.36)\dfrac{s}{\sqrt{N}} = 62.58 \pm .88$

95% interval: 61.70 ———— 63.46 for Q_3

MEAN DEVIATION

M.D. $\pm (1.96)(.60)\dfrac{s}{2\sqrt{N}} = 2.46 \pm .27$

95% interval: 2.19 ———— 2.73

COEFFICIENT OF VARIATION

$V \pm 1.96 \dfrac{V}{\sqrt{2N}} = 5.06 \pm .76$

95% interval: 4.30 ———— 5.82

EXAMPLE 54. Table of confidence intervals and standard errors from Example 53.

PARAMETERS	ESTIMATE	STANDARD ERROR	95% CONFIDENCE LIMITS
μ	60.43	.33	59.77–61.09
σ^2	9.35	—	7.10–13.04
σ	3.06	—	2.66– 3.61
M.D.	2.46	.14	2.19– 2.73
V	5.06	.39	4.30– 5.82
Median	60.41	.41	59.61–61.21
Q_1	58.19	.45	57.31–59.07
Q_3	62.58	.45	61.70–63.46

An Empirical Test

In order to exemplify and clarify the relationships between estimates, parameters, and confidence intervals, an artificial experiment like Example 38 was made. The results are summed up in Example 55. The length of the second lower molar was measured on the available specimens, 61 in all, of a species of fossil mammals, and the standard data were calculated. For experimental purposes, it may be taken that there were only these 61 members in the population of this species that lived in the Bighorn Basin in the Lower Eocene. Thus, whatever figures are calculated from these individuals can be regarded for our purposes not as estimates but as parameters, although, of course, this is highly artificial, for these really represent only a sample from the much larger unknown population that did exist there. Now smaller samples can be drawn from this "population" whose "parameters" are known. The estimates from these samples could then be directly compared with the parameters of the known population.

The first table in Example 55 shows the basic information for the comparison. The second table shows the 95 per cent confidence limits for each parameter in each sample together with the "population parameters" for comparison.

The confidence limits for V are certainly grossly incorrect since the sample sizes of 5 and 10 are far below the level where 1.96 standard errors will give even a rough approach to a 95 per cent area for this estimate. The confidence limits given clearly cover the parameter in every case, however, and the error due to sample size is in the opposite direction from that introduced by the lack of independence of the mean and standard deviation.

The curious point about this example is the closeness with which the sample estimates approach the "population parameters." Even 50 per cent confidence limits fail to cover the parameter in only three cases—the mean, the standard deviation, and the coefficient of variation for sample C. On the average, one would expect seven or eight of the fifteen 50 per cent confidence limits not to include the true values. The fact that one sample stands out from the others which are among themselves consistently close to the true values probably means that the sampling technique was somehow in error.

EXAMPLE 55. Comparison of parameters of a population and samples therefrom. Length of M_2 in *Phenacodus primaevus* from the Gray Bull Formation, Bighorn Basin, Wyoming. All available specimens (61) and five small samples drawn from this. (Original data)

SAMPLE	N	O.R.	$\bar{X}$	s	V
"POPULATION"	61	10.6–13.8	12.12	.96	7.9
A	10	10.7–13.5	12.26	.95	7.7
B	10	10.8–13.6	12.07	.88	7.3
C	10	10.7–12.8	11.93	.71	5.9
D	5	10.9–13.8	12.18	1.10	9.1
E	5	10.9–13.2	11.98	.91	7.6

Weighted means for 5 small samples:

$\bar{X}$	s	V
12.09	.94	7.3

95 PER CENT CONFIDENCE INTERVALS

SAMPLE	$\bar{X}$	s^2	s	V
"POPULATION"	12.12	.92	.96	7.9
A	11.58–12.94	.43– 2.99	.66–1.73	4.4–11.0
B	11.44–12.70	.37– 2.58	.61–1.61	4.2–10.4
C	11.43–12.43	.24– 1.68	.49–1.30	3.4– 8.4
D	10.82–13.54	.44–10.00	.66–3.16	3.4–14.8
E	10.87–13.09	.30– 6.78	.55–2.60	2.9–12.3

CHAPTER TEN

Comparisons of Samples

Most of the discussion up to this point has been devoted to the study of individual samples and of their relationships to populations. Probably the most frequent zoological operation with numbers, however, is in the comparison of two samples. Even in the study of a single sample, the usual aim is to obtain and present data that will permit subsequent comparisons. Such comparisons can be made intelligently and with reasonable objective probability only after the characteristics and relationships of samples and of distributions based on them are thoroughly understood. Once these ideas and the numerical operations based on them have been grasped, their use in comparisons is relatively simple; but without such data and concepts as means, standard deviations, confidence intervals, and standard errors, comparisons are largely subjective, highly unreliable, and often erroneous or meaningless.

In comparing two samples, the primary operation is of course to see whether they are or are not different and what the degree of difference is. In practice two samples always are different, for the chance of two samples, even though drawn from exactly the same population, being identical in character is practically nil. The degree of difference may be slight or great, and the real purpose of the comparison is to judge from this difference whether the samples were or were not drawn from the same population.

The meaning of "the same population" must not be taken too literally. If by "population" is meant a particular group living at the same time and in the same place, then very few samples are taken from the same population, for samples compared are often deliberately chosen from populations in different localities, different geological horizons, and so on. The object of inquiry is not whether the samples are from the same population in the sense of populations of objects, but whether the parameter in question, such as the mean or standard deviation is the same in the two populations from which the samples are taken. A sample of *Mus musculus* from Ohio is obviously not taken from the same real population of objects as one from Capetown, but what is of interest is whether house mice have the same average tail length, let us say, whether they be taken in America or South Africa.

The zoological population exists at many levels not coextensive with a narrowly defined population of animals restricted in time and space. An entire subspecies or species is for some purposes a population, while for others only those individuals that form a reproductive community can be considered members of a population.

In the real world, no two populations of objects can have precisely equal means or standard deviations. *A difference between two such populations can always be established, provided the observer is willing to go to enough trouble to refine and increase the number of his measurements.* This is a point often misunderstood in the application of statistical techniques. One always knows long before he starts that both the samples which he measures and the populations from which they are drawn are different to some greater or lesser extent in the value of every parameter that can be imagined. It is therefore essential that some degree of difference be assumed to be trivial and not germane to the problem. To take an absurd example, one could, with the aid of the techniques to be described in this chapter and by taking large enough samples, show beyond a reasonable doubt that elephants in the Belgian Congo differed in weight from those in Rhodesia by, let us say, one pound. But of what conceivable use is such information except to show that the elephants in the Belgian Congo are not the elephants in Rhodesia, something that was known to begin with? The patent absurdity of the example ought not to obscure the fact that this is a problem in all comparisons of zoological populations. If the differences between two samples are so small as to be trivial, the mere fact that they can be shown to represent a real difference between populations is not important. Throughout the discussion in this chapter, the concept of the significance of a difference will be used. This is a technical term relating the difference between samples to the difference between populations. What the zoologist must be first concerned with is the *biological significance* of a demonstrated difference, and for this kind of significance there is no test but that of intuition, experience, and intelligence. In some sense, the decision about the biological significance of a difference is a decision about which physical populations are to be included within the zoologist's concept of a biological population.

The notion of a population is basic to any comparison, however poor and subjective may be the method of carrying it out. In typological systematics, now becoming outmoded but still too commonly used, a specimen is compared with a type, and decision is made subjectively as to whether it belongs to the same species. This sample comparison as it is so often carried out is a thoroughly unscientific procedure. It involves no definite criterion of significance, no idea of what the range of variation really is, no conception of the relationship of the type to the variation of the species as a whole, and no method of relating the specimen being identified to this specific distribution beyond a vague and subjective opinion that is shown by more reliable methods to be as often wrong as right. It is possible to obtain definite criteria on all these points, as will be shown in the present chapter. The first and most essential point is that the only meaningful comparisons are those between *populations*, not between *specimens*. The specimens, types or otherwise, even if only two specimens are available for comparison with each other, are important only as samples from which the characteristics of populations are to be estimated.

The basic logic of statistical comparisons is that which has already been outlined in the introduction to the material on confidence intervals. A hypothesis is first set up—for example, that two samples do come from the same population. From this hypothesis the probability of various events can be deduced. If the event actually observed has a very small probability, the hypothesis is rejected. The choice of how small this probability must be to force rejection of the hypothesis is arbitrary as in the case of confidence intervals, and again, as in the construction of confidence intervals, it should be adjusted to the specific needs of the situation. The almost universally used rejection value is 5 per cent. That is, if the probability of the observation is .05 or less, the hypothesis is rejected; if it is greater than 5 per cent, the hypothesis is considered acceptable. This does not mean that the hypothesis is disproved or proved but rather that the degree of confidence in the hypothesis is so low that it is regarded as disproved "beyond reasonable doubt." Since one man's concept of what constitutes "reasonable doubt" may vary from another's, it is a far better procedure not to indicate a hypothesis as simply being rejected or accepted along with the criterion used as is so often done, but rather to indicate precisely what the calculated probability of the observation is under the hypothesis and allow the reader to decide whether he considers it so low as to make the hypothesis incredible.

It not infrequently happens that an author, using the 5 per cent probability value as a criterion for rejection of a hypothesis, will reject one hypothesis under which the observations had a probability of 4 per cent while accepting another, the observations for which had a probability of 6 per cent, without disclosing these probability values. Having chosen the 5 per cent level for rejection, this procedure is perfectly correct logically, but abhorrent biologically. On such slender threads hang many theories!

All hypotheses that are tested are basically of one form called the *null hypothesis*. The null hypothesis states that there is no difference between the populations from which the samples are drawn. It is a matter of statistical convenience that this is the standard hypothesis, rather than of any particular logical necessity. Basically the null hypothesis method assumes that populations are the same until demonstrated to be different. The burden of proof rests on the zoologists, so to speak, to demonstrate that the difference between the samples is so large that it is no longer credible that they are drawn from the same population. The null hypothesis is like the assumption of innocence in a court of law, which assumption is maintained until the weight of evidence is overwhelmingly against it.

It is often said that a difference between two populations is *significant* in a technical sense. This simply means that the difference is so large that the null hypothesis has been rejected. Again, how small the probability of the observation must be before the difference is considered to be significant is a decision which the zoologist must make. Moreover, establishment of the significance of a difference between two samples is not in itself a zoological conclusion but only a datum that makes such a conclusion possible. The zoological conclusion is not numerical and cannot be reached mathematically. If two samples differ significantly, then it is accepted for practical purposes (i.e., it is likely to the degree chosen as significant) that they represent different populations but it is not demonstrated what the populations are or why they differ, or what the zoological meaning of the difference is. If two samples of zoological specimens are significantly different in a morphological character, they may belong to different species or other taxonomic groups, they may represent different sexes, they may be of different age groups, they may have been affected by different food, one may have been affected by disease—these and many other possibilities remain to be considered and to have a choice made between them on zoological, nonnumerical grounds. Correct numerical treatment does not assure a correct conclusion, but it makes such a conclusion possible. Incorrect numerical treatment makes a correct conclusion impossible except by blind luck.

The Test of a Null Hypothesis

Even if two samples are drawn from the same population, they will have different values for the sample mean, standard deviation, and so on. This was shown in the artificial sampling experiment in the last chapter. Sometimes the difference between, say, the mean of two samples from the same population will be large, other times small, and this difference is itself a variate with a probability distribution which is dependent on the sample size and on the parameters of the population. Most differences between the two samples will be small but a few will be large. There must be a point, however, at which this difference is so large that it is, for all practical purposes, impossible, or better, "unbelievable." If such an unbelievable difference between two samples is really observed, since the observations are a hard fact which can not be disputed, it is the null hypothesis which must be discarded. What is required for testing a null hypothesis is the probability distribution of the difference between two sample estimates from the same population. If the probability is very small that one will observe a difference equal to or greater than the one actually observed, the null hypothesis is rejected and the difference is said to be significant.

The test of any difference between samples must depend upon the dispersion in the population so that it is not the absolute size of the difference but its size relative to the standard error of the estimates which is implicitly or explicitly tested in all the following cases.

The Means of Two Samples

In order to test the difference between the means of two samples, not only the sample means but also the standard deviations must be used, as suggested in the last paragraph. If the mean, standard deviation, and sample size of one sample are denoted by $\bar{X}_1$, s_1, and N_1, respectively, while the same values for the other sample are $\bar{X}_2$, s_2, and N_2, then the quantity

$$\frac{(\bar{X}_1 - \bar{X}_2)\sqrt{\frac{N_1 N_2}{N_1 + N_2}}}{\sqrt{\frac{(N_1 - 1) s_1^2 + (N_2 - 1) s_2^2}{N_1 + N_2 - 2}}} = t$$

is distributed as Student's t with $N_1 + N_2 - 2$ degrees of freedom.

This quantity, which has been designated as t, is roughly the difference between the two sample means divided by a measure of the standard deviation of that difference. To find the probability that such a difference or a greater one would be observed if the two samples really were taken from populations with the same mean, it is only necessary to look in the table of t (Appendix Table II) which has already been used for finding confidence intervals. It may be remembered that the values at the top of this table are the probabilities that t will fall *within* the limits of plus and minus the corresponding value of t in the body of the table. Thus for 15 degrees of freedom, the probability that t will fall between $+ 2.131$ and $- 2.131$ is .95. The probability that it will fall outside these limits is then .05. For convenience, the probabilities of falling outside the limits are given at the bottom of the table, and it is these probabilities which are used in testing hypotheses. They are simply one minus the corresponding probabilities at the top of the table. This means that only 5 per cent of the time would an observed value of t be as large or larger than 2.131 or smaller than $- 2.131$. If 5 per cent were the level of significance chosen for a particular test, then a difference between $\bar{X}_1$ and $\bar{X}_2$ which was so large as to make t greater than 2.131 or less than − 2.131 would cause rejection of the hypothesis that there was no difference between the two populations, and the difference between $\bar{X}_1$ and $\bar{X}_2$ would be called significant.

EXAMPLE 56. Student's *t*-test for the significance of the difference of two sample means. Mandible lengths in samples of *Peromyscus maniculatus bairdii*. Sample A from Alexander, Iowa; Sample B from Grafton, N.D. (Data from Dice, 1932)

SAMPLE	N	$\bar{X}$	s^2
A	13	15.721	.43128
B	43	15.492	.16923

$$t = \frac{(15.721 - 15.492)\sqrt{\frac{(13)(43)}{13 + 43}}}{\sqrt{\frac{12(.43128) + 42(.16923)}{13 + 43 - 2}}} = 1.517$$

Degrees of freedom = 13 + 43 − 2 = 54

EXAMPLE 57. Student's *t*-test for the significance of the difference of two sample means. Ratio of distal width to length in the ulna of A, a fossil turkey *Parapavo californicus*; B, an allied living species *Meleagris gallopavo*. (Data from Howard and Frost; see Howard, 1927)

SAMPLE	N	$\bar{X}$	s^2
A	133	10.8	.14593
B	29	10.1	.17808

$$t = \frac{(10.8 - 10.1)\sqrt{\frac{(133)(29)}{133 + 29}}}{\sqrt{\frac{132(.14593) + 29(.17808)}{133 + 29 - 2}}} = 8.741$$

Degrees of freedom = 133 + 29 − 2 = 160

Examples 56 and 57 are two *t*-tests with quite different results. Appendix Table II shows that for 60 degrees of freedom the probability that *t* will be within the limits ± 1.296 is .80 and the probability that it will be within the limits ± 1.671 is .90. The observed value of *t* for Example 56 is 1.517

for 54 degrees of freedom, so that the probability of observing a value between ± 1.517 is between .80 and .90. Conversely, the probability is between .20 and .10 that t would fall *outside* the limits ± 1.517. Between 10 per cent and 20 per cent of the time, a value of t as large or larger than the one observed would be expected even if the populations had the same mean. As 10 per cent is a fairly high value, one would not want to reject the null hypothesis that the populations have the same mean. It might happen, of course, that under some circumstances 10 per cent is too low a value for belief, in which case the hypothesis would be rejected. Whether the hypothesis be rejected or accepted, the value of t which was calculated and the accompanying probability, in this case, $.10 - .20$, would be published. In Example 57, although the means do not differ by much more than in Example 56, the result is quite different. The table shows that the probability is much smaller than .001 that a value of t will fall outside of the limits $+ 8.741$ and $- 8.741$. The highest value of t shown on the table for 120 degrees of freedom is only 3.291. For all practical purposes, the observation is impossible under the null hypothesis, and even the most skeptical mind would be forced to conclude that the populations from which the samples were drawn are indeed different. In publishing the result, it is sufficient to observe that the probability of exceeding the observed value of t is much smaller than .001.

"**One-sided tests.**" In the previous two examples no account was taken of the sign of $\bar{X}_1 - \bar{X}_2$. It would make no difference if the larger value were subtracted from the smaller, as this would only reverse the sign of t, making it negative instead of positive. The method of testing the hypothesis has been set up in such a way that the conclusion is the same whether t is $- 8.750$ or $+ 8.750$, since a deviation in either direction of this magnitude has a very low probability. This symmetry of the test is a result of the particular null hypothesis, i.e., that there is no difference between the means of the populations. There are some cases where this statement of the null hypothesis does not really answer the zoological question asked. An example of such is Allen's rule which states that in closely related groups of animals those from arctic regions will have shorter appendages than will those from temperate regions. The question raised by Allen's rule is not whether the mean length of the pinna of the ear, say, is *different* in arctic and temperate groups, but more specifically whether it is *larger* in temperate than in arctic populations. No matter how large the difference between the sample means, if the mean for arctic forms should be larger than that for temperate forms, the difference is not significant with respect to the hypothesis being tested. It is only when the difference is in the direction implicit in the null hypothesis that rejection of the null hypothesis can be considered.

For a test of such a "one-sided hypothesis," t is calculated in the usual way, but the probability interpretation must be different. In the table of t, if the probability is given as .05 that a value of t will fall outside the limits $+3.132$ and -3.132, what is meant is that half of this probability, .025, falls above $+3.132$ and the other half below -3.132. In a one-sided test, it is only one or the other of these areas that is of interest, but not both. If the calculated value of t should be 3.132 in a specific case, and this deviation is in the direction assumed by the hypothesis, then the appropriate probability of the observations under the hypothesis is .025 and not .05. The actual significance level for a one-sided test is always one-half of that for a two-sided hypothesis. Conversely, if the 5 per cent level of probability is chosen as the criterion for the rejection of the null hypothesis in a one-sided test, the appropriate value of t would be equal to the 10 per cent value for a two-sided test. To avoid confusion, the probabilities for one-sided and two-sided tests have been indicated on separate lines at the bottom of our table of t. Notice that the probabilities for one-sided tests are simply one-half of those for two-sided tests. That this consideration is important can be seen from reexamining Example 56. In this case t was 1.517 for 54 degrees of freedom, the corresponding probability for the original two-sided test falling between .10 and .20. Interpolation in the table of t gives a value of .14 more exactly. Suppose for some reason that the question were not whether there was simply *some* difference between sample A and sample B but more specifically whether sample A represents a population whose mean is actually larger than that from which B has been drawn. Then the corresponding probability is .07 rather than .14. While 14 per cent is a probability which is generally considered to be too high for significance, 7 per cent is very close to the conventional 5 per cent level of rejection and a definite suspicion about the second null hypothesis might be entertained.

In Example 57 where the difference between sample A and sample B is highly significant for a two-tailed test, it would remain highly significant for a one-tailed test if the hypothesis were concerned with A being greater than B. On the other hand, B is certainly not significantly greater than A, since it is in fact smaller.

This difference in probability between the one-sided and two-sided test raises a logical problem. If the one-sided test always has half the probability of the two-sided test, then why not test all hypotheses in a one-sided manner? After all, if the null hypothesis that A is not larger than B is rejected, then *ipso facto* the two-sided hypothesis that A equals B is also rejected. If A is larger than B, certainly A is different from B. The answer to this problem lies in the order in which decisions are made. *The hypothesis must always be constructed before the data are examined.* The observed results may not be used as a basis for constructing the hypothesis, or in the long run there will be a bias in the testing procedure. If A is observed to be larger than B and *then* the one-sided hypothesis "A is not larger than B" is constructed as a consequence of the observation, there is certainly an increased chance of rejecting the hypothesis. If, however, there is some *a priori* reason for choosing the hypothesis "A is not larger than B," there will be no bias in the test. Although less of a deviation of A from B is required in one direction for significance, no deviation in the other direction, no matter how large, will result in rejection of the null hypothesis. In the long run, then, this one-sided *a priori* hypothesis will be rejected no more frequently than it should be.

This problem of choosing a hypothesis independent of the observation is integral to the entire logic of statistical testing and arises in other ways. For example, how does one choose the proper significance level for a test? A dishonest investigator can easily lie with statistics if he chooses his significance level after performing his test. Thus, if the observed t corresponds to the 7 per cent level and the zoologist would really like very much to reject the hypothesis, he will decide that 7 per cent is too low a probability for acceptance. In yet another test he may find 3 per cent to be the proper level, and so on. Obviously, by adjusting the significance level to fit every occasion, all hypotheses can be accepted or rejected at will. It is to avoid unconscious biases of this sort that we recommend publishing actual probability values corresponding to a test, rather than simply denoting various differences as significant or nonsignificant.

The failure to distinguish carefully between one-sided and two-sided hypotheses is common even among workers familiar with statistical usages and often leads to erroneous conclusions. One-sided hypotheses are a good deal more common than is usually supposed, and the zoologist ought to examine the question that he is asking of the data with great care. One-sided hypotheses usually occur in the testing of biological "laws" and "rules" like Allen's rule, and in verifying previous conclusions about the way in which different populations may be related.

Paired Comparisons

Often in experimental sciences and occasionally in zoology, the two populations compared are paired. Paired samples have equal numbers of observations in each, and every observation in one sample has some biological correspondence with an observation in the other sample. A case in point is when two different measurements are made on a set of animals, and it is desired to test whether the two measurements differ significantly. Example 58 gives some original data on the lengths of two lower molars, M_1 and M_2, in the fossil mammal *Phenacodus primaevus*. The pairing of the measurements arises from the fact that both molars were measured in all 26 individuals and recorded in pairs. Had M_1 been measured in one group of specimens and at a later date M_2 measured in a different group or in the same group, without properly matching the measurements specimen for specimen, the samples could not be regarded as paired.

EXAMPLE 58. Student's t-test for the difference between paired samples. Measurements are the length in mm. of M_1 and M_2 in *Phenacodus primaevus*. (Original data)

ORIGINAL MEASUREMENTS

M_2	M_1	d
9.8	10.2	− .4
10.5	10.7	− .2
10.5	10.7	− .2
10.8	10.8	0
11.0	11.0	0
11.1	11.4	− .3
11.1	12.1	−1.0
11.3	12.6	−1.3
11.4	12.8	−1.4
11.4	10.8	.6
11.4	12.6	−1.2
11.9	12.3	− .4
12.2	12.4	− .2
12.2	12.0	.2
12.3	13.7	−1.4
12.3	13.0	− .7
12.4	13.2	− .8
12.4	12.4	0
12.5	13.8	−1.3
12.7	13.5	− .8
12.8	13.3	− .5
13.0	12.7	.3
13.1	13.1	0
13.2	13.6	− .4
13.4	12.6	.8
13.5	13.5	0

$N = 26$

$\Sigma d = -10.6$

$\bar{d} = -.41$

$(\Sigma d)^2 = 112.36$

$\dfrac{(\Sigma d)^2}{N} = 4.32$

$\Sigma d^2 = 13.58$

$\Sigma d^2 - \dfrac{(\Sigma d)^2}{N} = 9.26$

$s_d^2 = .3704$

$s_{\bar{d}}^2 = \dfrac{.3704}{26} = .0142$

$s_{\bar{d}} = .1192$

$t = \dfrac{\bar{d}}{.1192} = \dfrac{-.41}{.1192} = -3.44$

d.f $= N - 1 = 25$

When such paired measurements are available, the t-test for the difference between the means takes the simple form

$$t = \frac{\bar{d}}{\sqrt{\dfrac{s_d^2}{N}}}$$

with $N - 1$ degrees of freedom.

Here $\bar{d}$ is the mean *difference* between paired measurements, s_d^2 is the variance of these differences, and N is the number of specimens (not the total of measurements which is, of course, $2N$).

The difference between this form of t and the one given previously is in the denominator. The numerator, $\bar{d}$, is exactly equal to $\bar{X}_1 - \bar{X}_2$ and could be calculated from the difference between the means of the two measurements. To find s_d^2, however, it is necessary first to subtract X_1 from X_2 for each specimen and then to find the variance of these new quantities, d. The calculation of t has been carried out in Example 58 for paired data. It is important to note that the mean, $\bar{d}$, will depend on the signs. The probability of observing a value of t greater than or equal to 3.44 for 25 degrees of freedom, from Appendix Table II, lies between .01 and .001 so that the difference between the length of M_1 and M_2 is significant under a fairly stringent criterion.

The same data used to calculate t from the formula for unpaired comparisons gives $t = 1.43$, which for 25 degrees of freedom has a probability of around .17. Such a probability is generally assumed to show no significant difference as it is fairly high.

Example 58 was deliberately chosen to illustrate the great difference in result from the application of the paired and unpaired test. It should not be assumed that this difference will always be in the direction shown for the *Phenacodus* data. Under other conditions the paired test may show no significant difference, while the usual test gives a t with a very low probability. As in the choice between one-sided and two-sided hypotheses, a correct conclusion cannot be reached unless a test appropriate to the occasion is used. In the case of *Phenacodus*, it is obvious from examining the observations themselves that M_1 is somewhat larger than M_2 not only on the averages but consistently, for in only 4 specimens of the 26 was M_2 the larger. If the data had not been collected in pairs this information would be sacrificed not only to casual observation but in the t-test itself, because the large value of t obtained using the paired comparison formula is in part a result of this consistency. The advantage of the correct testing procedure is that it extracts more pertinent information from the measurements than does the incorrect one.

Comparison of a Single Specimen with a Sample

The t-test for the difference between the means of two samples is completely general with respect to sample size. The only limitation in this respect is that one of the samples must contain more than a single observation. Two single specimens obviously cannot be compared by this test for two reasons. First, the number of degrees of freedom for a t-test is $N_1 + N_2 - 2$ which would be zero when each population is represented only by a single specimen. Moreover, the t-test requires some value of s^2

in the denominator which again cannot be calculated from a single specimen. There is nothing to prevent, however, the comparison of a single specimen with a larger sample, an operation which is common in systematics. If there are N specimens in the larger sample, the t-test will have $1 + N - 2 = N - 1$ degrees of freedom and the form of t will be

$$t = \frac{(\bar{X} - X)\sqrt{\frac{N}{N+1}}}{s}$$

where $\bar{X}$ is the mean of the sample, X is the value of the single observation, and s is the standard deviation of the sample. This radically simplified form of t is just the result of setting $N_2 = 1$ and $s_2 = 0$ in the usual formula.

Example 59 shows the use of this method of comparison. The reasonable conclusion from this test is that the single specimen is drawn from a population with the same parameters as the larger sample, and this conclusion is made more likely by the agreement between the specimen and the sample in four characters.

EXAMPLE 59. Comparison of dimensions of teeth in a single specimen of the fossil pig *Chleuastochoerus*, from the Pliocene of China, with the sample means for these same dimensions in a larger series of specimens of the same genus from a different locality. (Data from Pearson, 1928)

VARIATE	15-16 SPECIMENS FROM LOCALITY 49		SINGLE SPECIMEN FROM LOCALITY 30_2			
	$\bar{X}$	s	X	$\bar{X}-X$	t	P
Length M_1	13.6	.7	14.3	−.7	−.94	.3−.4
Width M_1	9.6	.5	9.4	.2	.38	.7
Length M_2	16.4	.8	16.7	−.3	−.35	.7−.8
Width M_2	11.9	.7	10.9	1.0	1.34	.2

Assumptions of the *t*-test

The *t*-test is the best approximation for testing the differences between two samples, but it is exact only under a restricted set of conditions which are never met in practice. The first of these, as has already been mentioned, is that the measurement be normally distributed in the populations from which the samples are drawn. No real population is exactly normal, but experiments on the *t*-test show that it is quite insensitive in practice to deviations from normality or, to use the technical term, it is a *robust* test.

When the *t*-test is employed as a criterion for the significance of means, another underlying assumption is that the variances of the two populations are equal. If the variances are not equal, *t* is no longer an exact test on the means but again empirical assays have shown it to be quite insensitive to this assumption.

Up until very recently texts on biometry have given two different forms of the *t*-test, one of which purported to test the difference between the means independent of any assumption about the variances. No such exact test really exists, unless the variances are known exactly and these older methods have no particular use. In practice, the *t*-test as given in this book does test only the difference between the means, and it requires a very large difference between the variances of the populations to have any effect. It is, of course, conceivable that some populations might have the same mean but such radically different variances as to produce a significant value of *t*. While conceivable, it is eminently unlikely in any comparison which makes biological sense. Snedecor (1956) gives a correction to the *t*-test which may be used when there is strong reason to believe that the variances are very different, but the best plan is to employ the standard *t*-test and exercise some caution in concluding anything when the probability is very close to the significance level chosen.

The Variance of Two Populations

Whereas some users of biometrical techniques assume that the variance or standard deviation of a sample is useful only as a tool for determining the error with which the mean is estimated, a measure of variability is of interest in itself especially in zoology. The fact that one population is more variable than another is biological information of just as great significance to the systematist, morphologist, or other students of evolution as the knowledge that they differ in mean value. The processes of evolution act on individuals, and this action is reflected not only in the average structure of these individuals but in the degree to which they differ among themselves within a natural unit. Moreover, the variance of a population conveys a kind of information about the relation of the group to other groups and to the forces of the environment, which cannot be extracted from the mean alone, even if this mean were known exactly.

The significance of the difference between two sample estimates of variance is in principle identical with test on the means. A null hypothesis is set up—either that there is no difference (two-sided test) or that one is not larger than the other (one-sided test) and the probability of the observations under this hypothesis is determined.

Unlike tests on the mean, the difference between the two sample variances cannot be used as a test criterion. The ratio of the two variances can, however, and this is quite as satisfactory, for the hypothesis that the difference between the variances is zero is identical with the hypothesis that their ratio is unity. The standard null hypothesis for testing the significance of the difference between two variances, then, is that the ratio

$$\frac{\sigma_1^2}{\sigma_2^2} = 1$$

The greater the deviation of the ratio of the sample variances from unity, the less the probability of the observation. The ratio between the sample variances is denoted by

$$F = \frac{s_1^2}{s_2^2}$$

and the distribution of this ratio under the null hypothesis is given in Appendix Table IV. Since variances are always positive quantities, the smallest value that F can take is 0 although it may be infinitely large. Moreover, the distribution of F like that for t and χ^2 is different for different numbers of degrees of freedom. There is the added complication that the degrees of freedom for both samples must be separately specified rather than being combined as in the t-test. This is taken account of in Appendix Table IV, the degrees of freedom for the numerator and denominator of F being separately tabulated. Across the top of the table are the degrees of freedom for the numerator, denoted by m, and along the left-hand side in the second column are the degrees of freedom for the denominator, labeled n. In the body of the table are various values of F, and corresponding to each of these along the left-hand margin are the probabilities to be used in "one-sided" and "two-sided" tests.

At first, the values in this table may not appear to conform with the fact that F can be as small as zero, since the smallest value of F given in the table is unity. This is simply a space-saving device based on the simple expedient of always putting the larger sample variance in the numerator of the fraction.

The data of Example 56 may be used as an illustration of this test:

$$F = \frac{s_1^2}{s_2^2} = \frac{.43128}{.16923} = 2.55$$

with 12 degrees of freedom in the numerator and 42 degrees of freedom in the denominator.

Although the F table does not give values for 42 degrees of freedom in the denominator, the probability can be read off from the table with fair accuracy. For 12 degrees of freedom in the numerator, the F value corresponding to the 5 per cent level of significance lies between 2.41 at 30 degrees of freedom in the denominator and 2.17 at 60 degrees of freedom in the denominator. Since the observed value of F is outside these limits, the probability is below the 5 per cent level. Trying now the 2 per cent level, F should lie between 2.50 (60 degrees of freedom in the denominator) and 2.84 (30 degrees of freedom in the denominator). The observed value of F, 2.55, does lie in this interval so that the corresponding probability is close to .02. This is a sufficiently refined estimate of the probability in question. The difference between the variances in sample A and sample B is then significant at the 2 per cent level and for most purposes this would be regarded as strong enough evidence for rejection of the null hypothesis, i.e., for concluding that the population variances are different.

The data in Example 57 give a quite different result. Here

$$F = \frac{.17808}{.14593} = 1.22$$

with 28 degrees of freedom in the numerator and 132 degrees of freedom in the denominator. Appendix Table IV shows that for 30 and 120 degrees of freedom, F must be equal to or greater than 1.41 to be significant even at the 20 per cent level. Since the observed F is smaller than 1.41, these two samples may be regarded as not differing significantly. It is impractical to tabulate the F table for probabilities greater than 20 per cent and not particularly useful, since there are very few if any cases where more than a 20 per cent probability would warrant rejection of the null hypothesis.

The two examples given assumed that a two-sided hypothesis was being tested, namely, that the variances were not different. One-sided tests for these same examples could also be made by using the one-sided probability values at the left of the table. Many tests on variances are one-sided, and Chapter 12 on the analysis of variance will make extensive use of such tests. Again, the zoologist must be clear on whether the particular hypothesis he is testing is one- or two-sided.

It should be obvious that to test the difference between standard deviations, it is only necessary to square them, thus converting them into their equivalent variances. Standard errors in like manner must first be squared and then multiplied by N, to convert them back into the variances from which they were calculated. The resulting variances are then tested using the F ratio.

Binomial Proportions

Aside from the mean and variance of continuous variates, the most commonly calculated sample quantity is the proportion of individuals possessing a specific trait. The test of the significance for binomial proportions makes use of a probability distribution already discussed in a quite different context, the χ^2 distribution. It is not obvious why this function, which was introduced as the distribution of the sample variance, should also come into play in testing hypotheses about binomial proportions, nor can the connection be explained except by recourse to considerable mathematics. While the logic of most of the operations introduced in this book has been explained, at least in heuristic terms, the reader will simply have to take as a matter of faith that the distribution of a sample variance is also applicable to testing differences between binomial proportions. While the χ^2 test is used here for this specific purpose, it has much wider applications directly related to binomial proportions, and Chapter 13 in its entirety is devoted to these cases.

Suppose that there are two samples, A and B, of sizes N_A and N_B, respectively, and that the proportion of individuals possessing the trait is $\bar{p}_A$ in the first sample and $\bar{p}_B$ in the second. To test whether or not $\bar{p}_A$ and $\bar{p}_B$ are really estimates of two different proportions, it is easiest to work directly with the observed numbers and not with these relative frequencies themselves. The simplest method of calculating χ^2 for this test is to place the observations in a table of the following sort:

	Trait present	Trait absent	
Sample A	a	b	$N_A = a + b$
Sample B	c	d	$N_B = c + d$
TOTAL	$a + c$	$b + d$	$N = a + b + c + d$

In each cell are the number of observations fitting a particular specification. Thus a is the number of individuals in sample A that possess the trait in question; b is the number of individuals in sample A without the trait; c is the number in sample B that show the trait; and so on. N is the total number of observations in both samples and is thus equal to $N_A + N_B$ and to $a + b + c + d$.

Having placed the observations in such a table (called a "contingency table"), the calculation of χ^2 for the test of significance is simply given by

$$\chi^2 = \frac{N(ad - bc)^2}{(a + b)(c + d)(a + c)(b + d)}$$

No matter what the value of N, this calculated value for a 2×2 contingency table is distributed as χ^2 with *one* degree of freedom. At the top of the χ^2 table are the probabilities of observing a χ^2 less than or equal to a given value, which probabilities are the appropriate ones for confidence interval estimation. For a χ^2 test of the difference between binomial proportions, however, it is the probability of exceeding a given value of

χ^2 that is required. These are given at the bottom of the table and are simply one minus the former values. Under the null hypothesis—that there is no difference between the populations from which the samples have been taken—the value of χ^2 will exceed 3.84 only 5 per cent of the time and it will exceed 6.63 only 1 per cent of the time. If .05 were the significance level adopted, a value of χ^2 as large or larger than 3.84 would be considered significant, the null hypothesis rejected and the populations considered different. As in previous tests, it is better to publish the actual probability level attained for a particular observed value of χ^2 than to state only that the difference is significant or nonsignificant. Example 60 illustrates this calculation for some observations on shell coloration in a land snail. Although the binomial proportion is usually thought of in terms of presence and absence of a character, the dichotomy of red as opposed to yellow shells in this example is pertinent, since the character can be thought of as presence or absence of yellow color.

EXAMPLE 60. Numbers of red- and yellow-shelled snails *Cepaea nemoralis* in samples from three European localities. (Data from Lamotte, 1951)

LOCALITY	YELLOW	RED	TOTAL	% YELLOW
Dilbeek, Belgium	22	24	46	47.8
Stockholm, Sweden	23	33	56	41.1
Niederbronn, France	50	15	65	76.9

χ^2 TESTS

A	YELLOW	NONYELLOW	TOTAL	
Dilbeek	22	24	46	$\bar{p}_A = .478$
Stockholm	23	33	56	$\bar{p}_B = .411$
Total	45	57	102	—

$a = 22, b = 24, c = 23, d = 33, N = 102$

$$\chi^2 = \frac{(ad - bc)^2 N}{(a + b)(c + d)(a + c)(b + d)} = \frac{(726 - 552)^2 (102)}{(45)(57)(56)(46)} = .467$$

B	YELLOW	NONYELLOW	TOTAL	
Dilbeek	22	24	46	$\bar{p}_A = .478$
Niederbronn	50	15	65	$\bar{p}_B = .769$
Total	72	39	111	—

$a = 22, b = 24, c = 50, d = 15, N = 111$

$$\chi^2 = \frac{(ad - bc)^2 N}{(a + b)(c + d)(a + c)(b + d)} = \frac{(330 - 1200)^2 (111)}{(72)(39)(65)(46)} = 10.00$$

The table of χ^2 for 1 degree of freedom shows that the probability corresponding to a χ^2 of .467 is roughly .50. The difference between the frequency of yellow shells found at Dilbeek and Stockholm is then not inconsistent with the null hypothesis, and the populations must be regarded as having equal frequencies. On the other hand, a χ^2 value of 10.00 is too large even to be tabulated for 1 degree of freedom, and the probability of this observation under the null hypothesis is less than .005, so that the difference between the Dilbeek and Niederbronn populations would certainly be regarded as significant. The difference between Niederbronn and Stockholm is still greater than that between Dilbeek and Niederbronn and so may be assumed to be highly significant without further calculation.

Small Samples

The quantity discussed in the last section for testing significance from a contingency table is only approximately distributed as χ^2, and in small samples this approximation is so poor as to produce quite misleading results. The error in the test is always in the same direction, that is, it tends to assign too low a probability to the observations so that the null hypothesis will be rejected more often than it should. The reason for this discrepancy is that the distribution of χ^2 is continuous while that of the frequencies in a contingency table is necessarily discontinuous. The χ^2 distribution is approached as a limit by these discontinuous data, and if the frequencies are not unduly low, the approach is sufficiently close to give a valid estimate of the probability from χ^2; but this is not reliable if the values of the table are determined largely by very low frequencies in it. Suppose, for example, that an observed frequency is 2 and that this is lower than the theoretical frequency. What the use of χ^2 measures for that particular cell is the probability that a frequency of 2, 1, or 0, that is, of 2 or less, would arise by chance if the variables were independent. The probability is measured in terms of the relative area of the curve for the distribution of χ^2 lying beyond the value of χ^2 resulting from a frequency of 2 in this cell. The next possible higher frequency is 3; but the curve for χ^2 is continuous, and there is an area between the points corresponding to frequencies 2 and 3. Moreover, for the same reason, there is no area corresponding exactly and only to the frequency 2 or 3; but each of these must be assumed to have a certain range.

These considerations have given rise to two methods for coping with small samples. The first, *Yates' correction for continuity*, treats the observations as if they were, in fact, continuous and uses a modified form of χ^2. Up until recently this has been the most widely used approach, but recent work has shown that in many cases it is worse than no adjustment at all.

The second approach is to accept the discontinuous nature of the observations, avoid χ^2 entirely, and use an exact calculation of the probability based upon the *multinomial distribution*. Such a test, discussed in Fisher (1950, page 93), is quite laborious and seldom, if ever, need be invoked. Yates' correction, discussed in the next section, should be used when the total frequency, N, is *greater* than 40 and when the class with the lowest observed frequency is 10 or *less*. If every class contains more than 10 observations, the uncorrected χ^2 is quite adequate. For those few cases where the total frequency, N, is 40 or less, neither the uncorrected nor the corrected χ^2 can be trusted by itself, since the former will overestimate the true probability while the latter will considerably underestimate it. In these cases, it often happens that the uncorrected χ^2 is a better approximation to the true value than is given by the correction for continuity. The best approach for such small samples is to calculate both the corrected and uncorrected χ^2 values. If both show the difference between the samples to be significant, or not significant, no problem arises. Should the uncorrected χ^2 give a probability below the chosen significance level while the corrected χ^2 gives a value above the chosen level, two courses of action are open. The first is to perform the exact test and the other is simply to regard the null hypothesis with skepticism and regard the case as a doubtful one. Since the choice of significance levels is arbitrary, and since the observation of a difference significant at the 5 per cent level is only a single datum which aids in making some biological decision, no great loss is entailed in such an admission of uncertainty.

Yates' correction for continuity. This method assumes that the observed frequencies are really midpoints of continuous ranges. Thus a frequency of 2 is given an upper limit of 2.5 and a lower limit of 1.5. If the observed frequency in a cell is below the theoretical frequency in that cell, the observation is increased by 0.5, whereas if the observed frequency is greater than the theoretical frequency, the observation is reduced by 0.5. The way in which theoretical frequencies can be calculated in a contingency table is explained in Chapter 13, but Yates' correction can be made without recourse to this calculation. The value of χ^2 corrected for continuity, symbolized by χ_c^2 is given by

$$\chi_c^2 = \frac{\left(ad - bc - \dfrac{N}{2}\right)^2 N}{(a+b)(c+d)(a+c)(b+d)}$$

when $(ad - bc)$ is positive, while

$$\chi_c^2 = \frac{\left(ad - bc + \dfrac{N}{2}\right)^2 N}{(a+b)(c+d)(a+c)(b+d)}$$

when $(ad - bc)$ is negative.

In Example 61 the calculation of χ_c^2 is illustrated and the result of this test is contrasted with the unadjusted χ^2. As expected, the correction results in a lower χ^2 value and a higher probability. Neither test, however, gives a probability low enough to warrant rejection of the hypothesis that both populations have the same binomial probability. For this set of observations, Fisher's exact test yields a probability of .741, which is very close to that obtained by Yates' corrected χ_c^2:

EXAMPLE 61. Comparison of unadjusted χ^2 and Yates' corrected χ_c^2 for determining the significance of the difference between two binomial proportions. Data show the frequency of banded and unbanded forms of *Cepaea nemoralis* from two French localities. (Data from Lamotte, 1951)

LOCALITY	BANDS	NO BANDS	TOTAL
Emerainville	32	2	34
Orsay	120	13	133
TOTAL	152	15	167

A. UNADJUSTED

$$\chi^2 = \frac{[(32)(13) - (120)(2)]^2 (167)}{(34)(133)(152)(15)} = .502$$

Probability = .45

B. ADJUSTED FOR CONTINUITY

$$\chi_c^2 = \frac{\left[(32)(13) - (120)(2) - \frac{167}{2}\right]^2 (167)}{(34)(135)(152)(15)} = .140$$

Probability = .70–.75

Other Parameters

While the mean, variance, and binomial proportion are the most frequently calculated sample constants, the other sample values which have been discussed such as the median, mode, and coefficient of variability have their uses, and at times it may be necessary to test the difference between two sample values of these constants. As exact sampling distributions for these parameters are not available, the only alternatives are *distribution-free* tests to which we have already referred (p. 158) or a normal approximation such as has been suggested for confidence interval estimation for these parameters. If sample sizes are over 100, the difference between two sample values of one of these measures can be tested by

$$\frac{K_1 - K_2}{\sqrt{s_{k_1}^2 + s_{k_2}^2}} = \tau$$

where K_1 and K_2 are the values of the estimates of some parameter in the two samples, and s_{k_1}, s_{k_2} are the corresponding standard errors of the estimate in each sample. Formulae for these standard errors in large samples have been given on page 166.

As defined here, τ is normally distributed so that for a two-sided test the difference $K_1 - K_2$ would be considered significant at the 5 per cent level if 5 per cent of the area of the normal curve lies outside the limits $\pm \tau$. The table of the normal distribution in this book (Appendix Table I) is such that the area between the mean and a value of τ is given. While the probability of falling outside the limits $+ \tau$ and $- \tau$ could be calculated from this table by the methods shown on page 121, the simplest procedure is to use the table of the t distribution for an infinite number of degrees of freedom, these values being exactly equal to the required areas under a normal distribution. For example, if

$$\frac{K_1 - K_2}{\sqrt{s_{k_1}^2 + s_{k_2}^2}}$$

should be equal to 1.96, the difference between K_1 and K_2 would be significant at the 5 per cent level, since the last line of the t-table shows 1.96 as the 5 per cent point. For a one-sided test, the appropriate one-sided probabilities at the bottom of the table must be used.

For samples much smaller than 100, tests on medians, coefficients of variation, quartiles, and so on are most unreliable, and this lack of adequate test procedures is another argument for avoiding them unless they present necessary information not made clear by means and variances.

Comparisons with Hypothetical Values

In the place of comparing two samples, it may be necessary to compare a single sample with some hypothetical population. Hypothetical populations, although common in genetics and some branches of population ecology, do not enter into systematic and morphological enquiries with any frequency. There are no mathematical theories which would predict that a certain mean or variance ought to take a specified value in the realm of zoology, as we understand it in this book, but occasionally the zoologist, dealing with some theoretical concept of population ecology or genetics, may want to test whether a given sample has been drawn from a hypothetical population with a definitely known mean, variance, or other parameter. For such cases, test procedures already outlined in the chapter on confidence intervals are appropriate. If the 95 per cent confidence interval constructed from an observed sample covers the hypothetical value of the parameter, then there is no significant difference between the sample and the hypothetical population at the 5 per cent level. If, on the other hand, the 95 per cent confidence interval does not include the hypothetical value of the parameter, the difference between sample and theory is significant at the 5 per cent level. The same holds true for any level of significance, the general rule being: To test the agreement of an observed sample value with some hypothetical value at the K per cent level, construct the $(1 - K)$ per cent confidence interval for the parameter. If the theoretical value falls outside this interval, reject the null hypothesis and assume the difference to be significant. It is important to note that this technique is not applicable to single specimens, since a confidence interval cannot be constructed from one observation.

As an example of such a test, consider the problem of the sex ratio in an animal population. It is well known that in most, if not all, species the ratio of males to females changes with age class and may even be different from 1:1 at parturition in mammals. Genetic theory, however, predicts that at conception equal numbers of male and female zygotes are formed, at least in those species with regular sexual reproduction. Suppose that a sample of animals is collected which is fairly homogeneous with respect to age and without collecting bias with respect to sex. It is desired to test whether the observed frequency of males is in accordance with Mendelian expectation. Example 62, taken from Davis' work (1951) on *Rattus norvegicus*, illustrates the use of confidence intervals for such a test. In this example, there is no significant deviation from a 1:1 sex ratio.

Sample Size

One way of increasing the discriminatory power of a test for a given significance level is to take larger samples. If it were possible to make samples infinitely large, the true parameters of the population would be perfectly known, and without any error at all different populations could be distinguished. Thus every opportunity to increase the number of observations makes for increased accuracy of decision. In paleontological sampling, the sample size is usually not under the zoologist's control. He must work with what specimens are available. This is true to some extent in faunal sampling as well, but with populations of living animals it is often possible to collect and measure as many as is desirable for a particular purpose. When the sample size is under control, the question arises as to how many observations are needed to detect a given difference between populations. The answer to the question is found by algebraically solving the equation of the appropriate statistical test for the sample size. The most common problem involves the difference between the means of two populations, so that this case will be used as an illustrative example.

EXAMPLE **62**. Sex ratio among rats (*Rattus norvegicus*) trapped in Baltimore, Maryland. (Data from Davis, 1951)

N	f MALES	f FEMALES	PROPORTION MALES $= \dfrac{f_M}{N}$
1109	539	570	.486

95 per cent confidence interval for proportion of males:

$$\frac{f_M}{N} \pm 1.96 \sqrt{\frac{\frac{f_M}{N}\left(1 - \frac{f_M}{N}\right)}{N}}$$

$= .486 \pm 1.96 \sqrt{(.486)(.514)/1109}$
$= .486 \pm (1.96)(.015)$
or from .457 to .515

Since this interval includes the hypothetical value of .50, there is no reason to reject the hypothesis that this population had the normal Mendelian sex ratio.

The test for the significance of the difference between two means has, as has already been discussed, the form

$$t = \frac{(\bar{X}_1 - \bar{X}_2)\sqrt{\dfrac{N_1 N_2}{N_1 + N_2}}}{\sqrt{\dfrac{(N_1 - 1)s_1^2 + (N_2 - 1)s_2^2}{N_1 + N_2 - 2}}}$$

To make the problem at hand easier of solution two simplifying assumptions can be made. The first is that the two sample sizes are equal. If the sample size is under the zoologist's control, this is possible and, incidentally, gives the best test. The second simplifying assumption is that the sample variances are roughly the same, that is, $s_1^2 = s_2^2 = s^2$. Then the expression for t reduces to

$$t = \frac{(\bar{X}_1 - \bar{X}_2)\sqrt{N}}{\sqrt{2 s^2}}$$

If the means of the two populations are the same, the value of t calculated from samples of these populations will have Student's t distribution tabulated in Appendix Table II. Only 5 per cent of the time will t fall outside the limits -1.960 to $+1.960$ and only 1 per cent of the time will it lie outside the interval -2.576 to $+2.576$ (assuming large samples).

If the 5 per cent level of significance is the one chosen, then 5 per cent of the time populations which do not really differ will be erroneously considered different. On the other hand, if the null hypothesis is *not* true, if $\mu_1 - \mu_2$ is not zero, t will not have the tabulated Student's t distribution but a *noncentral t distribution*, and the probability that t takes a value outside the limits ± 1.96 will be greater than .05. If the 5 per cent significance level has been chosen, the populations will be considered different more often than 5 per cent of the time, and, since the populations really are different, this is all to the good. In fact, the greater the probability that t falls outside these limits, the better, since one would like to make the proper decision about the populations as frequently as possible. The problem of choosing a sufficiently large sample is then the problem of increasing as much as possible the probability that t will fall outside the limits considered significant. Reference to the simplified form of t given above shows that the absolute value of t will increase with increasing N and with an increase in the absolute value of $\bar{X}_1 - \bar{X}_2$. The probability that $\bar{X}_1 - \bar{X}_2$ takes any value is a function of the true difference between the populations, and clearly if the populations differ by a large amount, $\bar{X}_1 - \bar{X}_2$ is likely to be large in absolute size. Although the probability that $\bar{X}_1 - \bar{X}_2$ takes a certain value is fixed by the difference between the populations, the probability that t exceeds a given value can be increased by increasing N.

All that is required, then, is the distribution of noncentral t. It would be impractical to tabulate this distribution since it is different for every value of $\mu_1 - \mu_2$, nor is such a tabulation necessary since noncentral t differs from Student's t only in its mean, which is shifted by an amount proportional to $\mu_1 - \mu_2$. Then the probability that noncentral t will exceed 1.960, say, is equal to the probability that Student's t will exceed

$$t = 1.960 - \frac{(\mu_1 - \mu_2)\sqrt{N}}{\sqrt{2}\, s^2}$$

Suppose that the zoologist wishes to be 90 per cent sure that his sample value of t will exceed 1.960. Appendix Table II shows that t will exceed the value -1.282 for 90 per cent of the time in large samples. Substituting this value in the equation above results in the following:

$$-1.282 = 1.960 - \frac{(\mu_1 - \mu_2)\sqrt{N}}{\sqrt{2}\, s^2}$$

which on rearrangement gives

$$N = \frac{2\,(3.243)^2\, s^2}{(\mu_1 - \mu_2)^2}$$

The denominator of this expression contains only the difference to be detected, but the numerator contains the variance of the observations, s^2.

This quantity is unknown, so that unless some value can be assumed for it the problem is insoluble. Some previous information about the variance is necessary and this might be obtained in two ways. First, a small sample could be drawn from the populations and the value of the variance in this sample used. A better procedure would be to establish 95 per cent confidence limits for the population variance from this small sample and then to insert the upper confidence limit into the expression. This procedure will tend to give a larger N than absolutely necessary, but it is a safe procedure. If a small sample is not available, the best that can be done is to substitute for s^2 the largest reasonable value derived from observations on similar material. While it is not suggested that a given measurement will have the same variance in all species or populations, the coefficient of variation of the measurement will tend to be roughly the same. By choosing the *largest* reasonable value of V, known from other material, a working estimate of s^2 can be derived for substitution in the above relationship.

To generalize and sum up the procedure for finding a sample size large enough to be x per cent sure of having a significant t test at the y per cent level of significance, use the expression

$$N = \frac{2 s^2 (t_x + t_y)^2}{(\mu_1 - \mu_2)^2}$$

where

(1) $\mu_1 - \mu_2 = $ the postulated difference between the population means.

(2) $s^2 = $ the largest reasonable variance for the variate in question; this may be either the upper confidence limit for σ^2 from a previous sample, or s^2 derived from known coefficients of variation for the character.

(3) $t_x = $ the value of Student's t corresponding to the desired assurance of significance. This is derived from the "one-sided" probabilities as tabulated in Appendix Table II for infinite degrees of freedom. These "one-sided" probabilities at the bottom of the table are 1 minus the desired probability value. Thus, $t_x = 1.645$ for a 95 per cent assurance and $t_x = 2.326$ for a 99 per cent assurance.

(4) $t_y = $ the value of Student's t corresponding to the significance level desired. Care should be taken to use either the "one-sided" or "two-sided" test probabilities, depending upon whether the alternative hypothesis about the true difference between μ_1 and μ_2 specifies or does not specify the sign of the difference. Thus, if one uses the 5 per cent significance level but does not specify whether μ_1 is larger or smaller than μ_2, the proper value of t_y is 1.960.

An example of this calculation can be taken from Dice's observations on *Peromyscus maniculatus*. The average tail length of this species varies between 55 and 66 mm. in Dice's data and the largest variance observed among seven samples was 19.75. Suppose that one wanted to be 97.5 per cent sure of detecting in the future a difference as small as 1 mm. in tail length between two populations of this animal, using the 5 per cent level of significance. Then

$$N = \frac{2 s^2 (t_x + t_y)^2}{(\mu_1 - \mu_2)^2} = \frac{2 (19.75) (1.96 + 1.96)^2}{1} = 606.9$$

To be 97.5 per cent sure that a difference of 1 mm. between two population means will result in a significant *t*-test at the 5 per cent level of significance, it is necessary to take samples from each population of about 607 individuals. These are extremely large samples, but they are sufficient to be virtually certain of detecting a difference which is less than 2 per cent of the average length. For a 95 per cent assurance, it would be necessary to have samples of only 325 from each population, and lower assurances will result in correspondingly lower necessary sample sizes.

Turning to population variances, a method similar in principle to that used for the difference between means can be used. The F distribution as tabulated in Appendix Table IV is the distribution of s_1^2/s_2^2 under the null hypothesis that $\sigma_1^2/\sigma_2^2 = 1$. What is required for the sample size problem is the distribution of s_1^2/s_2^2 under a different hypothesis, namely, that σ_1^2/σ_2^2 is greater than unity. Denoting the ratio s_1^2/s_2^2 under the alternative hypothesis as F', we wish to know the probability that F' will be greater than or equal to that value of F considered significant. Clearly it is not practical to tabulate the distribution of F' for various alternative hypotheses. Instead, F' is converted to F by dividing it by σ_1^2/σ_2^2. This adjusted value can now be referred to the usual F distribution to determine the probability that it will be equaled or exceeded. This probability of detection will increase with increasing N, so that by a trial and error procedure it is possible to find a value of N (assuming always that the two samples are of equal size) sufficient to make the probability of detection as large as may be desired.

Appendix Table IV is not in a particularly convenient form for the trial-and-error procedure of finding the required sample size. Table 2 has been specially constructed for this purpose. Along the left-hand side are various values for degrees of freedom, assumed to be equal in both samples. As usual, the degrees of freedom are one less than the sample size. The next two columns contain the values of F which are significant at the 5 per cent and 1 per cent levels. The last four columns contain the values of the adjusted F' which give 75, 90, 95, or 99 per cent assurance of detection. To use the table choose a significance level, say, 5 per cent, and a trial value

198 QUANTITATIVE ZOOLOGY

of N. Divide the corresponding value of F by σ_1^2/σ_2^2 and compare this quotient with the four entries under "Probability of Detection" on the same line. Try successive values of N in this way until the quotient corresponds approximately to the desired detection probability.

TABLE 2. Selected values of F for use in determining the sample size necessary to detect a given ratio of two population variances.

D.F.	SIGNIFICANCE LEVEL		PROBABILITY OF DETECTION			
	5%	1%	75%	90%	95%	99%
10	3.72	5.85	.645	.430	.336	.206
20	2.46	3.32	.736	.557	.471	.340
30	2.07	2.63	.780	.623	.543	.419
40	1.86	2.26	.807	.666	.592	.479
50	1.74	2.07	.826	.696	.626	.518
60	1.67	1.96	.840	.718	.654	.548
70	1.60	1.85	.851	.736	.673	.573
80	1.55	1.78	.860	.751	.691	.594
90	1.51	1.72	.867	.763	.705	.612
100	1.48	1.67	.873	.774	.718	.628
120	1.43	1.60	.884	.791	.741	.654
140	1.39	1.55	.892	.805	.756	.675
160	1.36	1.50	.898	.817	.769	.692
180	1.34	1.47	.904	.826	.781	.707
200	1.33	1.44	.908	.834	.791	.719
250	1.28	1.39	.917	.850	.811	.745
300	1.25	1.35	.925	.862	.826	.765
350	1.23	1.32	.930	.872	.838	.780
400	1.22	1.29	.935	.880	.848	.792
450	1.20	1.28	.938	.887	.855	.803
500	1.19	1.26	.941	.891	.862	.812

For example, let $\sigma_1^2/\sigma_2^2 = 1.5$ and set the significance level at 5 per cent. Choosing 100 as a trial value for degrees of freedom, the corresponding entry in the F column is seen to be 1.48. Dividing 1.48 by 1.5 yields .987 which is larger than the entry under the 75 per cent detection probability, .873. The probability of detection for a sample of size 101, then, is less than 75 per cent which is for most purposes too low. For 300 degrees of freedom the 5 per cent value of F is 1.25. Dividing this by 1.5 gives .836 which has a probability of detection between 90 per cent and 95 per cent. Then, if a sample of size approximately 300 is taken from each of two populations, a ratio of 1.5 between their variances will be detected at the 5 per cent level about 94 per cent of the time. A sample of approximately 450 from each population would be required to be 99 per cent sure of detecting such a difference. These are extremely large sample sizes for most zoological inquiries, but they are typical of those required when variances are being compared.

A final problem which is related to the question of sample size is that of binomial sampling limits. Supposing that a character occurs in a certain proportion of the individuals of a species, how large a sample would be necessary to be sure, for all practical purposes, to include at least one individual with this character? More generally, within what limits are the number of individuals possessing the character sure to fall? These questions can be answered by use of Romig's tables, or the Harvard tables, to which reference has already been made. For a given sample size and a given value of the proportion of individuals possessing the trait in the population, these tables will give the probability of observing fewer than X or more than Y individuals in the sample with the characteristic in question. To be "sure" of observing no fewer than X or no more than Y individuals, this probability should be set at a very low level, say, .001. Even this low probability does not guarantee that the observed number of cases will fall within the limits, but it makes the probability very high.

Table 2 is a condensation of this information for selected sample sizes and population probabilities taken from the more extensive binomial tables. In each cell of the table, the figure to the left is the lower limit and that to the right the upper limit of the number of individuals possessing the character.

TABLE 3. Sampling limits.

SAMPLE SIZE N	PER CENT OCCURRENCE OF CHARACTER IN POPULATION								
	10%	20%	30%	40%	50%	60%	70%	80%	90%
5	0– 3	0– 4	0– 5	0– 5	0– 5	0– 5	0– 5	1– 5	2– 5
10	0– 5	0– 6	0– 8	0– 9	1– 9	1–10	2–10	4–10	5–10
15	0– 6	0– 8	0–10	1–12	2–13	3–14	5–15	7–15	9–15
20	0– 7	0–10	1–13	2–15	3–17	5–18	7–19	10–20	13–20
25	0– 8	0–12	1–15	3–18	5–20	7–22	10–24	13–25	17–25
30	0– 8	0–13	2–17	4–20	7–23	10–26	13–28	17–30	22–30
35	0–10	1–15	3–19	6–23	9–26	12–29	16–32	20–34	25–35
40	0–11	1–16	4–21	7–26	10–30	14–33	19–36	24–39	29–40
45	0–12	2–18	5–23	8–28	12–33	17–37	22–40	27–43	33–45
50	0–12	2–19	6–25	10–31	14–36	19–40	25–44	31–48	38–50
55	0–13	2–20	6–26	10–32	15–38	23–45	29–49	35–53	42–55
60	0–13	3–21	7–28	12–35	17–41	25–48	32–53	39–57	48–60
65	0–14	3–23	8–30	13–37	19–44	28–52	35–57	42–62	51–65
70	0–15	4–24	9–32	15–40	21–47	30–55	38–61	46–66	55–70
75	0–16	4–25	10–34	16–42	23–50	33–59	41–65	50–71	59–75
80	0–16	5–27	11–36	18–45	25–53	35–62	44–69	53–75	64–80
85	0–17	6–28	12–38	20–47	27–56	38–65	47–73	57–79	68–85
90	1–18	6–29	13–40	21–50	29–59	40–69	50–77	61–84	72–89
95	1–19	7–31	14–42	23–52	32–61	43–72	53–81	64–88	76–94
100	1–19	8–32	16–44	24–54	34–64	46–76	56–84	68–92	81–99

Thus, for a sample of 20 and proportion of population occurrence 40 per cent, the limits are seen to be 2 and 15; in other words, in these conditions the sample would surely have the character represented twice and would not have it represented more than 15 times. Any conditions not covered by the table can easily be inserted in the formulae discussed on page 125.

With this table or the formulae, it is easy to answer questions like those posed at the beginning of this section. The following statements can be verified by the table:

> If a character occurs in 20 per cent of the population, a sample of 35 specimens is necessary to be certain to include one with this character (in the 20 per cent column, the number 1 first appears as lower limit in the row $N = 35$).
>
> If a character does not occur in a sample of 15 specimens, it does not occur in more than 40 per cent of the population (in the $N = 15$ row, the 40 per cent column is the first in which the lower limits is not 0).
>
> A character must occur in at least 30 per cent of the population to be surely represented in a sample of 20 specimens (in the $N = 20$ row, the 30 per cent column is the first with the lower limit above 0).
>
> If a character is present in all individuals in a sample of 25 specimens, it must have occurred in at least 80 per cent of the population (in the $N = 25$ row, the 80 per cent column is the lowest with the upper limit equal to N).

It may be reiterated that these are limits of probability so great as to amount to practical certainty, the only ones on which such positive and conclusive statements may safely be based. Possibility and probability near an even chance, far from certainty, are different. Thus, if a character occurs in 20 per cent of the population, it is possible for it to occur in any sample of 1 or more individuals, and it is more likely than not to appear in any sample of 5 or more individuals, but it is only certain (for all practical purposes) to appear in a sample of 35 or more individuals. It is practically impossible for it to occur in all the individuals of any sample larger than 3 and improbable for it to occur in the whole of a sample of any size.

For sample sizes in excess of 100, the close approximation of the binomial distribution to normal curve can be utilized. If the population probability is p, then samples of size n will have a mean number of individuals possessing the trait equal to np. The approximate normal distribution will then have a mean of np and a standard deviation of $\sqrt{npq}$ (see page 138). The table of the cumulative normal distribution shows that 99 per cent of all observations will fall within the limits

$$\mu \pm 3.291$$

so to extrapolate from Table 3, the upper and lower sampling limits in large samples will be given by

$$np + 3.291 \sqrt{npq}$$

and

$$np - 3.291 \sqrt{npq}$$

respectively. For example, in a sample of 400 from a population in which a trait is possessed by 50 per cent of the individuals, it is practically certain that between

$$(400)(.5) + 3.291 \sqrt{(400)(.5)(.5)} = 232$$

and

$$(400)(.5) - 3.291 \sqrt{(400)(.5)(.5)} = 168$$

individuals will have the trait.

Heterogeneity of Samples

When a collection is brought in from the field, it is usually highly heterogeneous. Except as a broad faunal sample, it is not a sample of a single population for any study purposes but is made up of a mixture of samples of many different populations. The first step in study is to separate the collection as nearly as possible into samples, each drawn from a single population. This is done in the first instance by setting up population specifications that can be met from the field data. It will, for instance, usually be necessary to specify populations by localities or areas or by geological horizons and to separate the collection, according to the field data, into samples representative of these. Even for studies not primarily taxonomic, it almost always becomes necessary eventually to set up taxonomic specifications and to separate the samples into subspecies, species, genera, etc., and this is, of course, the principal aim of strictly taxonomic studies. Sometimes these specifications can be met from the field data, which means that the taxonomic groups are so readily recognizable that part of what is essentially laboratory study could in fact be done in the field. More often this is not true, and taxonomic separation of samples must be done from the collection as a secondary operation and not simply by filling specifications from field data.

The problem of splitting a heterogeneous sample into relatively homogeneous parts on the basis of its own characteristics thus arises from most all field collecting. The same problem sometimes arises in other ways than taxonomic. It is, for instance, frequently desirable to learn from a sample whether it is essentially homogeneous or recognizably heterogeneous as to age groups or as to sex. Observations on habits or other nonmorphological characters may also prove to be heterogeneous. With laboratory observations and controlled experiments, such heterogeneity can and should usually be avoided beforehand by meeting careful specifications; but this cannot always be done in the laboratory and often cannot in the field.

The easiest and the only conclusive method of splitting heterogeneous samples is by frequency distributions that are plainly multiple and that do not overlap. If the frequencies are low and are irregular, this cannot be observed or determined. If, however, the distribution can be so grouped as to give a fairly regular sequence and if it then has two definite modes and a definite break, with frequencies zero somewhere between these modes, it is obvious that two distributions are really present and they can be separated by inspection. A special example of this sort is given in Chapter 15, page 387. Example 63 shows another such case.

It may happen, however, that there are two modes but that the ranges overlap extensively. Even though the sample is really heterogeneous, the modes may be so close together and the ranges so nearly coextensive that it is impossible in a given distribution to detect the heterogeneity or to be sure that an apparent bimodality reflects two populations and is not the chance result of random sampling of one population. In taxonomic studies and others in which the sample observations are derived from a series of specimens with several variates pertinent to the problem under consideration, the correct procedure is to make distributions for all such pertinent variates. If any of these distributions show a clear and definite separation into two or more, it is possible to separate the other observations according to the specimen groups, even though these observations do not themselves have surely bimodal distributions.

In practice, taxonomically heterogeneous collections of related animals that are sufficiently homogeneous as to locality and, for fossils, horizon usually have at least one variate the distribution of which falls decisively into two or more parts and reveals the taxonomic heterogeneity. With the specimens of Example 63, for instance, the first of the two groups visible did not clearly split into two in any other distribution, and therefore the suggestion of bimodality here seen (one apparent mode in group 2.6–2.8 and the other, more definite, in 3.2–3.4) was very probably the result only of chance. The second group, however, did plainly split into two distributions for several variates (e.g., number of cusps on M_1) and therefore is really heterogeneous even though there is no clear hint of this in the distribution given in the example.

This usual result in zoology whereby heterogeneous samples are most reliably split in practice is made more probable by the zoological law of

EXAMPLE 63. Splitting a heterogeneous sample. Distributions of length of last lower premolar in small numbers of the extinct mammalian group Ptilodontidae a heterogeneous sample from approximately the same horizon and locality—Lower Fort Union, Crazy Mountain Field, Montana. (Original data)

X	INTERVAL .1 MM.	INTERVAL .3 MM.
2.3	0	
2.4	0	1
2.5	1	
2.6	2	
2.7	0	3
2.8	1	
2.9	2	
3.0	0	2
3.1	0	
3.2	2	
3.3	1	6
3.4	3	
3.5	0	
3.6	1	2
3.7	1	
3.8	0	
3.9	0	0
4.0	0	
4.1	0	
4.2	0	0
4.3	0	
4.4	0	
4.5	0	0
4.6	0	
4.7	0	
4.8	0	1
4.9	1	
5.0	1	
5.1	1	3
5.2	1	
5.3	1	
5.4	2	3
5.5	0	
5.6	0	
5.7	0	0
5.8	0	

ecological incompatibility, which is, in its simplest form, that two or more distinct groups of closely related and closely similar animals do not usually live together in one environment at one time. This has nothing to do directly with mathematics, but like many zoological principles it is essential in interpreting the results of numerical analysis. It may be restated from the present point of view: Really separable taxonomic groups of animals all of whose variates have extensively overlapping distributions are very rarely found together in nature. Such completely intergrading groups, usually contiguous geographic races or successive geologic subspecies, do occur but not normally in full association with each other. Distributions of variates from samples of such populations are shown in Example 64.

EXAMPLE **64**.

A. Distributions of the discontinuous variate serration number for P_4 of two geographically and geologically separate samples of fossil mammals of the genus *Ptilodus*. (Original data)

		f	
X	SAMPLE FROM TORREJON, N.M.	SAMPLE FROM FORT UNION, MONT.	BOTH TOGETHER
12	5	0	5
13	1	8	9
14	0	19	19
15	0	2	2

B. Distributions of tail length in geographically separate samples of the deer mouse *Peromyscus maniculatus*. (Data from Dice, 1932)

X	SAMPLE FROM ALEXANDER, IOWA	SAMPLE FROM GRAFTON, N.D.	BOTH TOGETHER
52–53	1	0	1
54–55	3	1	4
56–57	11	2	13
58–59	18	2	20
60–61	21	3	24
62–63	20	8	28
64–65	9	25	34
66–67	2	11	13
68–69	1	10	11
70–71	0	5	5
72–73	0	7	7
74–75	0	2	2
76–77	0	2	2

If such a heterogeneous sample, composed of two or more overlapping populations, cannot be split at all, the zoological problem is usually insoluble. There are ways in which the heterogeneity can be detected, or at least suggested, but even in such cases there is usually nothing practical that can be done about it. It may, for instance, be possible to show that two overlapping populations are represented and even, in the best possible circumstances, to calculate what percentage of specimens in the region of overlap belongs to each; but still the samples could not be separated, and it could not be determined what actual observations were from each population.

There are other hints of heterogeneity that never amount to proof but may be suggestive. A distribution with an unduly large V, much larger than for the same variate in related samples, is likely to be heterogeneous, and so is one that is strongly platykurtic. For instance, the large V for the tenth variate in Example 34 (page 91) suggests that the sample was heterogeneous, which was the fact; and so does the evident platykurtosis of the combined distribution of Example 64B, which is also heterogeneous in fact. Some homogeneous distributions do, however, have high V's and some are platykurtic, so that neither is ever conclusive evidence of heterogeneity. A significant deflection, a high V, and strong platykurtosis are not more than hints that something may be wrong with the sample; and the best thing to do in practice is to reexamine the sample and to try to get better data permitting assurance of essential homogeneity by specification of population.

Single Observations

Perhaps the commonest of all procedures in zoology is the comparison of the linear dimensions of two specimens. In taxonomic work, for instance, especially in paleontology where large samples are exceptional, a species is often represented chiefly or solely by its type, and identification of other specimens or description of other species involves comparison with this one specimen. The methods hitherto discussed are obviously inapplicable, and much as they may be desired, the comparison must be made without their aid. The procedure, however, is fundamentally the same as if groups of specimens and not single examples were available and is much clarified and placed on a much sounder basis if the relationship to such groups is recognized.

A single dimension of one specimen is part of a distribution even though it is the only known part. It is in every case assumed, consciously or unconsciously, that in comparing these isolated observations they are really considered as members of populations. This involves certain assumptions:

1. That the observation in hand is in fact representative of a group of more or less similar observations possible in nature; in other words, that it belongs to a population.
2. That it is more or less characteristic of that group.
3. That the group has a certain variability, the extent of which is at least vaguely in mind.

The second of these assumptions involves the belief that the observed value of the variate is within a limited range about the unknown mean of the population. Now it has been seen that the probability that the specimen is within a given distance of the mean can be closely estimated if the value of σ (the standard deviation) is known. In a normal curve, a little over 95 per cent of the area is included between $\mu + 2\sigma$ and $\mu - 2\sigma$; or, in other words, the chances are somewhat more than 95 in 100 that any one random observation will be within a distance of 2σ of the mean. Study of actual distributions of variates of reasonably homogeneous samples will readily show that the observed range does in fact nearly coincide with these limits in most cases.

For present purposes, then, the theoretical or inferred range of a variate on which only one or a few observations are available may be taken to be from $\mu + 2\sigma$ to $\mu - 2\sigma$. This implies that the specimen at hand is within those limits or is not over 2σ from the mean, an assumption sufficiently probable to be acceptable as a tentative working basis, in default of better evidence.

The value of σ is unknown; but if, as is necessary for the drawing of most sorts of inferences from single observations, a value of V, the coefficient of variation, is assumed, then by direct calculation from the equation $V = 100\sigma/\mu$, it is possible to calculate a hypothetical value of σ corresponding with any position that the unique specimen may be assumed to have in the distribution, that is, with any value of its deviation from μ and the consequent assumed position of μ. By assuming a value for V and supposing the specimen to be first at $\mu + 2\sigma$ and second at $\mu - 2\sigma$, its extreme probable positions, the extreme probable position of μ and of σ for the otherwise unknown group of which the specimen is a member may be calculated.

In such biological inferences, the assumption of a value for V is readily guided by analogy with known calculated values from larger samples of other groups. It is, for instance, reasonable to assume that if V for a linear dimension of a tooth is known on fairly good data in samples from one or several species and appears to be fairly constant, it will have about the same value in another closely related species. In general, in dealing with linear dimensions of functional unreduced anatomical elements, when marked changes due to growth are not involved, V may be assumed probably not to exceed 10 and it is almost never so great as 15. Usually it is from 3 to 7, provided that the sample is reasonably homogeneous. Experience with particular groups of animals is helpful in making such judgment. For instance, V's for functional anatomical parts are usually lower in birds than in mammals.

The general concept here involved may be most readily represented by graphic means. In graphs, the range, real or, as in this case, theoretical or inferred, is represented by a horizontal distance. From the meaning and derivation of V, it is clear that the greater the V, the greater the linear distance representing the range, provided that μ remains the same. If, however, V is supposed to remain the same, the basis of the present procedure, but μ can take different values, then the greater the mean, the greater the range. It is not necessary to the solution of the problem but would certainly greatly facilitate and clarify things if the representation of the range could be made the same for any given value of V or retain the same proportion of the value of V, whatever the value of μ; in other words, if the ratio $\dfrac{(\mu + 2\sigma) - (\mu - 2\sigma)}{V}$ were constant. This will be true if instead of the direct values of the various figures involved, their logarithms are used or if (what amounts to the same thing) they are plotted on logarithmic graph paper.

An example, plotted by logarithms, is given in Fig. 17. In this figure, A represents estimates of parameters for the length of M_3 from a sample of the fossil mammalian species *Plesiadapis gidleyi*. The circle is the mean, the short vertical lines $\pm$ s, and the longer vertical lines $\pm$ 2s. The horizontal line is the actually observed range. The dots in the series B represent the actual length of a single M_3 of the species *P. tricuspidens*, a close ally of *P. gidleyi*. The assumption is made that V for this species is the same as for the latter. The corresponding σ values may then be shown by making the horizontal distances (plotted by logarithms) the same as the estimated values (s) in A. As in the latter, circles represent means, short vertical lines s, and longer vertical lines 2s.

In B_1 the specimen is assumed to be at the mean for its species, in B_2 at -1σ, in B_3 at $+1\sigma$, in B_4 at -2σ, and in B_5 at $+2\sigma$. The theoretical extreme probable range of the species is then the extreme distance between -2σ of B_5 and $+2\sigma$ of B_4 and is shown as a broken horizontal line. In practice, the result may be reached by simply marking off on each side of the actual position of the specimen a distance equal to that between the longer vertical lines of A. Any specimen that does not fall within this range probably differs significantly from *P. tricuspidens* in the length of

208 QUANTITATIVE ZOOLOGY

M_3. Any specimen within this range cannot be assumed with any probability to differ significantly, on these data, although of course it might prove to do so were fuller data available. The circles connected by a dotted line represent the extreme probable positions of the mean of the species to which B belongs.

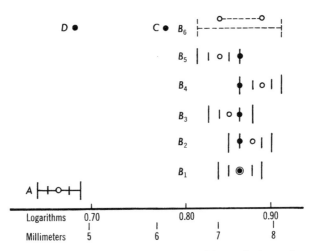

FIGURE 17. Graphic estimation of the probable variation of a species represented by a single specimen. The variate is length of the last lower molar in specimens of the extinct mammalian genus *Plesiadapis*. The horizontal scale is logarithmic; the vertical scale is not used. A complete explanation of the figure is given in the text (p. 207).

The points C and D represent two isolated specimens that one zoologist (Lemoine) believed to differ specifically from *P. tricuspidens* but that another (Teilhard) referred to that species. Both specimens are far outside the theoretical extreme range for the species, and the conclusion is that they almost certainly have significantly shorter last lower molars than *P. tricuspidens* and therefore probably are not of that species.

The various values involved may also and somewhat more usefully be computed. This computation is much simplified by using the logarithms of the values, known and hypothetical, of the variate. The theoretical range corresponding to a given value of V will then have the same value, regardless of the value of μ, in other words, of the absolute size of the

species. This theoretical range, on the distribution of logarithms, is then, by definition,
$$D = \log(\mu + 2\sigma) - \log(\mu - 2\sigma)$$
Since $V = \dfrac{100\sigma}{\mu}$,
$$\sigma = \frac{V\mu}{100}$$

Substituting this expression for σ in the formula for the theoretical range D,
$$D = \log\left(\mu + \frac{V\mu}{50}\right) - \log\left(\mu - \frac{V\mu}{50}\right)$$
By a series of eliminations, this becomes
$$D = \log(50 + V) - \log(50 - V)$$

Table 4 gives values of D corresponding to integral values of V up to 20 and also antilogarithms of D which represent the numerical value of $(\mu + 2\sigma)/(\mu - 2\sigma)$, the quotient of the upper theoretical limit divided by the lower, the use of which is explained below.

TABLE 4. Table of the quotient of upper limit of range divided by lower limit (antilog D) and of its logarithm (D) corresponding to a fixed value of the coefficient of variation.

V	D	ANTILOG D
1	.017	1.04
2	.035	1.08
3	.052	1.13
4	.070	1.17
5	.087	1.22
6	.105	1.27
7	.122	1.33
8	.140	1.38
9	.158	1.44
10	.176	1.50
11	.194	1.56
12	.213	1.63
13	.231	1.70
14	.250	1.78
15	.269	1.86
16	.288	1.94
17	.308	2.03
18	.327	2.13
19	.347	2.23
20	.368	2.33

The general graphic solution of the problem may be made from these figures without the preliminary plotting of a known variate (A in Fig. 17) and for any known value of V. On ordinary graph paper, place a dot at the point corresponding to the logarithm of the absolute value of the variate in question. Assume any desired value for V and mark off on each side of the dot the distance designated by the corresponding D. The total distance thus indicated (twice D, with the observed value of the variate in the middle) is the extreme likely range of the hypothetical species, and a second specimen must fall outside this to be significantly different as regards this variate.

Another form of comparison between two specimens possibly of one species is to make the likely assumption that the smaller is at or above $\mu - 2\sigma$, the larger at or below $\mu + 2\sigma$, and to calculate the least value of V for a species which would include both. Subtract the logarithm of the smaller measurement from that of the larger, set this as equal to log $(50 + V) - \log(50 - V)$, and solve for V, or with this as D, find the closest corresponding value of V in Table 4 (p. 209).

For instance, comparing B and D of the above example

Length M_3 of B—7.2 mm. . . . . . . log .85733
Length M_3 of D—4.8 mm. . . . . . . log .68124

Difference in logs $=$ .17809
Minimum value of $V =$ 10, from Table 4

Another exactly equivalent working of this type of problem is to divide the larger by the smaller number, when D will be the logarithm of the quotient (the subtraction of the logarithms, as above, being equivalent to a division and giving the logarithm of the quotient). The corresponding V may be found simply and with sufficient accuracy by considering this quotient as the antilog D and finding the nearest V in Table 4.

This relationship has an important application in a very common type of comparison, different in form but basically the same as the example just given. It is commonly said that a given dimension in one specimen is a certain percentage of, or percentage larger or smaller than, the same dimension in another specimen. This percentage is of course a quotient. If one value of a variate is 25 per cent greater than or 125 per cent of another, the quotient of the larger divided by the smaller is 1.25; and if the logarithm of this number is taken as D, the minimum reasonable V of a species including both can be read directly from the table. In this case, V is between 5 and 6. As before, the percentage, as a quotient, may be considered as antilog D and used to enter the table.

It is a commonly used criterion in some taxonomic work that a difference of 15 per cent in a linear dimension is in itself of probable specific value.

This is a traditional rule of thumb with no scientific basis. Clearly it implies some criterion as to variability, the most useful and exact measure of which (in dealing with analogous linear dimensions) is V. Now a difference of 15 per cent implies a quotient of 1.15, and the corresponding minimum likely value of V is between 3 and 4. This is a small value even for a pure sample of a single local race, and the conclusion is that the "15 per cent rule" will inevitably separate some specimens of a single species.

The nearest integral values of V corresponding to a given percentage of difference are given in Table 5. As before, V refers to the distribution of this variate in a hypothetical population to which both observations belong, on the assumption that one observation is at $\mu - 2\sigma$ and the other at $\mu + 2\sigma$.

In general, if no other criteria or characters are available, it is improper to assume that a difference in a linear dimensions may characterize different taxonomic groups unless it implies a minimum V of at least 10, since V's as high as 10 or even somewhat higher are known to occur in such characters of local races. Therefore, if a percentage rule of thumb is to be used at all in comparing isolated specimens, a difference of 50 per cent is necessary to indicate sufficient likelihood of a real taxonomic distinction. This is, of course, true only in dealing with a single observation on a single variate and when there is no valid reason to assume a value for V lower than 10.

TABLE 5. Percentage differences between two observations and corresponding minimum value of V in a population including both.

PERCENTAGE DIFFERENCE	V
5	1
10	2
15	3
20	5
25	6
30	7
40	8
50	10
60	12
70	13
80	14
90	16
100	17

It should be emphasized that the purpose and result of this procedure are to prove the likelihood of a negative: to show that it is highly likely that two observations do not belong to one population. It cannot prove that they *do* belong to one population. It is sound procedure for the burden of proof to be on the negative; and this is especially true in taxonomy, in which many inferences of this sort are involved. It must be assumed that any two comparable specimens are of one species unless it can be shown to be highly likely that they are not, and in general, it is usually proper to assume that any two observations are from one population until the contrary is proved.

The degree of accuracy and likelihood achieved and the exact meaning of what is demonstrated in these procedures are made clear by the following considerations. Consider the distribution of the variate in question for the theoretically infinite population, of which a single observation is available, as a single normal curve of definite constants, and consider how much of this normal curve is actually included within the limits given by the above procedure (within the range $2D$ when plotted by logarithms). If the unique observation happened to be at the actual mean, the limits here used would include 99.99 per cent of the real curve, or, for any practical purposes, 100 per cent. Any observation on the population would almost certainly be included in the theoretical limits. Now, the farther the observation really is from the real mean, the lower this figure will be. In approximately 955 cases out of 1000, however, which is a sufficiently high probability to serve as a practical criterion in most cases, it will fall within the limits of $\mu - 2\sigma$ and $\mu + 2\sigma$. Now even if the specimen was actually at one of these limits on the real curve, the theoretical limits will include 97.73 per cent of the original curve, so that 955 times out of 1000 the procedure here suggested will include within the theoretical limits at least 97.73 per cent (and generally much more—approaching 100 per cent) of all observations of the population. This high degree of probability is the basis for saying that if a second observation is not within these limits it is not from the same population as the first and may reasonably be separated. In order to reach this degree of likelihood for the negative conclusion, which is sought, the limits have necessarily been so placed that in every case they are likely to extend somewhat beyond the real (but unknown and indeterminable) limits of the population. For this reason, they may include observations that are in fact significantly different from those at hand but that cannot be shown to be so on these data. Therefore the positive proposition that all observations within the theoretical limits *do* belong to the same population is not true.

Finally, the general validity of the procedure as establishing a reasonable degree of likelihood is shown by the fact that in less than 3 cases in 1000 (0.27 per cent) will the unique specimen at hand deviate as much as 3σ from the unknown real mean in the distribution of the variate considered and that even in this extremely improbable case, the theoretical limits here used would include 84 per cent of the real distribution.

CHAPTER ELEVEN

Correlation and Regression

The idea that two variates may be related so that one tends to vary in such a way as to maintain a fairly constant relationship to the values of the other is fundamental and widespread not only in zoology and in kindred sciences but also in the whole field of human thought. Tall parents tend to have tall children and short parents short children. Children tend to weigh more as they become older. Within a given species, animals with larger heads tend also to have longer tails. The weather tends to be stormier when the sun has more spots. Increases in wages are generally accompanied by increases in living costs. A man is generally hungrier the longer he has gone without eating. These and thousands of other statements and ideas about every aspect of human life and of the universe are all examples of the correlation of two variates. All involve the idea that a change in one phenomenon or characteristic is usually accompanied by a change in some other.

It is not necessary to suppose that the relationship is constant and predictable in individual cases; indeed, it is seldom so in practice. Tall parents do sometimes have short children, children do sometimes lose weight, numerous animals with large heads do have short tails, the weather is sometimes calm when the sun has many spots, etc., but on the whole these relationships hold good, and it may confidently be stated from experience that they will always be true of averages if enough observations are made. It is obvious, however, that the relation may differ in intensity or in the accuracy of its predictions.

The two facets of the problem of relationship—the intensity and the form—are both of interest to the zoologist, although the latter is probably the more important. The critical point is that these two aspects of the problem are not interchangeable but are concerned with quite different practical questions. The aspect of relationship which has to do with its intensity is generally described as the *correlation* between the variables, while an expression of the form of this relation is the *regression* of one variable on another.

Correlation

A measure of the correlation between two variables is an index of the intensity of a relationship or the degree of accuracy with which the value of one variable may be predicted, given the value of the other.

Complete correlation or 100 per cent accuracy is almost confined to mechanical things and practically never occurs in biological data. That an automobile travels farther the more times its wheels turn around on the road is a positive correlation that is very nearly 100 per cent true; but that larger individuals have larger offspring may be true only two-thirds of the time, and that larger animals dig deeper burrows may be true only one-third of the time. The intensity of a supposed correlation may also prove to be zero, or, in other words, the supposed relationship may not exist. For instance, the old idea that some plants grow better if planted in the dark of the moon is shown by recent experiments to be a correlation with value about zero. A negative correlation is one in which an increase in the value of one variate is usually accompanied by a decrease in the value of another, and those also are common in nature. For instance, molar-tooth crown height and age are negatively correlated in most mammals; the older a horse becomes, the less is the distance from grinding surface to roots of his cheek teeth.

Most data of this sort can be reduced to numerical terms, and once this is done it is possible to measure the exact degree of correlation shown by a given series of observations and from this as a sample to estimate the probable degree in a population. The importance of these procedures for zoology is so great and so obvious that it hardly needs emphasis, and mention of a few types of problems will suffice to show how essential numerical correlation is in this field. Correlations between characters of parents and offspring show the effect and consistency of hereditary factors. Correlation between environmental conditions (such as temperature) and physiological variates (such as respiration rate) or morphological character (such as total length) show whether the latter are influenced by the former. Correlation of age with any physiological, psychological, or morphological characters reveals the phenomena of growth and maturation. Correlation of any two variates of an animal shows whether they are independent or linked together in some way, again throwing light on the phenomena of heredity and also on the function of growth, on physiological mechanisms, and on many other problems. It is also interesting in this connection that different taxonomic groups are often characterized by different intensities of correlations of this type.

Regression

Unlike correlation, regression does not measure directly the intensity of the relation between two variables, although it is not entirely independent of the correlation. The regression of one variable upon another is a measure of how much increase or decrease in one factor may be expected from a unit increase in the other. Regression serves primarily a predictive function and this is its great importance in science. It is not sufficient to know that animals with larger heads tend to have longer tails, or that an increase in wages will result in an increase in living costs. More often than not, the critical issue is that of how much longer the tail of a large-headed animal will be than that of a small-headed one, or how much the cost-of-living index will rise with each wage increase. The importance of such prognosis in economic affairs is obvious, but it may be equally important in zoology. If the regression of tail length on skull dimensions were well known for closely related animals, then the discovery of a fossilized skull belonging to this group would make possible fairly accurate estimates of tail length although the axial skeleton were missing. The popular misconception that a paleontologist can reconstruct the entire skeleton and flesh of an animal from a single tooth, while unfortunately incorrect, has some basis in truth. The paleontologist is constantly making use of the concept of regression when he reasons from parts of animals to whole animals. It is often said, for example, that the largest sauropod dinosaurs must have spent their lives in shallow water, because their legs could not possibly have supported the immense weight of their bodies except for the buoying effect of the water. This inference is based upon the fact that weight generally increases with the cube of linear dimensions, while load-bearing capacity of columnar elements like the legs increases only as the square of these dimensions. That there is a known correlation between linear dimensions and weight or load-bearing capacity is insufficient to make this judgment. What is required is a knowledge of the exact quantitative character of this dependence, namely, the cubic and quadratic proportionality of weight and strength to linear dimensions.

Regression is useful not only in prediction, where it serves a more or less exact function, but also in reasoning from observations to underlying mechanisms. Sir Francis Galton (1822-1911), who invented the statistical methodology of regression, used this technique in attempting to understand the general nature of heredity. Although he failed to infer the Mendelian laws which are now considered the basis of genetic observations, regression formulae are used today in the genetics of continuously varying characters. In morphology, the observation of a particular regression of one character on another is useful in making inferences about the growth relations of these organs. (See Chapter 15, Growth.)

Correlation or Regression ?

The zoologist may easily become confused on the question of whether correlation or regression is the proper technique to use, if he refers to the various works on statistics which are often in conflict with each other and with this book. One statement which appears in a recent book is: "In a correlation problem we sample from a population, observing two measurements on each individual in the sample. This contrasts with the purely regression problem where the sample is chosen with preassigned [values of one variable]." This is good mathematical statistics but may be poor biology.

The essential difference between correlation and regression is in the kind of information which they supply and not in the kind of sample to which they are applicable. The coefficient of correlation gives the intensity of the relationship only, while the quantitative relationship itself can be determined only from regression. Most zoological sampling is of the type referred to in the quotation as applicable to a correlation problem, for what the zoologist generally does is to sample a group of animals without reference to the value of either variable in which he is interested. To say that he is only entitled to estimate the intensity of the relation and not its quantitative nature is to render useless the entire procedure of observation. Despite statements to the contrary, the zoologist may with perfect confidence apply the methods of regression analysis suggested in this chapter to his material, whether or not one of his variables has been deliberately chosen to take certain preassigned values.

In a small number of situations, zoological problems do involve the arbitrary choice of values for one variable. Time series are obvious examples of this, in which case animals of specified ages may be measured and the relationship between age and the measurement sought. Again, measurements may be made on individuals chosen along a geographical, altitudinal, climatic, or geologic gradient. Thus, one might want to know the relation between a given character and the latitude in which the animal lives. Allen's, Bergmann's, and Gloger's rules are statements of such relationships. Mathematically speaking, regression is the only applicable concept in such cases, but again the zoologist would like to know the intensity of the relationship given by correlation. Unfortunately, the mathematics of the situation cannot be so lightly ignored here, and it must be admitted that not only is correlation illogical in such cases but that also the results of correlation analysis may be quite misleading.

In general, then, for the vast majority of situations in which animals are sampled at random with respect to the characters to be related, both correlation and regression analysis are applicable. In that small proportion of cases in which the sample is chosen deliberately with respect to one of the variables, only regression analysis should be used, the intensity of the relation being left in question.

Concepts of Regression

If two variates have some functional relation to each other, if one tends to increase or decrease in value with an increase or decrease in the other, this pattern of relationship will be fairly clear if the values are plotted on ordinary graph paper. Taking values of one variate as X and those of the other as Y, the scales of X and Y are laid out along the bottom and to the left, respectively, of the field of the proposed diagram. A dot is then placed in the field at a point corresponding to each pair of observed values of X and Y. The result is called a *scatter diagram* (Figs. 18 and 19). If there is a functional relationship between the two variates of a scatter diagram, the dots representing the observations will tend to be arranged along a line or in an elongate path across the diagram, as in Fig. 18. On the other hand, if the two variates are not related but take values completely independent of each other, the dots will be scattered over the surface of the diagram with no apparent pattern, as in Fig. 19. The pattern which the points take need not suggest a straight line but may show a definite curving trend. The methods of dealing with curvilinear regression may be more or less complex depending upon the shape of these curves, but fortunately the zoologist is seldom concerned with cases other than simple rectilinear relationships. There is a special part of zoological inquiry, that dealing with growth phenomena, which requires the analysis of curvilinear regressions and these will be taken up in Chapter 15 in that special context. The methods discussed in this chapter will assume that the pattern on the scatter diagram, if such a pattern exists, is the form of an elongated oval or ellipse so that the dots lie roughly in a straight path.

If the two variates are indeed related by a rectilinear function, this relationship can be most generally expressed by

$$Y = \alpha_y + \beta_{yx} X$$

where α_y and β_{yx} are the Y intercept and the slope, respectively, of the straight line relating Y to X. Here Y is called the *dependent* and X the *independent* variable and any straight line can be put in this form. It is really a prediction equation for Y, for if the values of α_y and β_{yx} are known, then the value of Y would be determined for any given value of X and could be predicted. The Y intercept, α_y, is the value of Y at which the line crosses the Y axis. That is, it is the value of Y when X is zero. The slope, β_{rx}, is the amount by which Y increases for a unit increase in X. Thus, if β_{rx} were 2, Y would increase 2 units for every increase of one unit in X, while if β_{rx} were -2, Y would decrease 2 units for each unit increase in X.

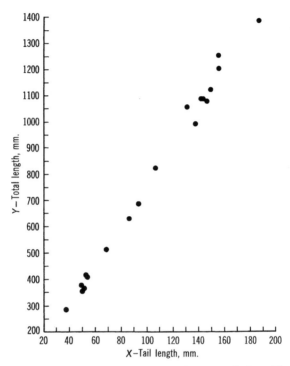

FIGURE 18. Scatter diagram of a nearly perfect correlation. Total length against tail length in females of the snake *Lampropeltis polyzona* (data of Example 65). The evident trend from lower left to upper right shows the regression of Y on X to be positive, as is the correlation, and the strong clustering around a straight line shows the correlation to be very strong.

Clearly, the function relating Y to X could be inverted to make X the dependent and Y the independent variable. Performing this operation gives

$$X = \frac{-\alpha_r}{\beta_{rx}} + \frac{1}{\beta_{rx}} Y$$

Here $\dfrac{-\alpha_r}{\beta_{rx}}$ is the X intercept, the value of X for which Y is zero, and $\dfrac{1}{\beta_{rx}}$

is the slope of this new line. This function can be put in terms of the symbols α_x and β_{XY} so that it appears as

$$X = \alpha_x + \beta_{XY}Y$$

where

$$\alpha_x = \frac{-\alpha_Y}{\beta_{YX}}$$

and

$$\beta_{XY} = \frac{1}{\beta_{YX}}$$

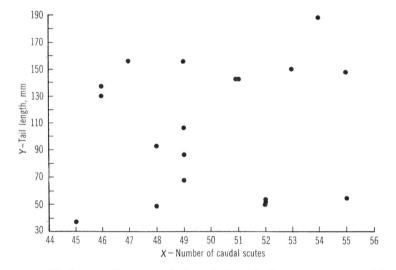

FIGURE 19. Scatter diagram of the relationship between two variables with no evident correlation. Tail length against number of caudal scutes in females of the snake *Lampropeltis polyzona*. The absence of evident trend and the fairly even scattering of observations indicates a regression of Y on X in the neighborhood of 0 and no correlation between the variates.

The following hypothetical series of values of X and Y illustrate a simple regression line:

X	Y
1	2
2	4
3	6

A unit increase in X obviously means an increase of 2 in Y, and they

maintain a constant ratio $Y/X = 2$ so that the regression equation for these points is
$$Y = 2X$$
In this equation, α_Y is zero and β_{YX} is 2. This same set of values could be represented by $X = Y/2$, and in accord with the previous relations,

$$\alpha_X = \frac{-\alpha_Y}{\beta_{YX}} = 0$$

and

$$\beta_{XY} = \frac{1}{\beta_{YX}} = 1/2$$

A slightly more complex case, in which the line does not pass through the origin, is given by the series of values:

X	Y
1	3
2	5
3	7

Here Y/X is not constant, but a unit increase in X is accompanied by an increase of 2 in Y. The equation representing this series of values is
$$Y = 1 + 2X$$
so that
$$\alpha_Y = 1 \text{ and } \beta_{YX} = 2$$
Similarly
$$X = -1/2 + Y/2$$
so that
$$\alpha_X = \frac{-\alpha_Y}{\beta_{YX}} = -1/2 \text{ and } \beta_{XY} = \frac{1}{\beta_{YX}} = 1/2.$$

If the points lie exactly on a straight line, there is no problem in determining the parameters α and β. In real situations, however, the points are never exactly colinear (in a single line) but are scattered over an area around the theoretical functional line. Even the extremely well clustered points of Fig. 18, which are unusually close to a perfect line for biological data, are not precisely colinear. The problem is then to estimate the unknown parameters α and β from the available information. An obvious procedure would be to draw a straight line through the points which seemed to the eye most closely to approximate the trend. Clearly such a subjective method is undesirable if some more objective approach is available, for different persons may disagree strongly as to which of many lines seems to fit the observations best. In line with the general philosophy of estimation, estimates have been derived for α and β which are unbiased and efficient. Denoting the estimate of β by b and that of α by a, these estimates are of the form

$$b_{YX} = \frac{\sum(X - \bar{X})(Y - \bar{Y})}{\sum(X - \bar{X})^2}$$

and

$$a_Y = \bar{Y} - b_{YX}\bar{X}$$

and in like manner

$$b_{XY} = \frac{\sum(X - \bar{X})(Y - \bar{Y})}{\sum(Y - \bar{Y})^2}$$

and

$$a_X = \bar{X} - b_{XY}\bar{X}$$

From the previous discussion it might be supposed that the estimates b_{YX} and b_{XY} would simply be reciprocals of each other since in the theoretical straight line β_{YX} is the reciprocal of β_{XY}. This is not the case, and although β_{XY} will be approximately the reciprocal of β_{YX}, this relation holds exactly only if all the points are strictly colinear. There is, of course, nothing to prevent using the reciprocal of β_{YX} as an estimate of β_{XY}, but it turns out that this estimate would not be unbiased. It is, however, true that if many samples were taken from the population, the average value of b_{YX} would be equal to the reciprocal of the average value of b_{XY}.

In any particular case, then, there are two regression lines, not one, representing the regression of Y on X and the regression of X on Y. These lines will coincide only if the points are strictly colinear, which in zoological practice never occurs.

To calculate b_{YX}, it is necessary to note that the numerator, called "the sum of cross-products," is obtained by multiplying the deviation from the mean of each observed Y by the *corresponding* deviation from its mean of each observed X and then summing these products. The denominator is simply the sum of squared X deviations which has been used before in estimating the variance. As a matter of fact, if the numerator and denominator of b_{YX} were multiplied by $1/(N - 1)$, the result would be

$$\frac{\frac{1}{N-1}\sum(X - \bar{X})(Y - \bar{Y})}{s_X^2}$$

where the numerator is called the "covariance" of Y and X and the denominator is the sample variance of X. The covariance is the two-variable analogue of the variance, and if X were substituted for Y in the expression of the covariance, the variance would result. In a sense, the variance of a quantity is simply the covariance of that quantity with itself.

EXAMPLE 65. Calculation of coefficients of regression from raw observational data on tail length and total length in females of the king snake *Lampropeltis polyzona*. (Data from Blanchard, 1921)

TAIL LENGTH (X)	TOTAL LENGTH (Y)	(X-$\bar{X}$)	(Y-$\bar{Y}$)	(X-$\bar{X}$)(Y-$\bar{Y}$)	(X-$\bar{X}$)^2	(Y-$\bar{Y}$)^2
37	284	−68	−511	34,748	4,624	261,121
49	375	−56	−420	23,520	3,136	176,400
50	353	−55	−442	24,310	3,025	195,364
51	366	−54	−429	23,166	2,916	184,041
53	418	−52	−377	19,604	2,704	142,129
54	408	−51	−387	19,737	2,601	149,769
68	510	−37	−285	10,545	1,369	81,225
86	627	−19	−168	3,192	361	28,224
93	683	−12	−112	1,344	144	12,544
106	820	+ 1	+ 25	25	1	625
130	1,056	+25	+261	6,525	625	68,121
137	986	+32	+191	6,112	1,024	36,481
142	1,086	+37	+291	10,767	1,369	84,681
142	1,086	+37	+291	10,767	1,369	84,681
146	1,078	+41	+283	11,603	1,681	80,089
149	1,122	+44	+327	14,388	1,936	106,929
155	1,254	+50	+459	22,950	2,500	210,681
156	1,202	+51	+407	20,757	2,601	165,649
187	1,387	+82	+592	48,544	6,724	350,464
TOTALS 1,991	15,101			312,604	40,710	2,419,218

$$\bar{X} = \frac{1,991}{19} = 105 \qquad \bar{Y} = \frac{15,101}{19} = 795$$

$$b_{YX} = \frac{\Sigma (X - \bar{X})(Y - \bar{Y})}{\Sigma (X - \bar{X})^2} = \frac{312,604}{40,710} = 7.68$$

$$b_{XY} = \frac{\Sigma (X - \bar{X})(Y - \bar{Y})}{\Sigma (Y - \bar{Y})^2} = \frac{312,604}{2,419,218} = .129$$

$$a_Y = \bar{Y} - b_{YX}\bar{X} = 795 - (7.68)(105) = -11.4$$

$$a_X = \bar{X} - b_{XY}\bar{Y} = 105 - (.129)(795) = 2.4$$

Example 65 shows the calculation of the two regression coefficients b_{YX} and b_{XY}, along with the intercept estimates a_Y and a_X. The results are as follows:

$$b_{YX} = 7.68$$
$$b_{XY} = .129$$
$$a_Y = -11.4$$
$$a_X = 2.4$$

The reciprocal of 7.68 is .1302 which is very close but not exactly equal to .129, showing that the regression lines of Y on X and X on Y will not exactly coincide. Moreover, a_x has a value somewhat different from $\dfrac{-a_y}{b_{rx}}$, this latter quantity being equal to 1.48. Hence the two regression equations are

$$Y = 7.68\ X - 11.4$$
$$X = .129\ Y + 2.4$$

From these, the regression lines can be easily plotted, noting that only two points are necessary to determine a straight line. Both lines pass through the point $\overline{X}, \overline{Y}$, so that one point for each is then established. The other point for each is most simply found by using the intercept values a_y and a_x. The regression line of Y on X crosses the Y axis at

$$Y = a_y = -11.4$$

and the regression line of X on Y crosses the X axis at

$$X = a_x = 2.4$$

If all the values of X and Y are far from zero, the X and Y intercepts are inconvenient points to plot. In such a case, a second point for the regression line of Y on X can be found by assigning any convenient value to X that lies within the range of measurements and solving the regression equation to find the equivalent value of Y. Similarly, for the regression of X on Y choose a convenient Y value and solve for the appropriate X. For the most accurate representation, these arbitrary points should be chosen as far from the means as possible, since the further apart are two points, the more accurately can a line be drawn between them. Figure 20 shows the two regression lines drawn for the data of Example 58.

If a mechanical calculator is available, the regression coefficients may be calculated more rapidly by a method analogous to that used for finding variances.

These calculating formulae are

$$b_{rx} = \frac{\sum XY - \dfrac{\sum X \sum Y}{N}}{\sum X^2 - \dfrac{(\sum X)^2}{N}}$$

and

$$b_{xr} = \frac{\sum XY - \dfrac{\sum X \sum Y}{N}}{\sum Y^2 - \dfrac{(\sum Y)^2}{N}}$$

where $\sum X$ and $\sum Y$ are the totals respectively of the X observations and the Y observations and the N is the sample size (not the total number of measurements, which would be $2N$). The quantities $\sum X$, $\sum Y$, N, $\sum X^2$, and $\sum Y^2$ have already been used in finding the variances of X and Y. The only new quantity is the sum of the cross-products of the observations which, it should be remembered, is the sum of the products of each X by its corresponding Y. The calculation of the regression coefficients by this method can be put in the form of a short table. This is done in Example 66 for the data of Example 65.

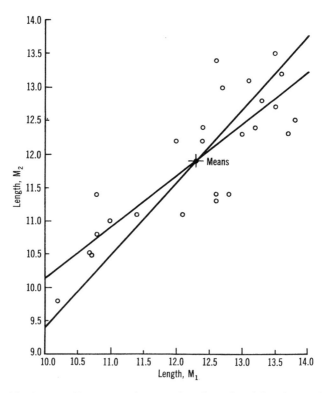

FIGURE 20. Scatter diagram and regression lines fitted by the method of least squares. Lengths of the first and second lower premolars of the fossil mammal *Phenacodus primaevus*. (Data of Example 58)

Confidence Interval for α and β

The estimates a_Y, a_x, b_{YX}, and b_{XY}, like estimates of any other parameters, are subject to sampling errors, and what is really wanted is some statement about the true values, α and β, based upon these estimates. To find confidence limits for these parameters, it is necessary to use not only the sample variances s_x^2 and s_r^2 but also a new quantity, the variance of estimate. The sample value of this quantity will be symbolized as s_{rx}^2 and s_{xr}^2, for the regressions of Y on X and X on Y respectively. By definition, these quantities are

$$s_{rx}^2 = \frac{N-1}{N-2}(s_r^2 - b_{rx}^2 s_x^2)$$

and by symmetry,

$$s_{xr}^2 = \frac{N-1}{N-2}(s_x^2 - b_{xr}^2 s_r^2)$$

where all the notation has been previously defined. It is again important to remember that N is the number of individuals measured, not the total number of measurements which is twice as great.

EXAMPLE 66. Machine method for calculating regression coefficients using the data of Example 65.

N	$\sum X$	$\sum Y$	$\sum X^2$	$\sum XY$	$\sum Y^2$
19	1,991	15,101	249,345	1,895,029	14,421,333

$$\sum XY - \frac{\sum X \sum Y}{N} = 1,895,029 - \frac{30,066,091}{19} = 1,895,029 - 1,582,426$$
$$= 312,603$$

$$\sum X^2 - \frac{(\sum X)^2}{N} = 249,345 - \frac{3,964,081}{19} = 249,345 - 208,636 = 40,709$$

$$\sum Y^2 - \frac{(\sum Y)^2}{N} = 14,419,847 - \frac{228,040,201}{19} = 14,421,333 - 12,002,116$$
$$= 2,419,217$$

$$b_{rx} = \frac{312,603}{40,709} = 7.68 \qquad b_{xr} = \frac{312,603}{2,419,217} = .129$$

The method of estimating β is such that s_{rx}^2 and s_{xr}^2 are the smallest possible values for the given observations. They are the mean square deviations of the observations from the regression line, and the principle of making them as small as possible is called the *method of least squares*.

It is sometimes erroneously stated that the estimates of β which are used have been derived on the principle of making this error variance a minimum. This is not true; b_{rx} and b_{xr} are simply unbiased, efficient estimates of β, and the fact that they happen to conform to the principle of least squares is incidental, though clearly a desirable property. Any other value of b except the ones estimated by our procedure will lead to larger standard errors of estimate. This means in a sense that the observed points lie, on the average, closer to the theoretical regression line calculated using b_{rx} or b_{xr} than to a line calculated in some other way. While there is no absolute requirement that one should choose the line from which the observations have the smallest mean square deviation, in preference to some other criterion, this is a universally employed principle and intuitively, at least, seems justifiable.

This principle of least squares is illustrated in Fig. 21, along with the difference between b_{xr} and b_{rx} implicit in the least squares method of estimation. Only four points are shown in this artificial example in order to make the illustration clearer. The solid line is the one calculated from b_{rx}, the regression of Y on X, while the broken line results from the calculation of b_{xr}, the regression of X on Y. The distances denoted by d_r are the differences between the Y value of the observed point and the Y value of the regression line $Y = a_r + b_{rx} X$. The distance d_x is the deviation in the X value of the observed point from the line $X = a_x + b_{xr} Y$. The two lines are such that for the first $\sum d_r^2$ is minimized, while for the second, $\sum d_x^2$ is minimized.

It is not possible, as we have already stated, to find a single line which meets both of these criteria simultaneously.

The confidence interval for β takes the simple form:

$$b_{rx} \pm \frac{t\, s_{rx}}{s_x \sqrt{N-1}}$$

or

$$b_{xr} \pm \frac{t\, s_{xr}}{s_r \sqrt{N-1}}$$

depending upon which estimate of β is used. The quantity t is the familiar Student's t with $N-2$ degrees of freedom. If, for example, the 95 per cent confidence limits for β are desired, the t-table is entered with $N-2$ degrees of freedom and the entry in the table in the .95 column is used. For 30 degrees of freedom, the 95 per cent value of t from Appendix Table II would be 2.042, or for a very large number of degrees of freedom, over 120, the 95 per cent value is 1.960.

For the X and Y intercepts, the parameters α_Y and α_X have the confidence intervals given by

$$a_Y \pm \frac{t\, s_{YX}}{\sqrt{N}} \text{ for } \alpha_Y$$

and

$$a_X \pm \frac{t\, s_{XY}}{\sqrt{N}} \text{ for } \alpha_X$$

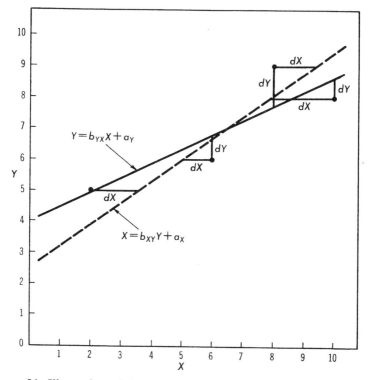

FIGURE 21. Illustration of the principle of least squares with hypothetical data. The solid line is the least squares regression line of Y on X such that the sum of squares of the deviations d_Y is minimized. The broken line is the least squares regression line of X on Y such that the sum of squares of the deviations d_X is minimized.

Again, t has $N - 2$ degrees of freedom. As an example of the calculation of these confidence intervals, the results of Example 65 may be used, all of the pertinent information having already been derived from the raw observations in calculating the estimates b_{XY}, b_{YX}, a_X, and a_Y. The calculation of these intervals is illustrated in Example 67.

EXAMPLE 67. Calculation of confidence intervals for α and β, using the results of Example 66.

$$b_{YX} = 7.68$$
$$b_{XY} = .129$$
$$a_Y = -11.4$$
$$a_X = +2.4$$

$$s_X^2 = \frac{1}{N-1}\left[\sum X^2 - \frac{(\sum X)^2}{N}\right] = \frac{1}{18}(40,709) = 2,261.61$$

$s_X = \sqrt{2,261.61} = 47.56$

$$s_Y^2 = \frac{1}{N-1}\left[\sum Y^2 - \frac{(\sum Y)^2}{N}\right] = \frac{2,419,217}{18} = 134,400.94$$

$s_Y = \sqrt{134,400.94} = 366.61$

$$s_{YX}^2 = \frac{N-1}{N-2}(s_Y^2 - b_{YX}^2 s_X^2) = \frac{18}{17}[134,400.94 - (58.98)(2,261.61)] = 1,070.66$$

$s_{YX} = \sqrt{1,070.66} = 32.72$

$$s_{XY}^2 = \frac{N-1}{N-2}(s_X^2 - b_{XY}^2 s_Y^2) = \frac{18}{17}[2,261.61 - (.0166)(134,400.94)] = 32.36$$

$s_{XY} = \sqrt{32.36} = 5.69$

For 17 degrees of freedom, $t = 2.11$.

$$b_{YX} \pm \frac{(2.11)(32.72)}{47.56\sqrt{18}} = 7.68 \pm .34$$

so that the 95% confidence interval for $\beta_{YX} = 7.34$–8.02.

$$b_{XY} \pm \frac{(2.11)(5.69)}{(366.61)\sqrt{18}} = .129 \pm .0077$$

so that the 95% confidence interval for $\beta_{XY} = .121$–$.137$.

$$a_Y \pm \frac{(2.11)(32.72)}{\sqrt{19}} = -11.4 \pm 15.8$$

so that the 95% confidence interval for $\alpha_Y = -27.2$–4.4.

$$a_X \pm \frac{(2.11)(5.696)}{\sqrt{19}} = 2.4 \pm 2.8$$

so that the 95% confidence interval for $\alpha_X = -.4$–5.2.

Although the confidence intervals for the parameters α_x and α_y are large, they are relatively small when the magnitude of the variate is considered.

Tests of Regression Coefficients

In addition to the determination of confidence intervals, it is of interest to be able to test the difference between two regression coefficients, or between a regression coefficient and some hypothetical value. As for other estimates, this latter test is performed by using the confidence interval itself. If the 95 per cent confidence interval includes the hypothetical value, then there is no significant difference between the observed regression coefficient and the hypothetical one (assuming a .05 rejection criterion). If the confidence interval does not include the postulated value, then there is a significant difference. The most usual test of this sort is with the hypothetical value zero. Upon finding a positive or negative regression coefficient, the question which must be asked is whether there is really any relation between the two variables, or whether the observed regression is simply a random deviation from zero, a β of zero indicating total lack of a rectilinear relation between the variates. The answer to this question is found in the confidence interval, for if this interval includes zero it must be said that there is no significant regression of one variate on the other, that is, they are independent of each other. It is sometimes said that a regression coefficient is "significant." What this means is that the observed coefficient is significantly different from zero. If no significance level is indicated along with this statement, it is probable that the author has used a 5 per cent level for significance, but one can never be sure, especially in the older literature.

Since the confidence interval for β depends upon the t-distribution, the test which immediately suggests itself for the difference between two regression coefficients is the t-test similar to that for means. The t-test for the difference between two regression coefficients is in the same form exactly as that for means so that the test statistic is

$$t = \frac{(b_{YX_1} - b_{YX_2})\sqrt{\dfrac{(N_1-1)(N_2-1) s_{X_1}^2 s_{X_2}^2}{(N_1-1) s_{X_1}^2 + (N_2-1) s_{X_2}^2}}}{\sqrt{\dfrac{(N_1-2) s_{YX_1}^2 + (N_2-2) s_{YX_2}^2}{N_1+N_2-4}}}$$

which is distributed as t with $(N_1 + N_2 - 4)$ degrees of freedom. To test the difference between two regression coefficients of the form b_{XY}, instead of b_{YX}, simply exchange X and Y wherever they appear in the formula. Where s_X^2 appears it is replaced by s_Y^2 and where s_{YX}^2 appears it is replaced by $s_{xr}{}^2$. In the formula, N_1 and N_2 refer to the number of individuals in the two samples being compared, $s_{x_1}{}^2$ and $s_{rx_1}{}^2$ are the variances for the sample of size N_1, while $s_{x_2}{}^2$ and $s_{rx_2}{}^2$ are the variances for the sample of size N_2. In like manner b_{rx_1} and b_{rx_2} are the estimated regression coefficients for the first and second samples respectively. Having obtained t, the probability associated with it is found from the table in the same way as for a test on means, and this probability has the same meaning. Such a test has been performed in Example 68.

EXAMPLE 68. Significance of the difference between two regression coefficients. Regression of tail length (X) on total length (Y) in males and females of *Lampropeltis polyzona*. (Raw data from Blanchard, 1921)

MALES	FEMALES
$N = 24$	$N = 19$
$b_{XY} = .153$	$b_{XY} = .129$
$s_Y^2 = 98{,}917$	$s_Y^2 = 134{,}401$
$s_{XY}^2 = 79.45$	$s_{XY}^2 = 32.36$

$$t = \frac{(.153 - .129)\sqrt{\dfrac{(23)(18)(98{,}917)(134{,}401)}{(23)(98{,}917) + (18)(134{,}401)}}}{\sqrt{\dfrac{22(79.45) + 17(32.36)}{39}}}$$

$t = (.024)\sqrt{19{,}898.16} = (.024)(141.06)$

$t = 3.39$

Student's t of 3.39 with 39 degrees of freedom has a probability of .001, so that the difference between the regressions is certainly significant.

The Choice of Regression Coefficients

Two regression coefficients b_{xr} and b_{rx} can be calculated from a sample, and the problem arises as to which of these ought to be used for a specific problem. The choice, when one exists, is entirely dependent upon the use to which these estimates are to be put and the nature of the sample from which they are calculated. The expressions

$$Y = \alpha_r + \beta_{rx} \quad \text{and} \quad X = \alpha_x + \beta_{xr}Y$$

which were assumed to be in some sense idealized representations of the functional relationship between the variables, have different meanings.

First, they may be regarded as *prediction equations*. If the value of Y is to be predicted from a known value of X, the procedure is to substitute in the first of these two equations the estimates a_Y and b_{YX} together with the value of X. The result is a predicted value for Y. If, on the other hand, X is to be predicted from a known value of Y, it is the second prediction equation which is pertinent and estimates a_X and b_{XY} are used in that expression. While it is true that the theoretical value β_{XY} is simply the reciprocal of the parameter β_{YX}, this is not exactly true of the estimates in any particular case, and if an unbiased prediction procedure is to be used, the calculated estimate b_{XY}, and not the reciprocal of b_{YX}, should be substituted in the second expression. When regression is used for prediction, the choice between b_{YX} and b_{XY} is really a choice between the two prediction equations and depends entirely upon which variable is to be predicted from which.

In a small number of cases such as time series, altitudinal gradients, and others, where the sample has been taken with definite values of one of the variates, there is no choice of estimates. Denoting the deliberately chosen variate as X and the associated random variate as Y, only the regression of Y and X, estimated by b_{YX}, has any validity. Certainly there is some regression of X on Y, but the method of sampling is such as to make the estimate b_{XY} a misleading and incorrect one. The only correct prediction equation is then

$$Y = \alpha_Y + \beta_{YX} X$$

As an example, suppose the total length of a number of fish of different age classes has been measured and the relation between age and length is desired. The ages have been deliberately, not randomly chosen, whereas the fish have not been selected for their length, but this length is a random variate. The dependent variate in this case is clearly length and the independent one, age. The only regression coefficient that can correctly be estimated is b_{YX}, the regression of length (Y) on age (X). The prediction equation

$$Y = a_Y + b_{YX} X$$

will allow a prediction of length for any new specimen, provided that the age is known. It may, of course, happen that only the value of the dependent variable is known and the independent one is to be predicted. For the hypothetical case of length and age, as a matter of fact, it is far more likely that a specimen's length will be known and its age unknown but needs to be predicted. In such a case, the important prediction problem is really the opposite of the valid prediction equation.

It is obvious, of course, that biologically age cannot be said to depend upon length, since it is the age which determines the length. Nevertheless, in the restricted sense of prediction, age can depend upon length and may be regarded as the dependent variable. Since b_{xr} cannot be correctly calculated because of the nature of the sample, the only procedure possible is to use the reciprocal of b_{rx} as an estimate of β_{xr}. The prediction equation for X with a known Y then becomes

$$X = \frac{-a_r}{b_{rx}} + \frac{1}{b_{rx}} Y$$

While $-a_r/b_{rx}$ and $1/b_{rx}$ are not unbiased or efficient estimates of α_x and β_{xr}, they are the only ones available from this sort of sample and so must be used. *This procedure for predicting X from Y should never be used unless the nature of the sample makes the calculation of the unbiased efficient estimates invalid.* Fortunately this sort of reverse prediction does not occur too frequently in zoology.

The other use of regression, which is logically distinct from that of prediction, is to find the straight line which best fits the observed data. This so-called "structural" problem assumes that there is some straight line relationship,

$$Y = \alpha + \beta X$$

and that α and β are to be estimated. Now the regression procedure that has been outlined provides not one but two estimates for both α and β. Thus b_{rx} is an estimate of β but so is $1/b_{rx}$. For any given set of observations these are not identical nor is there anything about them which would lead one to prefer the first to the second. In prediction problems, it is clear that b_{rx} is to be used for predicting Y from X and b_{xr} for predicting X from Y, but in finding the one line that best fits the data, neither the regression line of Y on X nor that of X on Y is wholly satisfactory.

For the prediction problem, the "best fit" line is the one which satisfies the least squares criterion. For the structural problem, however, some new criterion must be invoked since we have shown that no single regression line can simultaneously satisfy the least squares criterion for both X and Y deviations.

This problem of finding a linear representation of a function when both X and Y have been sampled at random has been attacked by many statisticians with somewhat varying results. For our purpose, perhaps the most desirable method has been suggested by Bartlett, an English statistician, partly because it is so simple an operation. This method will provide estimates of α and β that are unbiased and whose efficiency is fairly high. To find the "best fit" straight line by Bartlett's method:

1. Arrange the observations in ascending order with respect to one of the variates. A slightly different estimate of α and β will result, depending upon which of the variables is ordered, but both systems of ordering give unbiased estimates of α and β, so that in the long run it does not matter

which is chosen. The observations in Example 69 have been listed in this way.

2. Divide the ordered observations into three groups of equal size. If the sample size is not divisible by three, the lowest group and highest group should be made equal in size.

3. Calculate the sample means $\bar{X}$ and $\bar{Y}$ for the lowest and highest groups. These will be denoted as $\bar{X}_1$ and $\bar{Y}_1$ and $\bar{X}_3$ and $\bar{Y}_3$ for the lowest and highest group respectively.

4. Calculate the mean of X and Y for the whole sample, denoted by $\bar{X}$ and $\bar{Y}$. The slope of Bartlett's "best fit" line is given by the simple relationship:

$$B = \frac{\bar{Y}_3 - \bar{Y}_1}{\bar{X}_3 - \bar{X}_1}$$

The line also passes through the point corresponding to the grand means of the variates $\bar{X}$ and $\bar{Y}$. Since the slope and one point are sufficient to define a straight line, the problem of fitting is solved. The Y-intercept, α, is estimated by the expression

$$A = \bar{Y} - B\bar{X}$$

Example 69 uses this procedure to fit a straight line to the data of Example 65.

It is interesting to compare B calculated in this way with b_{yx} and b_{xy} from the previous method. b_{yx} is 7.679 and the reciprocal of b_{xy} is 1/.129 or 7.739, so that Bartlett's B of 7.684 lies between these two values although it is obviously not their arithmetic average.

In samples where only one regression coefficient, b_{yx}, can be validly computed, this problem of the best estimate of β does not arise. There is only one value of b, and this is the estimate of β whether a prediction problem or a line fitting problem is considered.

Confidence Intervals of Bartlett's β

Calculation of a confidence interval for β is somewhat involved, but if an estimate is to be of any use, a method for computing this interval is really essential. The quantity

$$t = \frac{(\bar{X}_3 - \bar{X}_1)(B - \beta)1/2K}{\sqrt{s_Y^2 - 2\beta s_{XY} + \beta^2 s_X^2}}$$

has the familiar Student's t distribution with $N - 3$ degrees of freedom.

234 QUANTITATIVE ZOOLOGY

The upper and lower confidence limits for β can then be found as the two roots of the quadratic equation

$$t^2 = \frac{(\bar{X}_3 - \bar{X}_1)^2 (B - \beta)^2 K/2}{s_Y^2 - 2\beta s_{XY} + \beta^2 s_X^2}$$

EXAMPLE 69. Calculation of Bartlett's "best fit" line for the data of Example 65.

	TAIL LENGTH	TOTAL LENGTH	
Group 1 (6 individuals)	37 49 50 51 53 54	284 375 353 366 418 408	$\bar{X}_1 = 49.00$ $\bar{Y}_1 = 367.33$
Group 2 (7 individuals)	68 86 93 106 130 137 142	510 627 683 820 1056 986 1086	$\bar{X}_2 = 108.85$ $\bar{Y}_2 = 824.00$
Group 3 (6 individuals)	142 146 149 155 156 187	1086 1078 1122 1254 1202 1387	$\bar{X}_3 = 155.83$ $\bar{Y}_3 = 1188.17$

$$\bar{X} = 104.79 \qquad \bar{X}_1 = 49.00 \qquad \bar{X}_3 = 155.83$$
$$\bar{Y} = 794.79 \qquad \bar{Y}_1 = 367.33 \qquad \bar{Y}_3 = 1188.17$$

$$B = \frac{\bar{Y}_3 - \bar{Y}_1}{\bar{X}_3 - \bar{X}_1} = \frac{1188.17 - 367.33}{155.83 - 49.00} = \frac{820.84}{106.83} = 7.684$$

$$A = \bar{Y} - B\bar{X} = 794.79 - (7.684)(104.79) = -10.42$$

The equation of Bartlett's "best fit" line is then $Y = -10.42 + 7.684 X$ or, what is equivalent, $X = 1.19 + .136 Y$

After all the numerical quantities have been substituted in this quadratic expression, it can be rearranged as in Example 70 to produce an equation of the form

$$a\beta^2 + b\beta + c = 0$$

whose solution is

$$\beta = \frac{-b \pm \sqrt{b^2 - 4ac}}{2a}$$

EXAMPLE 70. 95 per cent confidence intervals for β from the results of Example 69.

$\bar{X}_1 = 49.00$ $\bar{Y}_1 = 367.33$
$\bar{X}_2 = 108.85$ $\bar{Y}_2 = 824.00$
$\bar{X}_3 = 155.83$ $\bar{Y}_3 = 1188.17$
$N = 19$ $t_{.95} = 2.120$
$K = 6$
$B = 7.68$

	$\Sigma(X - \bar{X})^2$	$\Sigma(X - \bar{X})(Y - \bar{Y})$	$\Sigma(Y - \bar{Y})^2$
Group 1	190.0	1,390.0	11,446.1
Group 2	4,799.7	37,768.2	306,014.0
Group 3	1,313.1	9,113.8	85,217.3
	6,302.8	48,272.0	402,677.4

$s_X^2 = \dfrac{1}{16} \; 6{,}302.8 \;=\; 393.9$

$s_{XY} = \dfrac{1}{16} \; 48{,}272.0 \;=\; 3{,}017.0$

$s_Y^2 = \dfrac{1}{16} \; 388{,}459.9 \;=\; 25{,}167.3$

$t^2 = \dfrac{(\bar{X}_3 - \bar{X}_1)^2 (B - \beta)^2 K/2}{s_Y^2 - 2\beta s_{XY} + \beta^2 s_X^2}$

$2.120^2 = \dfrac{(155.83 - 49.00)^2 (7.68 - \beta)^2 \, 6/2}{25{,}167.3 - 2\beta(3{,}017.0) + \beta^2 \, 393.9}$

Rearranging terms, we find $\beta^2 - 15.36\beta + 58.71 = 0$

The values $a = 1$, $b = -15.36$, and $c = 58.71$ may now be substituted in the quadratic solution $\dfrac{-b \pm \sqrt{b^2 - 4ac}}{2a}$, and the two roots of this equation are

$\beta = 7.68 + .52$
and $\beta = 7.68 - .52$

which are the upper and lower 95 per cent confidence limits.

The quantities in the quadratic expression for B are defined as follows: t is the appropriate value from the table of Student's t for $N-3$ degrees of freedom which gives the desired confidence coefficient. For example, if N were 7, there would be 4 degrees of freedom, and a 95 per cent confidence interval would correspond to a t of 2.776.

K is the number of individuals in the highest (or lowest) group.

$\bar{X}_1$ and $\bar{X}_3$ are the means of X for the lowest and highest groups respectively, and B is the estimate calculated for β.

The quantity s_x^2 is the pooled within-group variance of X. That is,

$$s_x^2 = \frac{1}{N-3}\left[\Sigma(X-\bar{X}_1)^2 + \Sigma(X-\bar{X}_2)^2 + \Sigma(X-\bar{X}_3)^2\right]$$

where $\Sigma(X-\bar{X}_1)^2$ is the sum of the squared deviations from the mean of the first group of the observations in the first group only, and so on. In a similar manner,

$$s_{xy} = \frac{1}{N-3}\left[\Sigma(X-\bar{X}_1)(Y-\bar{Y}_1) + \Sigma(X-\bar{X}_2)(Y-\bar{Y}_2)\right.$$
$$\left. + \Sigma(X-\bar{X}_3)(Y-\bar{Y}_3)\right]$$

and

$$s_y^2 = \frac{1}{N-3}\left[\Sigma(Y-\bar{Y}_1)^2 + \Sigma(Y-\bar{Y}_2)^2 + \Sigma(Y-\bar{Y}_3)^2\right]$$

In Example 70, the 95 per cent confidence limits have been calculated for the data on tail length and total length of *Lampropeltis* for which B was calculated in the last example.

A similar procedure can be used to find confidence limits for the Y intercept, α, although these limits are not as exact as for β. The upper and lower 95 per cent limits for α are given by

$$A \pm \frac{t}{\sqrt{N}}\sqrt{s_y^2 - 2Bs_{xy} + B^2 s_x^2}$$

where t is the value of Student's t with $N-3$ degrees of freedom. The term under the square root sign is identical with the denominator of the confidence limit expression for β used above *except* that the observed value of the slope, B, is used instead of β which is, of course, unknown. It is this substitution which makes the confidence interval somewhat inexact but not to an important extent. Using the results already obtained in Example 70, the confidence limits for α in the *Lampropeltis* data are calculated in Example 71. The rather broad limits, from -32.49 to 11.65, are not surprising in view of the very steep slope of the regression line and the fairly large confidence interval for that slope, from 7.16 to 8.20.

EXAMPLE 71. Calculation of 95 per cent confidence limits for α estimated by Bartlett's "best fit" method. Data are from Examples 69 and 70.

$$\begin{aligned} N &= 19 \\ s_Y^2 &= 25{,}167.3 \\ s_{XY} &= 3{,}017.0 \\ s_X^2 &= 393.9 \\ A &= -10.42 \\ B &= 7.68 \\ t_{.95} &= 2.12 \end{aligned}$$

$$A \pm \frac{t}{\sqrt{N}} \sqrt{s_Y^2 - 2B\, s_{XY} + B^2\, s_X^2}$$

$$= -10.42 \pm \frac{2.12}{\sqrt{19}} \sqrt{25{,}167.3 - 46{,}341.1 + 23{,}232.2}$$

$$= -10.42 \pm (2.12)(10.41) = -10.42 \pm 22.07$$

The 95 per cent confidence limit is -32.49–11.65

It is also possible to test the significance of the difference between two values of A or B calculated by Bartlett's "best fit" method. The tests are essentially Student's t-tests, and while they do not give *exact* results, they are quite adequate in practice. To test the difference between two slopes, B and B', use the expression

$$t = \frac{(\bar{X}_3 - \bar{X}_1 - \bar{X}_3' + \bar{X}_1')(B - B')}{\sqrt{\dfrac{2 s_B^2}{K} + \dfrac{2 s_{B'}^2}{K'}}}$$

where t has $(N + N' - 6)$ degrees of freedom.

In this formula, the primed symbols refer to one sample, the unprimed symbols to the other sample and

$$s_B^2 = s_Y^2 - 2B s_{XY} + B^2 s_X^2$$

All the other quantities have been defined in the discussion of confidence intervals.

For the test of significance involving the two Y intercepts, A and A', the appropriate expression is

$$t = \frac{A - A'}{\sqrt{\dfrac{s_B^2}{N} + \dfrac{s_{B'}^2}{N'}}}$$

with all the quantities defined as before.

Security of Predictions

The predicted value of a variate Y from a regression equation is only an estimate and is subject to error. It is clearly necessary to be able to construct confidence limits for such a prediction, just as for any estimate of a parameter. The confidence limits for a predicted value of Y usually turn out to be quite large when the particular value of X is very far from the mean—in some cases so large as to make prediction nearly useless. No attempt at prediction should be made without specifying the confidence limits, for a completely erroneous impression may arise.

There are really two predictions that can be made from the regression equation. The first concerns the mean value of Y for any given X, and the other the value of Y for an individual whose X value is known. Whereas the predicted value is the same for both of these problems, the confidence limits are different. The average value of Y for a given X will clearly have a smaller confidence interval than the individual value of Y, means always being more certain than single measurements. If the predicted value of Y for a given X is denoted by Y_x, then the confidence interval for the *mean value of* Y_x is given by

$$Y_x \pm t\, s_{yx} \sqrt{\frac{1}{N} + \frac{(X - \bar{X})^2}{(N-1)\,s_x^2}}$$

and the confidence interval for an *individual* predicted value of Y_x is

$$Y_x \pm t\, s_{yx} \sqrt{1 + \frac{1}{N} + \frac{(X - \bar{X})^2}{(N-1)\,s_x^2}}$$

The symbols have all been defined before and t has $N - 2$ degrees of freedom.

As predicted the confidence interval for the mean of Y_x is smaller than that for an individual Y_x. In addition, the confidence interval is seen to be larger the greater the deviation of X from $\bar{X}$.

Example 72 demonstrates the calculation of these confidence intervals for the prediction of total length from a known tail length using the prediction equation derived in Example 65. This is a problem which often confronts the paleontologist, for example.

The Correlation Coefficient

Several means of measuring correlation numerically have been devised. They are usually based on a conception of perfect correlation in which an increase in the value of one variate is always accompanied by an exactly proportionate change (increase in positive and decrease in negative correlation) in the other. Thus, in these series of paired observations,

X	Y	X	Y
1	1	1	8
2	2	2	6
4	4	4	2

both are perfect correlations—the first positive, because every time that X increases, the corresponding Y increases by an equal amount, and the second negative, because every time X increases, Y decreases by an amount twice as great as the increase in X. Perfect correlation involves only the same relative change, not the same absolute change. The following series all have perfect positive correlation:

X	Y	X	Y	X	Y
1	.02	1	1.0	1	50
2	.04	2	1.5	2	100
4	.08	4	2.5	4	200

EXAMPLE 72. Confidence intervals for the prediction of total length from tail length in *Lampropeltis*. (Data from Example 65)

$b_{YX} = 7.68$ $\qquad N = 19$
$a_Y = -11.4$ $\qquad \bar{X} = 104.78$
$s_{YX} = 32.72$ $\qquad s_X^2 = 2,261.61$

The prediction equation is then

$$Y = -11.4 + 7.68\ X$$

Choosing an X arbitrarily, say, 150, the predicted value of Y is

$$Y = -11.4 + (7.68)(150) = 1,140.60$$

The 95 per cent confidence interval for the *mean* Y corresponding to an X of 150 is

$$1{,}140.60 \pm (2.11)(32.72)\sqrt{\frac{1}{19} + \frac{(150 - 104.78)^2}{(18)(2{,}261.61)}} = 1{,}140.60 \pm 69.04\ \sqrt{.10286}$$

$$= 1{,}140.60 \pm 22.16$$

or from 1,118.44 to 1,162.76

The 95 per cent confidence interval for an *individual* predicted Y value, when that individual has an X of 150, is

$$1{,}140.60 \pm 69.04\ \sqrt{1.10286} = 1{,}140.60 \pm 72.49$$

or from 1,068.11 to 1,213.09

The measure of correlation is then a measure of how near the actual observations come to this relationship. The best measure would be analogous to the measure of dispersion for one variate by its variance or standard deviation but would involve a relationship between the variances of two variates as arranged in paired observations, their variances together, or covariance. Such a measure, devised by Pearson, which has been found to have very useful properties, is the ratio of the mean product of corresponding deviations of the two variates to the geometric mean of their separate variances. This measure is always symbolized by r, and the term "coefficient of correlation," if not qualified, generally refers to r, the formula for which is

$$r = \frac{\sum(X - \bar{X})(Y - \bar{Y})}{\sqrt{\sum(X - \bar{X})^2 \sum(Y - \bar{Y})^2}}$$

The formula is exactly equivalent to

$$r = \frac{\sum(X - \bar{X})(Y - \bar{Y})}{(N - 1) s_X s_Y}$$

in which N is the number of pairs of observations and s_X and s_Y are the standard deviations of X and Y respectively. The parameter of which this is an estimate is usually denoted by ρ.

This definition of r is suggestive of the formulae for estimating b_{YX} and b_{XY} and, as a matter of fact, the reader can easily verify for himself that

$$r = \sqrt{b_{XY} b_{YX}}$$

For cases in which the regression coefficients are to be calculated, which includes most situations, the sample correlation coefficient is then easily calculated by the use of this relation. For example, the two regression coefficients for tail length and total length in the data cited on *Lampropeltis* were calculated to be

$$b_{YX} = 7.68 \text{ and } b_{XY} = .129$$

The correlation between these two variates is then simply

$$r = \sqrt{(7.68)(.129)} = +.99$$

a very high value. This formula would give the value $+1$ for perfect positive and -1 for perfect negative correlation, 0 for no correlation, and other values between these limits. In order for r to be $+1$, b_{YX} and b_{XY} must be exact reciprocals of each other so that the product $b_{YX} b_{XY}$ is equal to unity. This result relates to the discussion of regression in which it was pointed out that if the points lay on a perfect straight line, one regression coefficient would be the exact reciprocal of the other. Points which lie exactly on a straight line are then perfectly correlated.

There is one difficulty about calculating the correlation coefficient from the square root of the product of the regression coefficients. No matter whether the regression coefficients are positive or negative, the algebraic sign of r calculated in this way may be either. Negative regression coefficients, however, mean a negative correlation. It must be remembered, then, that the *sign of the correlation coefficient calculated from the expression*

$$r = \sqrt{b_{xr} b_{rx}}$$

is always the same as the sign of the regression coefficients. (Unless some error has been made in calculating the two regression coefficients, they will both have the same sign.)

It was pointed out that the correlation coefficient cannot be validly calculated for samples in which one of the variates takes only values not chosen at random as for time series, altitudinal transects, and the like. The reason for this should now be clear. If r is defined as the geometric mean of b_{rx} and b_{xr}, then both of these quantities must be calculable from the sample. But it was argued that b_{xr} is not a valid estimate of β_{xr} when the sample has been taken deliberately with respect to the X variate. It then follows that r cannot be validly calculated from such a sample.

It is not necessary to calculate r from the regression coefficients, but it may be found by direct application of the first formula given. The important point is that the quantities

$$\Sigma(X - \bar{X})^2, \ \Sigma(Y - \bar{Y})^2, \text{ and } \Sigma(X - \bar{X})(Y - \bar{Y})$$

are the only pieces of information necessary for finding r, b_{xr}, and b_{rx}, so that once these basic quantities have been obtained from the data, both the correlation and the regression coefficients follow quite simply. There is no reason, then, to publish only the correlation coefficient, since no more work is required to indicate the regressions as well. Whereas the value of r can always be found if b_{rx} and b_{xr} are given, the reverse is not true.

Correlation Tables

It is often very helpful in making a preliminary judgment of the data if they are put in a tabular form, as shown in Examples 73A and 73B. The two variates are grouped in any convenient number of divisions, and the X variate groups are listed across the top in ascending order from left to right, while the Y variate groups are listed at the left-hand end of the table in descending order from top to bottom. In each cell is the number of individuals whose X and Y measurements fall in the corresponding X and Y groups. This is a sort of numerical graph which makes clear whether or not there is an obvious correlation between the variates. Thus, the variates in Example 73A are more highly correlated than in Example 73B, since the individuals are more closely clustered around the diagonal set of cells. If these same data are put on a scatter diagram as in Fig. 18, the same trend is obvious. Calculation of the correlation coefficients from the raw data used to construct these two tables shows, in fact, that for Example 73A, $r = +.99$ while for Example 73B it is only $+.82$. Correlation tables are not substitutes for correlation coefficients but only a way to visualize the relationship more clearly. With experience, it may also be possible to tell at a glance that the correlation is not significant and hence not, as a rule, worth the trouble of calculation.

Significance of Correlation

Even if the sample correlation coefficient is different from zero, there may be no correlation in the population. After all, r is only a sample estimate of ρ and, subject to sampling error as it is, it may show a correlation where none exists. Whereas all sorts of tests of hypotheses might be made concerning ρ, the question most often of importance is whether ρ is really different from zero. To test the null hypothesis, that ρ is not different from zero, it is necessary to know the probability distribution of the sample correlation coefficient, r, given that ρ is really zero. If r is so large that the probability of its occurrence in a sample from a population with $\rho = 0$ is very small, the null hypothesis would be rejected and it would be necessary to believe that some real correlation exists between the variates. The distribution of the sample correlation coefficient, r, is given in Appendix Table V for various degrees of freedom. In order to find the significance of an observed r, enter the table with $N - 2$ degrees of freedom. The sign of r is ignored, the probabilities in the table representing the area outside the limits $+r$ and $-r$ in the distribution. These are then probabilities for a "two-sided test." If, for some reason, it is definitely known that a correlation must be positive or negative, a "one-sided" test would be needed, the one-sided probabilities being equal to one-half those in the table.

It can be seen from the table that the correlation between tail length and total length in *Lampropeltis* which was calculated to be $+.99$ is highly significant, since for 17 degrees of freedom ($N = 19$) an r as low as .6932 has a probability of only .001. For the same set of specimens, the correlation between scute number and tail length, the scatter diagram for which is Fig. 19, has a positive correlation of .10. For 17 degrees of freedom, this is far from significant at the 5 per cent or any other level usually employed. The smallest value in the table for 17 degrees of freedom is .3887, corresponding to a probability of .1. A value of r as small as .10 would have even a higher probability of occurrence when ρ was zero.

EXAMPLE 73. Correlation tables.

A. Tail length and total length in females of the king snake *Lampropeltis polyzona*. (Data from Blanchard, 1921)

Y–TOTAL LENGTH	X–TAIL LENGTH							
	30–49	50–69	70–89	90–109	110–129	130–149	150–169	170–189
1350–1449	—	—	—	—	—	—	—	1
1250–1349	—	—	—	—	—	—	1	—
1150–1249	—	—	—	—	—	—	1	—
1050–1149	—	—	—	—	—	5	—	—
950–1049	—	—	—	—	—	1	—	—
850– 949	—	—	—	—	—	—	—	—
750– 849	—	—	—	1	—	—	—	—
650– 749	—	—	—	1	—	—	—	—
550– 649	—	—	1	—	—	—	—	—
450– 549	—	1	—	—	—	—	—	—
350– 449	1	4	—	—	—	—	—	—
250– 349	1	—	—	—	—	—	—	—

Calculated $r = +.99$

B. Lengths of M_1 and M_2 in the fossil mammal *Phenacodus primaevus* from the Gray Bull Formation of the Bighorn Basin, Wyoming. (Original data)

Y–LENGTH OF M_2	X–LENGTH OF M_1							
	10.0–10.4	10.5–10.9	11.0–11.4	11.5–11.9	12.0–12.4	12.5–12.9	13.0–13.4	13.5–13.9
13.5–13.9	—	—	—	—	—	—	—	1
13.0–13.4	—	—	—	—	—	2	1	1
12.5–12.9	—	—	—	—	—	—	1	2
12.0–12.4	—	—	—	—	2	1	2	1
11.5–11.9	—	—	—	—	1	—	—	—
11.0–11.4	—	1	2	—	1	3	—	—
10.5–10.9	—	3	—	—	—	—	—	—
10.0–10.4	—	—	—	—	—	—	—	—
9.5– 9.9	1	—	—	—	—	—	—	—

Calculated $r = +.82$

Because of the close relationship between correlation and regression, the significance of r can be judged from the regression coefficients alone. If the regression coefficients are not different from zero—that is, if the confidence interval for β overlaps 0—then the correlation is not significantly different from zero. Conversely, a significant regression means a significant correlation as well. From this consideration it should be clear, if it is not so already, that by "correlation" we have meant "rectilinear correlation."

The absence of a significant *r* does not mean that the variables have no dependence on each other, but rather that there is no rectilinear dependence. Two variables may be exact functions of each other, yet show no significant correlation on the basis of the parameters β and ρ, when the graphic relationship is a curve and not a straight line.

Of all the statistics in common use, the correlation coefficient is the one whose significance is most often ignored. It is virtually impossible to draw samples which do not show some value of *r* different from zero, but very often this *r* is not significant. A correlation coefficient is meaningless without a measure of the probability that it really estimates a ρ which is not zero.

Transformation of *r* to *z*

Fisher, whose first major contribution to statistics was the distribution of the sample correlation coefficient, has presented a measure of correlation denoted by *z*, which is algebraically related to *r* in the following way:

$$z = \frac{\log_e (1 + r) - \log_e (1 - r)}{2}$$

where $\log_e$ is the *natural logarithm*, or logarithm to the base *e*. Tables of natural logarithms are readily available, usually included in books of mathematical tables in common use. If logarithms to the base 10 are the only ones available, use may be made of the relation

$$\log_e X = 2.30259 \log_{10} X$$

The measure *z* has two important attributes. First, it is roughly normally distributed and for sample sizes above 100 may be regarded as a very close approximation to a normal variate, and, second, the variance of this normal variate is simply

$$s_z^2 = \frac{1}{N - 3}$$

These facts make *z* useful for testing the significance of a correlation coefficient when the sample size is in excess of 100 and for testing the difference between two correlation coefficients.

To test whether *r* is significantly different from zero, use the expression

$$z\sqrt{N - 3}$$

which is a standardized normal deviate. If the probability of the observed value of *z* is very small, as judged from the table of the standardized normal distribution, the null hypothesis is rejected and r is significant. Example 74 shows this calculation for the small sample of *Lampropeltis* as an illustration, although z is only needed for large samples.

EXAMPLE 74. Calculation of s_z and z/s_z

A. Data of Figure 19

$N = 19$ $\quad s_z = \dfrac{1}{\sqrt{N-3}} = \dfrac{1}{\sqrt{16}} = \dfrac{1}{4} = .25$

$r = .10$

$z = .10$ $\quad \dfrac{z}{s_z} = \dfrac{.10}{.25} = .40$

The table of t shows that P is greater than .6 (it is in fact between .68 and .69), hence the value of z is not significant.

B. Data of Example 65 and Figure 18

$N = 19$ $\quad s_z = \dfrac{1}{\sqrt{16}} = .25$

$r = .99$

$z = 2.30$ $\quad \dfrac{z}{s_z} = \dfrac{2.30}{.25} = 9.2$

This z is decisively significant. Note that the value of s_z is the same as in A, although the value of z is nearly 30 times greater.

To test the significance of the difference between two correlation coefficients, use the expression

$$\dfrac{z_1 - z_2}{\sqrt{\dfrac{1}{N_1 - 3} + \dfrac{1}{N_2 - 3}}}$$

which is an approximate standardized normal deviate even for small values of N. Example 75 shows the calculation and test of differences for two sets of observations.

EXAMPLE 75. Significance of the difference between two correlation coefficients using the transformed correlation coefficient z.

A. Total length and tail length in males and females of *Lampropeltis polyzona*. (Raw data from Blanchard, 1921)

	N	r	z
Males	24	.974	2.16
Females	19	.988	2.56

$$\frac{z_1 - z_2}{\sqrt{\frac{1}{N_1 - 3} + \frac{1}{N_2 - 3}}} = \frac{2.56 - 2.16}{\sqrt{\frac{1}{21} + \frac{1}{16}}} = \frac{.40}{.33} = 1.21$$

A standardized normal deviate of 1.21 corresponds to a probability of .2 to .3, so that the difference would not be considered significant.

B. Total length and tail length in females of *Lampropeltis polyzona* and of *Lampropeltis elapsoides elapsoides*. (Data from Blanchard, 1921)

	N	r	z
L. polyzona	19	.988	2.56
L. elapsoides	25	.899	1.47

$$\frac{z_1 - z_2}{\sqrt{\frac{1}{N_1 - 3} + \frac{1}{N_2 - 3}}} = \frac{2.56 - 1.47}{\sqrt{\frac{1}{16} + \frac{1}{22}}} = \frac{1.09}{.33} = 3.3$$

This value of the standardized normal deviate corresponds to a probability of less than .001 so the difference is certainly significant.

Cause and Effect and Spurious Correlation

Measures of correlation and test of their significance demonstrate only that two variates are shown or are not shown by a given sample to vary in such a way as to tend to maintain a definite relation to each other. They tell nothing about the cause of the relationship but reveal only its existence or the lack of reliable evidence for it. There is a danger in passing from these numerical results to biological conclusions that the relationship may be misunderstood or may, without due consideration, be assumed to represent cause and effect. The fallacy is that most common in vulgar thought, *post hoc, ergo propter hoc* (or *cum hoc propter hoc*), and is among the foundations of much of the untrue "natural wisdom" of the unlettered and of astrology, alchemy, and the other false sciences. On these discredited levels, the fallacy is easy enough to detect, and numerical correlation itself disposes of many logically obscure cases by showing that the supposed or apparent correlation is not significant.

This fact, however, increases the danger of supposing that a causal relationship does exist when the correlation is found to be numerically significant, although even on this level common sense is an adequate safeguard against the more egregious blunders. It may be shown that rodents are more abundant in an area when crop prices are lower—in other words, that number of rodents and crop price show significant negative correlation—but no one would conclude that rodents tend to lower crop prices or that cheap crops tend to increase rodent families. It is obvious that the two phenomena are really quite independent except as both may depend on a third factor, crop abundance or, still more remotely, favorable weather. The relations may in other cases be so obscure that considerable analysis is necessary to differentiate cause and effect or to distinguish a real correlation from one that is spurious—being caused not by any true relationship between the variates considered but by their relationship to a third variate that has been omitted from the problem.

It is not true that any correlation between variates not truly related as cause and effect is to be considered spurious. Some of the most useful correlations are between two variates, both of which are affected by the same unmeasured cause. In fact, such correlations may reveal the existence of any important variate or characteristic that cannot be directly observed and may serve to measure it indirectly by measuring its effects. Such correlations are common in zoology, and that between lengths of two adjacent teeth (Example 73B) is a good example. Obviously these are not cause and effect; but they are, or more properly their relationship is, the result of some cause that cannot otherwise be detected or measured. The data do not show what this cause is, whether genetic, environmental, or otherwise. Its existence having been detected in this way, its nature remains to be determined, if possible, by further observation, calculation, and experimentation.

There is no general rule for differentiating true and spurious correlation beyond the application of logic and testing of other possible correlations of the variates in question. It is, however, usually true that the correlation technique is properly applied to two variates that may be related as cause and effect or that may be analogous effects of a cause not directly observable.

Aside from purely spurious correlation, a true correlation may have its value falsified by the existence of another distinct correlation not excluded from the data and tending to either increase or decrease the correlation that is sought. For instance, correlation between thymus size and body size of a species of mammal may correctly be sought, and its result may throw important light on the biology of the species; but a good sample of the whole species will not give the correct value of the correlation. Thymus size will probably be found to be positively associated with body size in animals of the same age, but negatively correlated with body size if age is not a constant because it shows a stronger negative correlation with age than positive correlation with body size.

Correlation either with age or with body size is real and significant, but correlation with either that does not exclude the effect of the other gives a false value. Such relationships may become very complex, but their disentanglement can often be accomplished in one of two ways. It may be possible to select a sample such that all but two of the variates involved, or possibly involved, in the correlation are made as nearly constant as possible, thus making possible a true observation of the correlation of those two variates. Sometimes, however, this will reduce the available sample so that it is too small to detect a significant correlation, and even with abundant data it may be impossible to keep all but two variates approximately constant. In such cases, the only practical solution is, if possible, to measure all the variates involved in the correlation and to derive a corrected true correlation value for any two of them from the interrelationships of all.

Partial Correlation

The process just mentioned is called *partial correlation*. When a measurable correlation involves more than two factors or variates, partial correlation is a technique for finding how much of the correlation value results from the relationship between any two of these variates. A variate the value of which depends on that of two or more other variates is said to show *multiple correlation*; and partial correlation, defined in another way, is the measurement of the relationship between the value of the given variate and that of any one of the other variates on which it depends. Partial correlation is used when it is not practical to keep all but two variates involved in correlation constant, and it gives a numerical result not significantly different from what would have been obtained had this been possible.

Partial correlation is possible only when corresponding values of the three or more variates involved can be obtained from the same body of data. When this is possible, the process is surprisingly simple in concept, although the calculation involved may be laborious, sometimes prohibitively so if there are more than three variates. With three variates the formula is

$$r_{12 \cdot 3} = \frac{r_{12} - r_{13} r_{23}}{\sqrt{(1 - r_{13}^2)(1 - r_{23}^2)}}$$

The subscripts 1, 2, and 3 refer to the three variates. The first step is to correlate these in pairs from the original data, correlating the first two and obtaining a coefficient r_{12}, the first and third, giving r_{13}, and the second and third, giving r_{23}. These are total correlations, because in obtaining them nothing was excluded—the value of r_{12} may still be influenced by that of r_{13} and of r_{23}. The symbol $r_{12 \cdot 3}$ represents the partial correlation of variates 1 and 2 with the effect of the variation of 3 on either of them eliminated. Similarly, a partial correlation $r_{13 \cdot 2}$ (of 1 and 3 with 2 eliminated) may be obtained by transposing 2 and 3 throughout the formula and a partial correlation $r_{23 \cdot 1}$, by transposing 1 and 3.

If there are four or more variates, it is necessary to eliminate them one at a time. Thus, with four variates it is first necessary to obtain $r_{12 \cdot 4}$, $r_{13 \cdot 4}$, and $r_{23 \cdot 4}$ by applying the formula three times and then to use these values, with the effect of 4 eliminated, in the formula to obtain $r_{12 \cdot 34}$, the correlation of 1 and 2 with the effects of the variability of both 3 and 4 eliminated, as follows: r_{12}, r_{13}, r_{14}, r_{23}, and r_{34} are first calculated, and then

$$r_{12 \cdot 4} = \frac{r_{12} - r_{14} r_{24}}{\sqrt{(1 - r_{14}^2)(1 - r_{24}^2)}}$$

$$r_{13 \cdot 4} = \frac{r_{13} - r_{14} r_{34}}{\sqrt{(1 - r_{14}^2)(1 - r_{34}^2)}}$$

$$r_{23 \cdot 4} = \frac{r_{23} - r_{24} r_{34}}{\sqrt{(1 - r_{24}^2)(1 - r_{34}^2)}}$$

$$r_{12 \cdot 34} = \frac{r_{12 \cdot 4} - r_{13 \cdot 4} r_{23 \cdot 4}}{\sqrt{(1 - r_{13 \cdot 4})^2 (1 - r_{23 \cdot 4}^2)}}$$

With four variates, the rather laborious operation summarized in the formula thus has to be performed four times, and this labor increases more rapidly than does the number of variates: for five variates the formula is calculated ten times, for six, twenty times, and for ten, two hundred and twenty times. In some sociological, medical, psychological, and kindred studies, partial correlation with large numbers of variates is used. For most zoologists, the use of three variates or at most four is adequate. The process of obtaining $r_{12 \cdot 3}$ is shown in Example 76. The significance of the correlation coefficients in Example 76 is found in the same way as ordinary correlations, except that the number of degrees of freedom is different. A partial correlation coefficient has a number of degrees of freedom equal to

$$N - 2 - K$$

where K is the number of variates held constant in the particular partial correlation. Thus $r_{12 \cdot 3}$, $r_{13 \cdot 2}$, and $r_{23 \cdot 1}$ all have one variable held constant and would each have $N - 3$ degrees of freedom. A partial correlation like $r_{12 \cdot 34}$ would have $N - 4$ degrees of freedom.

EXAMPLE 76. Computations of partial correlations with three variates. Maximum length, width, and height of crowns of last upper molars of the extinct mammal *Acropithecus rigidus*. (Original data)

VARIATES
    1. Length
    2. Width $N = 28$
    3. Height

TOTAL CORRELATIONS
    $r_{12} = .355$
    $r_{13} = .795$
    $r_{23} = -.046$

PARTIAL CORRELATIONS

$$r_{12 \cdot 3} = \frac{r_{12} - r_{13} r_{23}}{\sqrt{(1 - r_{13}^2)(1 - r_{23}^2)}} = \frac{.355 - .795(-.046)}{\sqrt{(1 - .795^2)(1 - .046^2)}} = \frac{.355 + .037}{\sqrt{.368 \times .998}}$$

$$= \frac{.392}{.605} = +.648$$

$$r_{13 \cdot 2} = \frac{r_{13} - r_{23} r_{12}}{\sqrt{(1 - r_{23}^2)(1 - r_{12}^2)}} = \frac{.811}{.933} = +.867$$

$$r_{23 \cdot 1} = \frac{r_{23} - r_{12} r_{13}}{\sqrt{(1 - r_{12}^2)(1 - r_{13}^2)}} = \frac{-.328}{.562} = -.584$$

TESTS OF SIGNIFICANCE

	r	Degrees of Freedom	P
r_{12}	.355	26	.05–.10
r_{13}	.795	26	.001
r_{23}	−.046	26	.50
$r_{12 \cdot 3}$	.648	25	.001
$r_{13 \cdot 2}$	.867	25	.001
$r_{23 \cdot 1}$	−.584	25	.01–.001

r_{12} might or might not be considered different from zero, depending upon the criterion used. The values of r_{13}, $r_{12 \cdot 3}$, $r_{13 \cdot 2}$, and $r_{23 \cdot 1}$ would all be considered significant even under a very stringent criterion, while r_{23} would not be considered different from zero unless a very weak criterion of significance were employed.

The problem chosen as an example of partial correlation, although relatively simple arithmetically, has somewhat complex zoological connotations, and the partial correlations give information not visible or directly deducible from the original data and not so well shown by any other method of analysis. The unexpected nature of some of these results and their hardly obvious zoological meaning exemplify very clearly (1) the need for numerical analysis in such problems and (2) the fact that the results of such analysis are not zoological conclusions but must be interpreted logically and with care, in the light of the real meaning of each operation, starting with the gathering of the original measurements. The zoological consideration of this problem and of its numerical conclusions will therefore be given in more detail than for most previous examples, since the reasoning involved is analogous to that for any problem of partial correlation and cannot be taken for granted.

The measurement of length was taken on the wearing surface of these teeth, and so it changes or may change with degree of wear. This is in this case almost invariably the maximum length of a tooth as preserved. It would be preferable to take a length measurement not affected by wear, but this was not done in this case for three reasons: (1) because it had not usually been done in samples of allied animals, and hence the measurement as taken is more nearly comparable with data already available; (2) because it is desirable to find out what influence wear does have on this length measurement; and (3) because length on the wear surface can be accurately measured in all cases and other lengths can be accurately measured only on isolated teeth out of the jaw. The maximum width is at the base of the tooth and had not been visibly affected by wear in any specimen measured. The maximum crown height would doubtless vary in unworn teeth, but all these teeth were in fact worn to various degrees. The differences in height caused by wear were obviously very much greater than any such differences could have been on the unworn teeth, so that this measurement is for all practical purposes a measure of degrees of wear rather than of genetic differences in tooth heights. It is thus indirectly an approximate but useful measure of individual ages. Age cannot, of course, be measured directly for fossils; but it is certain that if it could, it would show a very high positive correlation with molar wear and hence an equally high negative correlation with molar height. It can thus be reasonably assumed that molar height is a good approximate inverse indirect measure of age.

In order to understand and interpret this sample correctly, it is necessary to know and measure two things that are not apparent from the original data: the relationship between length and width and the relationship between length and age (or wear) or molar height. We want, in other words, reliable measures of the correlation of 1 and 2 and of the correlation of 1 and 3. The total correlation r_{12} gives a result that is not significant, but evidently this may not be reliable; for if either 1 or 2 is also correlated with 3, which is known from the method of measurement to be probable for 1, then 3 also influences the value of r_{12}. Similarly, it cannot be assumed that r_{13} is a reliable measure, for 2 may well be correlated with 1 and hence may influence this value. The only way to get the information sought is therefore somehow to eliminate any influence of 3 on the correlation of 1 and 2 and any influence of 2 on the correlation of 1 and 3.

One solution would be to select specimens in which height, 3, is the same for all and to obtain a value of r_{12} from them, then to select specimens of the same width, 2, and obtain a value of r_{13}. If this were done, r_{12} and r_{13} would be reliable measures of the relationships being investigated; but in this case, it could not be done because the largest available samples with nearly constant height or width would (as the original measurements show) have included only three or four specimens, too small to give a useful value of r. Recourse is therefore had to partial correlation, which produces the same result but uses all the observations made. $r_{12 \cdot 3}$ and $r_{13 \cdot 2}$ are reliable measures of the two relationships sought.

The measure r_{12} is of doubtful significance, but $r_{12 \cdot 3}$ shows a significant positive correlation. The zoological conclusion is that for any given height —hence, in any given wear stage, any given period of life, or for unworn teeth—greater length tends to be accompanied by greater width, and vice versa. It then follows, zoologically, that length and width are affected by some common influence, almost surely genetic, and further that they would tend to have an approximately constant ratio in teeth unaffected by wear. These are important conclusions, essential to intelligent study of the specimens involved, and they could not be reached by any more direct consideration of the actual measurements available.

The value of r_{13} differs significantly from 0 and that of $r_{13 \cdot 2}$ still more so, showing that a correlation existing between 1 and 2 or between 2 and 3 tended to reduce and, to that extent, to falsify the apparent association between 1 and 3. For a constant width, there is a strong positive correlation between length and height. There are really two factors in this, both tending in the same direction, so that correlation of length with only one of them would be less; but they cannot be separated on these data, and their separation is not really necessary. One factor is variation in original unworn height and the other is variation in height caused by wear; but the former is certainly very slight compared with the latter (as known on zoological grounds, although not numerically demonstrable from these data), and it may safely be concluded that there is a significant negative correlation between length and wear and, beyond that, between length and age. We have, then, demonstrated beyond reasonable doubt that in these teeth, the length of the grinding surface tends to become less as the animal grows older. Aside from the importance of this fact in itself, it also has an important practical bearing, for it shows that comparisons of lengths of these teeth with any others will not be valid unless height is also taken into account and its influence eliminated or discounted.

In this case, the value $r_{23 \cdot 1}$ does not measure any relationship of zoological importance or answer any question naturally suggested by the data. In the actual research, there was no reason to calculate or discuss this correlation, but it is given in our example because the value will probably be found surprising and because it well illustrates the danger of hasty judgment or of faulty nonnumerical reasoning in dealing with numerical procedures. The value of r_{23} is very small and certainly does not differ significantly from 0, but the value of $r_{23 \cdot 1}$ shows a fairly strong, surely significant negative correlation between width and height if the effect of length on these is discounted. On the face of it, this seems to mean that these teeth become wider (at the base, beyond the reach of the actual wear) as they are worn down, which might be considered either as a sensational discovery or as a manifest absurdity.

In fact, this is not at all what the negative partial correlation means. It means only that in teeth of a given length those that are narrower will tend to be less worn. We know, from the value of $r_{13 \cdot 2}$, that length and height are positively correlated. It can then logically be seen that a tooth in which the unworn length was greater will have to be worn down to some degree before its length becomes equal to that of a smaller unworn tooth, and hence that when two teeth have the same length, the tooth that was longer when unworn will tend at this stage to have the lesser height. But we also know from the value of $r_{12 \cdot 3}$ that the tooth longer when unworn will, at any wear stage, tend to have the greater width. Hence it follows inevitably that in a miscellaneous sample with all stages of wear represented greater width will be associated with lesser height, in other words, that width and height will be negatively correlated. This is what the value of $r_{23 \cdot 1}$ means, and it has no particular zoological significance, certainly not that suggested at first sight. It is simply a corollary of the high positive values of $r_{13 \cdot 2}$ and $r_{12 \cdot 3}$ and is seen to follow inevitably when the real meaning of these is kept in mind.

Rank Correlation

Aside from r, the only directly calculated measure of correlation in general use is Spearman's measure of rank correlation, usually designated by ρ. As this symbol has already been used for the population values of Pearson's correlation coefficient, we will employ the symbol C for Spearman's rank correlation. This method arranges the two series of values in the order of their magnitude and gives each a rank according to its position in the series. Thus the largest value of X is ranked as 1, the largest value of Y also as 1, and the next value of each 2, etc. The differences in rank between corresponding values of X and Y are then recorded (none being regarded as negative) and the measure calculated as follows:

$$C = 1 - \frac{6 \sum d^2}{N(N^2 - 1)}$$

where d = a difference in rank

N = the number of pairs of observations.

If two or more observations have the same value and hence are tied for rank, they are best given the middle value of the ranks that they would occupy if different, and the next higher value is given its correct serial ranking. The value of C is generally about the same as that of r if the original distributions were approximately normal; and C is somewhat easier to calculate than is r, but this is almost its only advantage. The use of C in zoology is not recommended except as a relatively quick way of finding whether r is likely to be significant, or when rank can be determined but accurate and equally spaced absolute values cannot, or when accuracy and exact tests of significance are not required. The calculation is shown in Example 77, as well as a demonstration of some of the relationships between C and r

From their respective formulae, it can be seen that C shows perfect correlation, $+1$, if all the corresponding ranks of X and Y are in the same order and hence if Y increases or decreases with every increase of X (or vice versa), regardless of the amount of increase or decrease, while r does not show perfect correlation unless this condition is fulfilled and also the condition that the ratio between increments of X and Y is constant. For this reason, for certain types of curvilinear correlation, C will give higher and perhaps more truly representative values than will r; but this relationship is usually unreliable in practical use, and C may overestimate or (less often) underestimate the true value of the correlation.

EXAMPLE **77**. The use of C.

A. Calculation of C from data of Example 73B.

ORIGINAL MEASUREMENTS		RANKS		DIFFERENCES IN RANK	
X	Y	X	Y	d	d^2
9.8	10.2	1	1	0	0
10.5	10.7	2.5	2.5	0	0
10.5	10.7	2.5	2.5	0	0
10.8	10.8	4	4.5	0.5	0.25
11.0	11.0	5	6	1	1
11.1	11.4	6.5	7	0.5	0.25
11.1	12.1	6.5	9	2.5	6.25
11.3	12.6	8	14	6	36
11.4	12.8	10	17	7	49
11.4	10.8	10	4.5	5.5	30.25
11.4	12.6	10	14	4	16
11.9	12.3	12	10	2	4
12.2	12.4	12.5	11.5	2	4
12.2	12.0	12.5	8	5.5	30.25
12.3	13.7	15.5	25	9.5	90.25
12.3	13.0	15.5	18	2.5	6.25
12.4	13.2	17.5	20	2.5	6.25
12.4	12.4	17.5	11.5	6	36
12.5	13.8	19	26	7	49
12.7	13.5	20	22.5	2.5	6.25
12.8	13.3	21	21	0	0
13.0	12.7	22	16	6	36
13.1	13.1	23	19	4	16
13.2	13.6	24	24	0	0
13.4	12.6	25	14	11	121
13.5	13.5	26	22.5	3.5	12.25
					$\Sigma d^2 = 556.50$

$N = 26$

$$C = 1 - \frac{6 \Sigma d^2}{N(N^2 - 1)} = \frac{6 \times 556.50}{26(26^2 - 1)} = \frac{3{,}339}{17{,}550} = 1 - .19 = .81$$

r for these data is .82, a very close agreement

EXAMPLE 77. *continued*

B. Demonstration of some relationships between C and r

1. The following hypothetical distribution has $r = +1$

RANKS		RANKS		DIFFERENCES IN RANK	
X	Y	X	Y	d	d^2
1	1	1	1	0	0
2	2	2	2	0	0
4	4	3	3	0	0

$$\sum d^2 = 0$$

$$C = 1 - \frac{6 \times 0}{3(9-1)} = 1 - 0 = 1$$

When $r = +1$, $C = +1$.

2. The following has $r = -1$

RANKS		RANKS		DIFFERENCES IN RANK	
X	Y	X	Y	d	d^2
2	6	1	3	2	4
4	5	2	2	0	0
6	4	3	1	2	4

$$\sum d^2 = 8$$

$$C = 1 - \frac{6 \times 8}{3(9-1)} = 1 - 2 = -1$$

When $r = -1$, $C = -1$.

3. The following has $r = 0$

RANKS		RANKS		DIFFERENCES IN RANK	
X	Y	X	Y	d	d^2
1	1	1.5	1.5	0	0
1	3	1.5	3.5	2	4
3	1	3.5	1.5	2	4
3	3	3.5	3.5	0	0

$$\sum d^2 = 8$$

$$C = 1 - \frac{6 \times 8}{4(16-1)} = 1 - .8 = .2$$

When $r = 0$, C is small but is not necessarily 0.

EXAMPLE 77. *continued*

4. The following has $C = +1$

X	Y	d_X	d_Y	$d_X d_Y$	d_X^2	d_Y^2
1	1	−1.33	−1	1.33	1.7689	1
2	2	−0.33	0	0	0	0
4	3	1.66	1	1.66	2.7556	1
				$\sum d_X d_Y = 2.99$	$\sum d_X^2 = 4.5245$	$2 = \sum d_Y^2$

$\overline{X} = 2.33$

$\overline{Y} = 2.00$

$r = 2.99/\sqrt{4.5245 \times 2} = .99$

When $C = +1$, r is large but is not necessarily $+1$. (When $C = -1$, r has a large negative value but is not necessarily -1.)

CHAPTER TWELVE

The Analysis of Variance

In the preceding chapters we have developed what might be called the basic statistical repertory for a zoologist. It is possible with these methods to estimate various parameters of populations, to make a judicious guess as to the range in which these parameters are likely to lie, to find the dependence of one measurement upon another, and to test whether two populations differ in a more than trivial way in the value of a particular parameter. While these matters cover much of the gamut of problems in which zoologists are interested, they have two restrictions that make them inadequate for solution of further problems of frequent occurrence and great importance in zoology. First, the methods so far discussed deal only with the comparison of two populations, whereas in practice there are often three or more groups to be compared. Second, the assumption is inherent that the two populations compared are homogeneous within themselves in respects pertinent to the problem, and that they differ from each other only in a single factor.

It is more usual that a number of factors affecting the measured individuals are simultaneously varying. A sample of animals may be derived from several geographic localities and at the same time contain both sexes, animals of different ages, specimens collected at different seasons, and so on. Each of these factors may contribute to the differences among the specimens, and some measure of their relative importance is obviously desirable.

What is required, then, is a technique that is capable of testing simultaneously the differences among several populations and of assigning to the various factors which affect the measured character their relative roles. The *analysis of variance* is such a technique. No new statistical concepts are required; the basic repertory already discussed supplies the various tests and concepts contained in the analysis of variance. This technique is simply a new way of looking at the old problems so that more generality is introduced.

The Meaning of the Analysis of Variance

The value of any measurement of an animal or group of animals is a unique result of the joint action of many environmental factors with each other and with the heredity of the organism. In this sense, every organism is unique, for it is inconceivable that exactly the same interplay of heredity and environment could be duplicated in nature. Not only is the result of any combination of forces on the animal unique, but it is unpredictable as well. All other things being equal, a rise of ten degrees in temperature may raise the metabolic rate of an animal by a given amount; but all other things are never equal and the effect of a ten-degree rise in temperature may be quite different when combined with other variable forces acting on the animal. This principle of unpredictability or *uniqueness of the environment*, if pursued to its obvious end, prevents the biologist from assigning to any single factor in the environment, a relative role in the determination of observable phenomena. In this sense, all factors are equally important and equally confounded with each other in determining the morphology and physiology of an animal, and any attempt to separate the effects of these forces is biologically unrealistic.

In a more restricted sense, however, various determining factors do affect an organism to different degrees, and the relative importance of each variable is a valid object of biological inquiry. The observed variation among animals is due both to variation in their genes and in the environments in which the organisms have developed. Great variation in some factors of the environment will result in great morphological diversity, while just as great variation in some other environmental variables may have no detectable effect upon the differences among animals.

The difference between these two senses of "the relative importance of factors" in determining the morphology of an animal can be easily illustrated by a hypothetical situation. Suppose that snout-to-vent length has been measured on ten live anesthetized frogs, and it is found that, within the limits of accuracy of the measuring instrument, all ten frogs have the same length. Each specimen is now killed and preserved in a different way; one in alcohol, one in formaldehyde, one pinned and dried, and so on. If, after a period of preservation, each specimen is again measured, there will be distinct differences in length from frog to frog as a result of the preservation processes. It is an obvious absurdity to say that the snout-to-vent length of any given specimen is due entirely to the method of preservation since this length has been determined during the development of the frog and only later has it been slightly altered by the zoologist. On the other hand, it is true that the observed *variation* in length from frog to frog is due to variation in preservation methods since the frogs were all of the same length when alive.

The purpose of the analysis of variance, as its name implies, is to determine how much of the variation among observations is due to *variation* in each factor influencing the character being studied. Even greater care must be exercised in the interpretation of results of the analysis of variance than for most other statistical techniques. The fact that a given factor like temperature cannot be shown to be the major source of variation in a particular sample is not in itself evidence that temperature plays no role in the determination of the character studied. It may show that, but it may show equally that the range of temperatures involved in the analysis was insufficient to produce as much variation in the character as did some other factor. As with any statistical methodology, the results of any analysis of variance are specific facts to be used cautiously as a basis for inductive generalization.

Effects and Interactions

On the supposition that a given value of a variate is the result of an analyzable set of factors, a series of simple models can be set up. Suppose that a sample of animals is measured and that this sample is derived from 10 different localities. The amount by which any one measurement deviates from the true mean of all measurements, may be symbolized as

$$(X - \mu) = A + e$$

X is a particular value of the variate.

μ is true mean of all the measurements in the population.

A is that portion of the deviation associated with the particular locality from which the individual came and hence presumably due to conditions peculiar to that locality,

and

e is the remainder of the deviation, not explained by the factor of locality.

Whereas the simple model is written in terms of deviations of the observed values from the population mean, μ, this mean is not known and the actual statistical manipulations must be done in terms of the sample mean, $\bar{X}$. In statistical terminology, the contribution of the locality to the deviation of the observations from the mean is known as the *effect* of locality. It must be remembered that under our model this effect is defined in terms of *deviations*. The term e, called the *error* deviation, is simply a potpourri of all the uncontrolled and undefined factors influencing the character other than the defined effect—locality, in this example. It is assumed that these errors occur in individuals from all localities without association with those localities, that is, that they are random with respect to locality.

A more complex case than the simple one-factor model is that in which the entire sample is heterogeneous with respect to two components. Suppose that the sample referred to above has been collected not only in different localities, but in different months of the year as well, each locality having been sampled every month. Then the individual measurement would be resolved as follows:

$$(X - \mu) = A + B + AB + e$$

where A and B are the effects of locality and month, respectively, while AB is that part of the deviation due to *interaction* between locality and month.

This new term, the interaction, appears when more than one factor is varying. It is the amount added to or subtracted from the basic value, arising from the particular and unique interaction of a given month with a given locality. For example, locality 5 may on the average have longer individuals than the other localities, and individuals collected in February might be larger on the average than those in other months, but it is entirely possible that individuals collected in February from locality 5 may be smaller than the average of other members of the sample. This would presumably be due to a unique interaction of the particular locality with the particular conditions during February. Some artificial examples will make this concept of interaction as opposed to main effect somewhat clearer. Examples 78A and 78B show hypothetical measurements on individuals taken from three localities at three different times of year, each locality having been sampled at each season. The entries in the body of the table are hypothetical values of some measurement, and it is assumed for the sake of simplicity that there is no random effect attached to them ($e = 0$). At the right and lower margins of each table are the means of the measurements for a given row or column. In Example 78A, all of these means are equal. On the average, there is no *effect* of season because the average measurements in each season as shown by the row means are equal to each other and to the grand mean. The same applies to locality, since the average measurement is equal for the three localities. There is a great difference among the individual entries in the table, however, cells AI and CIII having the lowest measurements, CI and AIII the highest. Thus, the differences are due entirely to the interaction of a given locality with a given season.

Example 78A is more than an artificial arithmetic trick, for it often happens that neither of two factors can be shown to have a main effect, yet a significant interaction will be present. This means simply that the two factors are not meaningful for the problem in hand. Suppose, for example, that the critical environmental variable in the determination of a character were rainfall. In temperate regions, precipitation is an extremely localized phenomenon and while two localities may have the same average yearly rainfall, during any particular part of the year the pattern of rainfall may differ markedly between localities. In such a case, there will be no average effect either of locality or time, but there will be a strong interaction of the two. Each locality at each time is a different environment and there is as much difference from locality to locality at a given time as there is from time to time in a given locality.

EXAMPLE 78. Hypothetical measurements on individuals from three different localities collected in three seasons, illustrating the difference between interaction effects (*A*) and main effects (*B*).

A SEASON	LOCALITY A	B	C	$\bar{X}$	*B* SEASON	LOCALITY A	B	C	$\bar{X}$
I	1	2	3	2	I	1	2	3	2
II	2	2	2	2	II	4	5	6	5
III	3	2	1	2	III	7	8	9	8
$\bar{X}$	2	2	2	2	$\bar{X}$	4	5	6	5

Example 78B is in direct contrast to this picture. There is a difference among the three seasons as well as among the three localities as shown by the marginal means. Moreover there is no interaction. The lowest measurement is in cell AI which coincides with the season and locality having the lowest means. The highest cell value is CIII which coincides with the highest locality and season. Not only are the rough magnitudes of the individual measurements in exact accord with the magnitudes of the locality and season from which they were samples, but also the numerical values are completely explained by the main effects. Thus cell AI has a measurement of 1, which deviates from the grand mean of all the measurements by 4 units. This can be entirely attributed to the first season's deviation of 3 from the grand mean, plus the first locality's deviation of 1. In symbolic form,

$$(X - \mu) = A + B + AB$$
$$4 = 3 + 1 + 0$$

so that the main effect of season I is 3 and that of locality A is 1 with no interaction (*AB*) between seasons and locality.

It is inconceivable that in nature the value of a variate can be entirely attributed to the average effect of two factors like locality and season, but there are many cases in which this is so nearly true as to be an acceptable working picture.

If there are three factors affecting a variate—say, locality, season, and year—then the symbolic representation of a measurement would be

$$(X - \mu) = A + B + C + AB + BC + AC + ABC + e$$

where A, B, and C are the *main effects* of the factors; AB, AC, and BC are the *first-order interactions* taken two at a time; and ABC is the *second-order* interaction denoting that part of the measurement which is ascribable to the interaction of all three factors acting jointly. This scheme may be generalized for any number of factors with similar results. For example, a four-factor analysis would contain the four main effects (A, B, C, and D), six first-order interactions comprising all possible combinations of the four factors taken two at a time, three second-order interactions made up of all the combinations of the four factors taken three at a time, and finally one third-order interaction term representing the joint effects of all four factors.

In discussing the one-, two-, and three-factor analyses, no mention was made of the number of classes represented in each factor. For example, there may be 2 or 20 localities, 2 or 12 months, 5 or 50 years represented. The type of analysis is in no way governed by the number of these classes but only by the number of factors. In the jargon of the analysis of variance, the classes are often spoken of as *levels*. Thus, January, February, and March are 3 levels of the factor months. This term is a carry-over from the agricultural origin of the analysis of variance, where varying amounts of fertilizer were applied to some crop. As the agronomist often talks about "levels of fertilization," the term "levels" has become generally synonymous with classes in a factor.

Another such carry-over from agronomy, one that we will not use in this book but which is often encountered in other texts, is the use of *treatments* as synonymous with factors. Thus, in agricultural literature, phosphate fertilizer and insecticide application would be two treatments (factors) and in each there might be several levels.

Level is in such common usage that we will adopt it for our discussion, but it must be understood as used in a technical sense, so that five different species are 5 levels of the factor species, just as 10, 20, 30, 40, and 50 lbs. of phosphate would be levels of fertilizer to an agronomist.

Random and Fixed Factors

The two problems to be solved by the analysis of variance are: first, does a given factor like locality or temperature contribute to the observed variability of the observations, and, second, if a factor does contribute to this variability, how important is it relative to other factors? The first of these questions is a problem in testing null hypotheses. If a factor has no effect, then all of the deviations symbolized by A or B in the model

$$(X - \mu) = A + B + AB + e$$

will be zero. The null hypothesis is basically, then, that $A_1 = A_2 = A_3 = \cdots A_n = 0$ for n levels of the factor.

The second question asked, on the other hand, is one of estimation. If all the A's are not zero, then the variation among them contributes some amount to the total variation, an amount which one would like to estimate.

The precise methodology of testing the null hypothesis and the possibility of estimating the contribution of the factors to the total variation depend to some extent upon the way in which the various levels of a factor are related to each other. One possibility is that the particular classes or levels chosen for the analysis are fixed by the problem itself. For example, if four species were being compared, it is these four entities which are the object of inquiry and not simply any group of four species. In such a case, the null hypothesis is that the effects $A_1 = A_2 = A_3 = A_4 = 0$. On the other hand, in some analyses the identity of the levels within a factor are not of primary concern. They are simply a random sample of all the possible levels, random at least with respect to the character measured. This is usually the case with years, where the particular years in which the sampling takes place are chosen by convenience rather than any criterion directly related to the problem in hand. In this case, the question asked is not whether 1954, 1955, and 1956, specifically, have different effects, but whether years *in general* make a difference in the character measured. A is then an estimate of year effect derived from random samples from some probability distribution with a mean α, say, and a variance σ_A^2. The null hypothesis, that there is no differential effect of years, is equivalent to the hypothesis $\sigma_A^2 = 0$, since if $\sigma_A^2 = 0$ all the A's must be equal. Moreover, if the null hypothesis is rejected and there is indeed some effect of years, it is possible to estimate σ_A^2 and compare it with σ_B^2, the variance associated with some other factor in the analysis. In this way the relative importance of the various factors affecting the organisms can be measured.

The first situation in which the identity of the levels is of prime importance is the *fixed model* of the analysis of variance, and a factor whose levels are related in this way is a *fixed factor*. The second situation is the *random model* based upon *random factors*. It may occur, and often does, that an analysis involves both a random factor and a fixed factor—for example, if five species were collected in each of three years. This *mixed model* will not introduce any serious complication into the method.

It is difficult to make a rule of thumb for deciding whether a factor is fixed or random, but the closest approximation might be to ask the question, "If the observations were to be made several times, would the biology of the problem force the same choice of levels each time, or would any set of levels answer?" If the choice of levels is fixed by the problem, the factor is a fixed one; if the choice of levels is arbitrary, it is a random factor.

The sorts of factors of interest to a zoologist are fairly restricted and they can be easily catalogued with respect to the type of model involved.

Time. 1. YEARS. Usually a random factor, chosen mostly at the convenience of the investigator rather than for any reason inherent in the investigation. Occasionally there may be some peculiarity of a set of years, such as unusual climatic or biotic conditions not found at other times, but even in such case any set of years in which these occur would suffice for the problem. Actually, if years are chosen with respect to such variables, it is not time but these others which are the factors, and they are fixed factors. The zoologist must always be wary of testing differences due to one factor under the guise of another extraneous variable.

2. MONTHS. Usually fixed as are other cyclic time factors. While it is conceivable that for some purpose one sample of months, say, January, March, and July, might be an equivalent random sample to another set of three, in most instances the differences in animal populations from one month to another are reflective of cyclical seasonal variation unique for each month. The pertinent factor is then seasonal, and months are used only as formal subdivisions of seasons. The seasons, as such, may of course also be used and are fixed factors.

3. SMALLER TIME INTERVALS. Random factors when the number of levels is small in comparison to the available number of levels within larger cyclical time units. For example, any three days in a given month may be regarded as equivalent random samples with respect to cyclical temperature changes, but the 60 days from January to March are not random levels from the same population as 60 days from March to May, for a wild population in a temperate region.

4. GEOLOGICAL HORIZONS. Clearly a fixed-level factor.

Locality. Geographical localities are random levels, if the choice of localities from which to take samples is only a matter of convenience and representativeness. In such circumstance, many possible sets of localities can be chosen which are equally representative. In testing altitudinal or geographical clines, however, localities are at fixed levels, the concern being not with a general effect of different localities but with effect due to clinal change or regression on geographical position. This is again a case of obscuring the real factor, altitude or distance, with a superficial one, locality.

Environment. Environmental factors are almost always fixed-level factors. Those involved in various zoological problems are extremely diverse: climate, chemical composition (e.g., of environmental soil or water), exposure to light, color or texture of substrate, and many others. Apparent random-level effects of locality are usually really due to fixed-level effects of environmental factors, and it is usually much more enlightening to analyze variances in terms of the latter rather than in terms of locality alone. Random-level analysis by locality may indicate whether there are likely to be fixed-level environmental effects and may thus be a useful preliminary step in attempting to identify pertinent environmental factors.

Physiological factors. Age, sex, and other morphological and physiological factors which may be specified are all fixed factors. That sex is fixed is obvious. Age, in general, is of interest from the standpoint of regression, and any factor whose levels are serially arranged as in a regression analysis must be considered fixed.

Taxonomic categories. In nearly every case, various taxonomic units are at fixed levels. The question is generally whether a number of given genera, species, or subspecies differ with respect to a measurement. The identity of the groups is of prime concern.

Techniques. It is sometimes of value to determine whether different observers vary in their measurements. Observers are random factors since one set of observers will differ in their measurements about as much as any other set. Sampling methods, on the other hand, are clearly fixed-level factors, since the interest lies in whether there is a difference among several specific techniques.

This is not an exhaustive list of the possible factors which might enter into a zoological sample to which the analysis of variance is applied, but it provides an idea of what is meant by fixed and random factors.

The Fixed Model

One-factor designs. Example 79 illustrates the simplest set of observations for which the analysis of variance is useful. There is only a single factor which appears at a number of levels, and within each level a number of specimens has been measured. For concreteness, suppose the factor to be species and the number of levels c. Within each of the c species, the tail length of n_i animals has been measured. It is not necessary that the number of observations, n_i, be the same for all species, but the analysis is somewhat easier to compute when this equality holds. Each entry in the body of the table is a single measurement, and it is denoted by X_{ij}. The subscripts refer to the level and the observation in the level, respectively. Thus X_{13} is the value of the third observation in the first species, X_{36} the sixth observation in the third species, and so on, so that in general X_{ij} is the jth observation in the ith level. At the bottom of the table are listed the means for each level denoted by $\bar{X}_i$. The last symbol to be introduced is the grand mean of all the observations, $\bar{X}$.

EXAMPLE 79. The form of observations for a one-factor analysis of variance. The factor is species and each entry in the body of the table is a single measured individual.

		SPECIES		
1	2	3	4 ... c	
X_{11}	X_{21}	X_{31}	X_{41}	
X_{12}	X_{22}	X_{32}	X_{42}	
X_{13}	X_{23}	X_{33}	X_{43}	
X_{14}	X_{24}	X_{34}		
	X_{25}	X_{35}		
	X_{26}	X_{36}		
$\bar{X}_1$	$\bar{X}_2$	$\bar{X}_3$	$\bar{X}_4 \ldots \bar{X}_c$	$\bar{X}$

The sample means, $\bar{X}_i$, are estimates of the true means μ_i of each species or level, and the problem to be solved by the analysis is whether there is any real difference between these true means. Even if the μ_i are all equal, the $\bar{X}_i$ will differ because of sampling deviations. Under the null hypothesis that $\mu_1 = \mu_2 = \mu_3 = \ldots$, all of the levels are really samples from a single population with some mean μ. If one chooses a sample of size N from a population, the mean of that sample will not be equal exactly to the true mean, and if several samples are taken, the means of the samples will differ from each other and from the true mean because of simple sampling deviations. We have discussed this point in great detail in Chapter 7. The problem of the analysis of variance, then, is to determine whether the observed variation in the $\bar{X}_i$ could be entirely ascribed to random sampling deviations or whether it is reflective of a real difference in the means μ_i of each level. This problem is exactly analogous to that of testing the equality of two means by Student's t-test.

The method used for determining whether the differences among the $\bar{X}_i$ are due solely to random sampling deviations is basically the same for all analyses of variance, whether one-factor or five-factor, fixed or random model.

The total variation among the observations is broken up into several parts, each arising from a different source of variability. The partitioning is so arranged that one of the components of variation is that due only to random sampling error within each level or combination of levels. The other components are such that they will be equal to the random sampling component if the null hypothesis is true but larger if there is a real effect of the factor or factors. The entire analysis of variance, then, reduces to the calculation of statistics which should be equal under the null hypothesis of no effect. The test of the null hypothesis becomes a test of the equality of two calculated quantities. If these quantities are chosen judiciously, procedures will already exist for these tests. In a sense the analysis of variance reduces the problem of testing the equality of a *number* of means to the problem of testing the equality of *two* quantities.

In every analysis, the total variation of the observations is measured by the *total sum of squares*, which is simply the sum of the squared deviations of all the observations from the grand mean. In symbolic form:

$$\text{Total sum of squares} = \sum_{ij}(X_{ij} - \bar{X})^2$$

for the one-factor case. The subscripts i and j under the summation sign mean that the deviations are taken for every i and every j, or, in other words, for every observation. This total sum of squares is then broken up into a number of components such that one of them is the sum of squares of deviations to be ascribed to random variation and the others are sums of squares which will be dependent upon the null hypothesis.

For the simple one-factor case, the partition is obviously

total sum of squares = sum of squares associated with the main effect of factor + sum of squares due to random deviations,

which algebraically is

$$\sum_{ij}(X_{ij} - \bar{X})^2 = \sum_{i} n_i (\bar{X}_i - \bar{X})^2 + \sum_{i}\left[\sum_{j}(X_{ij} - \bar{X}_i)^2\right]$$

The first term on the right-hand side is simply the sum of the squared deviations of each level mean from the grand mean, each squared deviation being multiplied by the number of measurements taken in that level. The second term is calculated by first summing up the squared deviations of each measurement from its own level mean and then summing these sums over all levels. The resulting values are usually put in a tabular form illustrated by Table 6.

Each component should be calculated by the appropriate formula, or an equivalent calculating formula which will be given shortly, and entered into the second column of the table. A check on the correctness of calculation is automatic because the values calculated for the main effects and deviations must add up to that calculated for the total sum of squares. One should never calculate just the main effects and totals, hoping to find the deviation sum of squares by subtraction of the first from the second. This will make it impossible to check the accuracy of the calculation. It is best to make use of all self-correcting devices available in a complex calculation.

TABLE 6. Analysis of variance table for a one-factor design.

SOURCE	SUM OF SQUARES	DEGREES OF FREEDOM	MEAN SQUARE	F
Main effect	$\sum_i n_i(\bar{X}_i - \bar{X})^2$	$c-1$	$\dfrac{1}{c-1}\sum_i n_i(\bar{X}_i - \bar{X})^2$	$\dfrac{\text{M.S.}_1}{\text{M.S.}_2}$
Sampling deviations	$\sum_i\left[\sum_j(X_{ij}-\bar{X}_i)^2\right]$	$N-c$	$\dfrac{1}{N-c}\sum_i\left[\sum_j(X_{ij}-\bar{X}_i)^2\right]$	
TOTAL	$\sum_{ij}(X_{ij}-\bar{X})^2$	$N-1$		

The third column of the table contains the degrees of freedom for each sum of squares. If there is a total of N observations in the entire analysis, the degrees of freedom for the total sum of squares is $N-1$. Denoting the number of levels of the main effect as c, there are $c-1$ degrees of freedom for the main effect sum of squares. At each level there are $n_i - 1$ degrees of freedom for the random deviations, so that the total degrees of freedom for this component is the sum of these, or $N-c$. Like the sums of squares, the degrees of freedom for the two components must add to the number for the total sum of squares.

The fourth column of the table for the analysis contains the *mean squares* for each component, which are simply calculated by dividing each sum of squares by its appropriate degrees of freedom. These mean squares are analogous to variances since they are the sum of squared deviations divided by degrees of freedom. In a rough sense, the deviation mean square is an estimate of the variance due to random deviations, while the main effects mean square estimates the variability contributed both by true differences between the means and by random deviations.

If there is no real effect due to the factor, then the mean square for main effect should be equal to the mean square due to deviations. If there is some effect of the factor, this main effect square will be larger than the deviation mean square. The test for the existence of a main effect of the factor is, then, simply a test of the equality of two mean squares which are both a kind of variance under the null hypothesis. In Chapter 10 it was shown that the equality of two variances is tested by their ratio,

$$\frac{s_1^2}{s_2^2} = F$$

with n_1 degrees of freedom in the numerator and n_2 in the denominator.

To test for the existence of a real effect of the factor, the ratio

$$F = \frac{\text{Mean square for main effect}}{\text{Mean square for deviations}}$$

is calculated, and the result referred to the table of the F distribution with $(c - 1)$ degrees of freedom in the numerator and $(N - c)$ in the denominator. If F is quite large, the probability in the table will be small. If 5 per cent has been chosen as the required level of significance, an F so large as to have a probability of less than .05 requires the rejection of the null hypothesis. One would then conclude that the means for the various levels are not equal, or that there is a real effect due to the factor under investigation. On the other hand, should F be close to unity, the corresponding probability in the table will be high, the null hypothesis will not be rejected, and there will be no evidence for an effect of the factor.

The appropriate F ratio for testing the main effect is shown in the last column of the analysis of variance table.

While the sum of squares and the degrees of freedom for the sums of squares must add up to those for the total, this is not true for the mean squares. The mean squares for main effects plus the mean square for deviations will not in general add up to the mean square for total deviations, and as this last mean square does not enter into the analysis, there is no purpose in calculating it.

Calculation. The calculation of the sums of squares is similar to that for variances, in that it involves squaring deviations from a mean and summing these squared deviations.

Precisely the same simplification of calculation procedure can be made for the analysis of variance as for any variance—by eliminating deviations entirely and only operating on the raw observations. The calculating formulae for the three basic quantities in the one-factor analysis are as follows:

$$\text{Total sum of squares} = \sum_{ij} X_{ij}^2 - \frac{\left(\sum_{ij} X_{ij}\right)^2}{N}$$

$$\text{Main effect sum of squares} = \sum_i \frac{T_i^2}{n_i} - \frac{\left(\sum_{ij} X_{ij}\right)^2}{N}$$

Deviation sum of squares $= \sum_{ij} X_{ij}^2 - \sum_i \frac{T_i^2}{n_i}$

where

$T_i =$ total value of observations in the ith level
$n_i =$ number of observations in the ith level
$N =$ grand total number of observations

and the other symbols are defined in the usual way.

These calculation formulae have the advantage of completely eliminating deviations in actual computation, thus saving a step which is not only time consuming but also conducive to error.

To demonstrate the calculation of a one-factor analysis of variance, Example 80 has been constructed from Marien's observations on starlings from India, illustrating a common use of the analysis of variance—the comparison of measurements from a number of taxonomic entities. It is not implied that the lack of a significant difference among these entities in regard to a given character is a demonstration that they should be lumped into a single species. Neither would the demonstration of a difference force one to maintain them as distinct taxonomic units. That populations differ with respect to a given variate is a single datum which contributes to a decision in systematics, but it is not a substitute for that decision.

By inspection of the means of the three species it would appear that *Sturnus ginginiamus* and *S. fuscus* are nearly identical, while *S. contra* is considerably smaller than either. The analysis of variance is not capable of showing which differences are significant but only that the three species differ, and this is one of the drawbacks of the entire methodology. While it will test the homogeneity of a number of groups it will not, in itself, permit any inference about which group or groups may be out of line with the rest. This can only be accomplished by individual t-tests for the differences between pairs of populations, although often, as in this case, the main or perhaps the sole source of the statistical significance may be obvious on inspection.

The Analysis of Variance in Regression

The regression problem and the analysis of variance for a single fixed factor are very closely related. To say that some variate Y shows a significant regression on a deliberately chosen variate X is another way of saying that the various levels of X in some way effect the value of Y. If there is a rectilinear regression of Y on X, there is an effect of X on Y in the sense of the analysis of variance. The reverse is not true, however. An effect of the level of X on the value of Y does not imply a linear regression

EXAMPLE 80. Wing length in males of three species of starling *Sturnus*, from India. Measurements in millimeters. (Data from Marien, 1950)

Sturnus contra	*Sturnus ginginiamus*	*Sturnus fuscus*
120	123	122
120	124	122
121	125	125
122	125	127
122	126	127
122	127	127
123	127	128
125	127	129
125	128	
126	128	
126	129	
	129	

$T_1 = 1{,}352$ $T_2 = 1{,}518$ $T_3 = 1{,}007$
$n_1 = 11$ $n_2 = 12$ $n_3 = 8$
$\bar{X}_1 = 122.9$ $\bar{X}_2 = 126.5$ $\bar{X}_3 = 125.9$
$N = n_1 + n_2 + n_3 = 31$ $\sum_{ij} X_{ij} = T_1 + T_2 + T_3 = 3{,}877$

TOTAL SUM OF SQUARES

$$\sum_{ij} X_{ij}^2 - \frac{\left(\sum_{ij} X_{ij}\right)^2}{N} = 120^2 + \cdots + 218^2 + 129^2 - \frac{3{,}877^2}{31} = 485{,}097 - 484{,}875 = 222$$

MAIN EFFECT

$$\sum_i \frac{T_i^2}{n_i} - \frac{\left(\sum_{ij} X_{ij}\right)^2}{N} = \frac{1{,}352^2}{11} + \frac{1{,}518^2}{12} + \frac{1{,}007^2}{8} - \frac{3{,}877^2}{31}$$
$$= 484{,}956 - 484{,}875 = 81$$

DEVIATIONS

$$\sum_{ij} X_{ij}^2 - \sum_i \frac{T_i^2}{n_i} = 485{,}097 - 484{,}956 = 141$$

ANALYSIS OF VARIANCE

SOURCE	SUM OF SQUARES	D.F.	MEAN SQUARE	F
MAIN EFFECT	81	2	40.5	$\frac{40.5}{5.0} = 8.1$
DEVIATIONS	141	28	5.0	
TOTAL	222	30		

$F = 8.1$ with 2 degrees of freedom in the numerator and 28 in the denominator. The probability is less than .005.

because this effect may be of many sorts. Thus as X increases, Y may increase for a time, but at higher levels of X, Y may in fact decrease. The regression of Y on X is not rectilinear, but there is an effect of the factor X. Moreover, if the factor is a taxonomic category or a method of sampling, the concept of regression becomes absurd. What does it mean to say that tail length shows a regression on species? Regression analysis, then, is a special case of the analysis of variance, and a significant regression coefficient means that there is a special kind of effect of the independent factor.

The total variation of Y, the dependent variate in regression, is due to a number of causes each with its associated sum of squares. First, the values of Y for a given X will vary around their mean. This is simply the random deviation sum of squares analogous to that in a single-factor analysis. Second, the mean values of Y for each X will vary from the grand mean of Y because of the regression itself. If the regression is positive, values of Y associated with small values of X will be less than the average Y, and values of Y associated with larger value of X will be greater than the average Y. Finally, the average Y for each X may deviate from the theoretical regression line, because of either random deviations or the fact that the points do not really have a rectilinear regression. Symbolically,

> Total sum of squares of Y = sum of squares due to random deviations of individual Y values in each level of X + deviation of each level mean from grand mean due to the regression + deviations of each level mean from the expected regression line.

Or

$$\sum_{ij}(Y_{ij} - \bar{Y})^2 = \sum_i \left[\sum_j (Y_{ij} - \bar{Y}_i)^2\right] + b^2 \sum_i n_i (X_i - \bar{X})^2 + \sum_i n_i (\bar{Y}_i - a - b X_i)^2$$

where

Y_{ij} = jth observation at the ith level of X
$\bar{Y}$ = grand mean of all observations
$\bar{Y}_i$ = mean of the observations in the ith level of X
b = estimated regression coefficient
a = estimated Y intercept of regression line
X_i = ith value of X
$\bar{X}$ = mean of all X values
and n_i = number of observations in the ith level of the factor

The first term on the right-hand side of the expression is obviously the sum of squared deviations of each observed Y from the mean of Y at a particular level of X. It is the same formula as the random deviation component in the one-factor analysis. The second factor does not appear to contain Y explicitly but does so implicitly in b, the regression coefficient. It could be rewritten as

$$\sum_i (bX_i - b\bar{X})^2$$

and since bX_i is the theoretical value of Y_i and $b\bar{X}$ the theoretical value of $\bar{Y}$, if there is a true regression coefficient of value b, this term is simply the sum of the squared deviations of the predicted values of Y from the predicted mean. This is what is meant by the sum of squares due to regression. The last term is obviously derived from the deviations of each $\bar{Y}_i$ from what it should be if the points fitted the regression line $Y = a + bX$ perfectly.

These components may be put in the form of Table 7 with the degrees of freedom and mean squares.

TABLE 7. Analysis of variance in regression.

SOURCE	SUM OF SQUARES	DEGREES of FREEDOM	MEAN SQUARE
Within levels of X	$\sum_i \left[\sum_j (Y_{ij} - \bar{Y}_i)^2 \right] = $ S.S.$_1$	$N-k$	S.S.$_1/(N-k)$
Regression	$b^2 \sum n_i (X_i - \bar{X})^2 = $ S.S.$_2$	1	S.S.$_2$
Deviations from regression	$\sum_i n_i (\bar{Y}_i - a - bX_i)^2 = $ S.S.$_3$	$k-2$	S.S.$_3/(k-2)$
TOTAL	$\sum_{ij} (Y_{ij} - \bar{Y})^2$	$N-1$	

The regression sum of squares *always has one degree of freedom* no matter at how many levels X appears. The sum of squares within levels has $N - k$ degrees of freedom where N is the total number of observations and k the number of levels of the factor. This is the same as for the one-factor analysis of variance. The deviations from the regression have $k - 2$ degrees of freedom, which is one less than the equivalent component (main effect) in the one-factor analysis. This lost degree of freedom is the one used up in the regression sum of squares.

With the mean squares in Table 7, it is possible to test two hypotheses about the observations. The first concerns the rectilinear regression coefficient, b. If there is any upward or downward trend in the observations, b will be different from zero even if this trend is not exactly rectilinear. That is, even in a curvilinear regression, there usually is some rectilinear component, measured by b, indicating a generally rectilinear trend. If $b = 0$, there is no such trend, which means, in effect, that the sum of squares due to regression will be entirely the result of deviations from a regression line with coefficient zero. Then, the F ratio,

$$F = \frac{\text{Mean square due to regression}}{\text{Mean square due to deviations from regression}}$$

or

$$F = \frac{(k-2) \left[b^2 \sum_i n_i (X_i - \bar{X})^2 \right]}{\sum_i n_i (\bar{Y}_i - a - bX_i)^2}$$

with 1 degree of freedom in the numerator and $k - 2$ in the denominator, is a test for the existence of a regression coefficient b different from zero. If the F ratio is so large as to make its probability very small, we reject the null hypothesis that $b = 0$ and conclude that there is a significant regression.

The second hypothesis which is testable from this analysis of variance is that of rectilinearity. The problem of whether the regression is exactly of the form

$$Y = a + bX$$

is an entirely separate consideration from that of whether b is different from zero, and it can be tested by a different F ratio. The null hypothesis is that no systematic deviations exist from the rectilinear regression line. If there are no systematic deviations, then the mean square due to deviations from the line ought to be entirely accounted for by the random deviations within each level of X. If these random deviations within each level are not sufficient to account for the deviations of $\bar{Y}_i$ from the rectilinear regression line, the null hypothesis must be rejected and the conclusion made that there is a nonrectilinear trend. The F ratio to test rectilinearity is, from these considerations,

$$F = \frac{\text{Mean square due to deviations about regression}}{\text{Mean square within levels of } X}$$

or

$$F = \frac{\sum_i n_i (\bar{Y}_i - a - bX_i)^2 (N - k)}{\sum_{ij} (Y_{ij} - \bar{Y}_i)^2 (k - 2)}$$

with $k - 2$ degrees of freedom in the numerator and $N - k$ in the denominator. If the F is too large as judged by its probability, the null hypothesis is rejected and it is concluded that there is a significant nonrectilinear component in the regression line.

276 QUANTITATIVE ZOOLOGY

If there is only a single Y observation for each level of X, it is not possible to test for rectilinearity since the mean square within each level of X is zero. It is not necessary to rewrite the analysis of variance for this special case, the same formulae applying exactly. In such a case, the first entry in the table will be zero and n_i which appears in the other entries is equal to one. It is unfortunate and a little paradoxical that when only one observation is made for each value of X, even the most obviously curvilinear relation cannot be tested for this curvilinearity, but this is a consequence of the entire model of the analysis of variance.

Calculation.   All of the sum of squares components can be reduced to simple machine calculating formulae. These formulae are

WITHIN LEVELS:
$$\sum_{ij} Y_{ij}^2 - \sum_i \frac{T_{Y_i}^2}{n_i}$$

REGRESSION:
$$b^2 \sum_i n_i X_i^2 - b^2 \frac{\left(\sum_i n_i X_i\right)^2}{N}$$

DEVIATIONS FROM REGRESSION:
$$\sum_i \frac{T_{Y_i}^2}{n_i} - b^2 \sum_i n_i X_i^2 - \frac{T_Y^2}{N} + b^2 \frac{\left(\sum_i n_i X_i\right)^2}{N}$$

TOTAL:
$$\sum_{ij} Y_{ij}^2 - \frac{T_Y^2}{N}$$

The data of Example 81 are presented to illustrate the calculation of this analysis in a typical regression situation. The values in the table are the total lengths in millimeters of smallmouth bass in four age groups. The factor X is the age, which occurs at four fixed levels, and within each level there are a number of observations of Y, the total length. There is one year difference between successive age groups, starting with year-old individuals.

To test the significance of the regression coefficient, the appropriate F ratio is
$$F = \frac{18{,}959.54}{343.50} = 55.20$$

with 1 degree of freedom in the numerator and 2 in the denominator. By reference to the F table, this value of F is seen to correspond to a probability of about .01. The conclusion must be that the regression is significant.

The test for rectilinearity is made by F ratio, as follows:
$$F = \frac{343.50}{677.59} = .507$$

with 2 and 15 degrees of freedom in the numerator and denominator, respectively. Again, there is no need to consult the F table, since this ratio is smaller than unity. The only reason for rejection of the null hypothesis would be an F larger than one, whereas $F = .507$ corresponds to a probability greater than .50. There is then no evidence for a curvilinear regression in these data.

Two-factor designs. The presence of two variable factors introduces two new complications into the analysis of variance, one of which has already been discussed under the general topic of *interaction*. Not only will there be a sum of squares due to the main effects of each factor and the random variation within each level of each factor, but the unique interaction of level a of one factor with level b of the second factor will also contribute its own component to the sum of squares.

The second complication involves the necessity of understanding the principle of the *factorial design*. Simply because two factors are known to be varying in a given sample does not mean that the model of the two-factor design applies. What is essential in the factorial model of the analysis of variance is that *every level of one factor appear with every level of the other factor* in gathering the data. For example, if three localities are sampled in each of three years, then data must be gathered in each locality every year. On the other hand, if locality 1 were sampled in 1951, locality 2 in 1952, and locality 3 in 1953, it would be impossible to make a two-factor analysis of variance. Nor would the factorial model be any more applicable to sampling locality 1 in 1951, 1952, and 1953, locality 2 in 1954, 1955, and 1956, and locality 3 in 1957, 1958, and 1959. It is not the number of levels but their identity which determines the application of the model. Nonfactorial models are special cases of the one-factor analysis of variance and they will be discussed separately under hierarchical designs (see page 298). All the discussion of two- and three-factor designs below refers strictly to factorial models.

For a two-factor design, the total sum of squares is partitioned into four components.

Total sum of squares =

    Sum of squares due to main effect of factor A

    + Sum of squares due to main effect of factor B

    + Sum of squares due to interaction of A and B

    + Sum of squares due to random variation within each combination of factor A and factor B

The data for such an analysis can be put into the form presented in

EXAMPLE 81. Total length in millimeters of smallmouth bass in four age groups, from the Merrimac River, Missouri. A random sample of a more extensive set of original observations by E. M. Lowry.

AGE GROUPS (X)

LENGTH (Y)	1 YEAR	2 YEARS	3 YEARS	4 YEARS
	139.7	139.7	203.2	241.3
	127.0	215.9	241.3	247.7
	133.4	171.5	209.6	235.0
	177.8	152.4	215.9	
		228.6	190.5	
		190.5		
		149.2		
	$n_1 = 4$	$n_2 = 7$	$n_3 = 5$	$n_4 = 3$
	$T_1 = 577.9$	$T_2 = 1{,}247.8$	$T_3 = 1{,}060.5$	$T_4 = 724.0$
	$\bar{Y}_1 = 144.5$	$\bar{Y}_2 = 178.3$	$\bar{Y}_3 = 212.1$	$\bar{Y}_4 = 241.3$

To calculate the regression, the following formula is used:

$$b_{YX} = \frac{\sum XY - \dfrac{T_X T_Y}{N}}{\sum X^2 - \dfrac{T_X^2}{N}}$$

$N = n_1 + n_2 + n_3 + n_4 = 19$

$T_Y = T_1 + T_2 + T_3 + T_4 = 3{,}610.2$

$T_X = n_1 X_1 + n_2 X_2 + n_3 X_3 + n_4 X_4 = 45$

$$b_{YX} = \frac{9151.0 - \dfrac{(45)(3{,}610.2)}{19}}{125 - \dfrac{2{,}025}{19}} = \frac{590.6}{18.4} = 32.10$$

$$a_Y = \bar{Y} - b\bar{X} = \frac{3{,}610.2}{19} - 32.10\left(\frac{45}{19}\right) = 113.98$$

REGRESSION EQUATION

$Y = 113.98 + 32.10 X$

EXAMPLE **81.** *continued*

ANALYSIS OF VARIANCE FOR REGRESSION

WITHIN LEVELS

$$\sum_{ij} Y_{ij}^2 - \sum_i \frac{T_{Yi}^2}{n_i} = 139.7^2 + 127.0^2 + \ldots + 247.7^2 + 235.0^2$$

$$- \frac{577.9^2}{4} - \ldots - \frac{724.0^2}{3} = 715{,}786.38 - 705{,}622.54 = 10{,}163.8$$

REGRESSION

$$b^2 \sum_i n_i X_i^2 - b^2 \frac{\left(\sum_i n_i X_i\right)^2}{N} = 32.10^2 \left[4(1)^2 + \ldots + 3(4)^2\right] - \left[(4)(1) + \ldots + (3)(4)\right]^2 = 32.10^2 (18.4) = 18{,}959.54$$

DEVIATIONS FROM REGRESSION

$$\sum_i \frac{T_{Yi}^2}{n_i} - \frac{T_Y^2}{N} - b^2 \left[\sum_i n_i X_i^2 - \frac{\left(\sum_i n_i X_i\right)^2}{N}\right] = \frac{577.9^2}{4} + \ldots + \frac{724.0^2}{3}$$

$$- \frac{3{,}610.2^2}{19} - 18{,}959.54 = 705{,}622.54 - 685{,}976.00 - 18{,}959.54 = 687.00$$

TOTAL

$$\sum_{ji} Y_{ij}^2 - \frac{T_Y^2}{N} = 715{,}786.38 - 685{,}976.00 = 29{,}810.38$$

ANALYSIS OF VARIANCE TABLE

SOURCE	SUM OF SQUARES	DEGREES OF FREEDOM	MEAN SQUARE
WITHIN LEVELS	10,163.84	15	677.59
REGRESSION	18,959.54	1	18,959.54
DEVIATIONS	687.00	2	343.50
TOTAL	29,810.38	18	

Table 8. In the most general situation there are r levels of factor A and c levels of factor B, r standing for rows and c for columns in the table. In addition, within each cell there are n observations. The analysis can be made, and does not differ fundamentally, if n for each cell is different, but the added computational complication is great so that every effort should be made to have an equal number of measurements or observations for each cell. If most of the cells have the same number of observations and the numbers in the other cells differ only slightly, these differences may be disregarded, since the decrease in accuracy of the test is virtually nil while the simplification of computation is enormous. Should the zoologist be so foolhardy as to design his set of observations with radically

TABLE 8. Tabular form for observations in a two-factor analysis of variance.

FACTOR A LEVELS	FACTOR B LEVELS							
	1	2	3	.	.	.	c	
1	X_{111} X_{112} X_{113}	X_{121} X_{122} X_{123}	X_{131} X_{132} X_{133}	.	.	.	X_{1c1} X_{1c2} X_{1c3}	$\bar{X}_{1..}$
2	X_{211} X_{212} X_{213}	X_{221} X_{222} X_{223}	X_{231} X_{232} X_{233}	.	.	.		$\bar{X}_{2..}$
3	X_{311} X_{312} X_{313}	X_{321} X_{322} X_{323}	X_{331} X_{332} X_{333}	.	.	.		$\bar{X}_{3..}$
.	.	.	.	.	.	.	.	.
.	.	.	.	.	.	.	.	.
.	.	.	.	.	.	.	.	.
r	X_{r11} X_{r12} X_{r13}							$\bar{X}_{r..}$
	$\bar{X}_{.1.}$	$\bar{X}_{.2.}$	$\bar{X}_{.3.}$	.	.	.	$\bar{X}_{.r.}$	$\bar{X}_{...}$

unequal numbers in each cell, or so unfortunate as to have this situation thrust upon him by circumstances not under his control, he will find directions for an appropriate analysis of variance in Snedecor (1956). In most cases, a little foresight and planning will avoid this situation, but in paleontology, for example, where the number of measurable specimens is not ordinarily under reasonable control, a resort to more complex calculations may be necessary.

A given observation in the table is denoted by X_{ijk}, by which is meant the kth measurement at the jth level of factor B and the ith level of factor A. Following this notation,

$\bar{X}_{ij}$ = the mean of all measurements in cell ij

$\bar{X}_i$ = mean of all measurements in the ith row (level of A)

$\bar{X}_j$ = mean of all measurements in the jth column (level of B)

$\bar{X}$ = grand mean of all measurements

With this notation, the various sums of squares are

Total sum of squares = $\sum\limits_{ijk}(X_{ijk} - \bar{X})^2$

which is simply the sum of the squared deviations of all the observations from the grand mean.

Main effect of A: $nc \sum\limits_{i}(\bar{X}_i - \bar{X})^2$

Main effect of B: $nr \sum\limits_{j}(\bar{X}_j - \bar{X})^2$

Interaction of A and B: $n \sum\limits_{ij}(\bar{X}_{ij} - \bar{X}_i - \bar{X}_j + \bar{X})^2$

In this last component, the mean of each cell is corrected for the main effect of A and of B by subtracting the column and row means. That is, this component is equivalent to the form

$$n \sum\limits_{ij}\left[\bar{X}_{ij} - (\bar{X}_i - \bar{X}) - (\bar{X}_j - \bar{X}) - \bar{X}\right]^2$$

which is sensible intuitively.

Deviations: $\sum\limits_{ijk}(X_{ijk} - \bar{X}_{ij})^2$

which is the pooled sum of squares of the deviation of each observation from its own cell mean.

The analysis of variance is conventionally written out in the form of Table 9, with the degrees of freedom and mean squares appropriate for each component. The three components—A effect, B effect, and AB interaction—are each tested by an F ratio between their respective mean squares and the deviation mean square. Thus, to test for the presence of an effect of factor A, the F ratio is

$$F = \frac{\dfrac{1}{r-1} nc \sum_i (\bar{X}_i - \bar{X})^2}{\dfrac{1}{rc(n-1)} \sum_{ijk} (X_{ijk} - \bar{X}_{ij})^2}$$

with $r-1$ degrees of freedom in the numerator and $rc(n-1)$ in the denominator. As usual, an F so much larger than unity as to correspond to a very small probability will be considered as evidence of an effect. This form of the F test, with the deviation mean square in the denominator, is common to all fixed model analyses of variance. This method of testing does not exclude a comparison by inspection of main effects with each other or with the interaction mean square. This comparison may show that the interactions are far more important than main effects or vice versa. Such comparisons are important pieces of biological information, and although all three components may be significant, if one is very much larger than the other, this provides a basis for a zoological conclusion. Even if there is an effect of both locality and season on the size of some animal, if the locality season interaction is much greater as evidenced by its mean square than either of the main effect mean squares, then the character is for all practical purposes determined neither by season or locality but by a specific interaction of the two.

TABLE 9. Analysis of variance for a two-factor design.

SOURCE	SUM OF SQUARES	DEGREES OF FREEDOM	MEAN SQUARES
Main effect of A	$nc \sum_i (\bar{X}_i - \bar{X})^2$	$r-1$	$\dfrac{nc}{r-1} \sum_i (\bar{X}_i - \bar{X})^2$
Main effect of B	$nr \sum_j (\bar{X}_j - \bar{X})^2$	$c-1$	$\dfrac{nr}{c-1} \sum_j (\bar{X}_j - \bar{X})^2$
AB interaction	$n \sum_{ij} (\bar{X}_{ij} - \bar{X}_j - \bar{X}_i + \bar{X})^2$	$(r-1)(c-1)$	$\dfrac{n}{(r-1)(c-1)} \sum_{ij} (\bar{X}_{ij} - \bar{X}_j - \bar{X}_i + \bar{X})^2$
Deviations	$\sum_{ijk} (X_{ijk} - \bar{X}_{ij})^2$	$rc(n-1)$	$\dfrac{1}{rc(n-1)} \sum_{ijk} (X_{ijk} - \bar{X}_{ij})^2$
TOTAL	$\sum_{ijk} (X_{ijk} - \bar{X})^2$	$rcn-1$	

The two-factor analysis of variance presented here was based on many observations in each cell. It may be, however, that only a single observation is available for each locality at each season. In such a case, there will be no deviation mean square, and n will be equal to unity. Without a deviation mean square, it is the interaction term that is the only estimate of random deviation and the interaction mean square appears in the denominator of the F test for each main effect. There will no longer be a test possible for interaction, as it cannot be separated from the effect of random deviations. In any zoological application more than in the usual industrial uses of statistics, there is very likely to be an important interaction. Such being the case, more than a single observation should be made for each combination of levels wherever possible. An important biological fact may be obscured, an erroneous conclusion may even result from the failure to detect interactions between factors. If, in the analysis of such data, the main effects turn out to be nonsignificant, it may very well be due to a large real interaction between factors. While there is no main effect of the factors in a statistical sense, the existence of a strong interaction between two factors is a very real effect of these variables in a biological sense.

Calculation. The machine computing formulae for the sums of squares are as follows:

A effect: $\quad\dfrac{1}{nc}\sum_i T_i^2 - \dfrac{T^2}{nrc}$

B effect: $\quad\dfrac{1}{nr}\sum_j T_j^2 - \dfrac{T^2}{nrc}$

AB interaction: $\dfrac{1}{n}\sum_{ij} T_{ij}^2 - \dfrac{1}{nc}\sum_i T_i^2 - \dfrac{1}{nr}\sum_j T_j^2 + \dfrac{T^2}{nrc}$

Deviations: $\quad\sum_{ijk} X_{ijk} - \dfrac{1}{n}\sum_{ij} T_{ij}^2$

TOTAL: $\quad\sum_{ijk} X_{ijk}^2 - \dfrac{T^2}{nrc}$

where

T_{ij} = total of observations in the cell at the ith level of factor A and jth level of factor B

T_i = total for ith row (A level)

T_j = total for jth column (B level)

T = grand total for all observations

The following example of a two-factor analysis of variance has been deliberately chosen to show the flexibility of this technique. While the analysis has been described in terms of measurements of individual specimens, it can be as easily applied to other variables, like the total number of organisms collected under certain conditions. The data in Example 82 are the total numbers of aquatic insects collected in two streams in North Carolina in each of four months. Each cell contains six values, which are the results of performing the sampling technique six times in each stream in each month. Sampling was done by means of a standard square-foot bottom sampler, so that the six entries in each cell can be regarded as six measurements exactly analogous to some continuous measurement on six specimens.

EXAMPLE 82. Number of aquatic insects taken by a standard sampling instrument in Shope Creek and Ball Creek, North Carolina, in the months of December 1952, March 1953, June 1953, and September 1953. Each cell with six replications. (Original data of W. Hassler)

MONTH	SHOPE CREEK			BALL CREEK				
December	7 19 18	9 1 15	$T_{11} =$ 69 $\bar{X}_{11} =$ 11.5	25 16 10	9 28 14	$T_{12} =$ 102 $\bar{X}_{12} =$ 17.0	$T_{1\cdot} =$ $\bar{X}_{1\cdot} =$	171 14.3
March	29 114 24	37 49 64	$T_{21} =$ 317 $\bar{X}_{21} =$ 52.8	35 22 18	45 29 27	$T_{22} =$ 176 $\bar{X}_{22} =$ 29.3	$T_{2\cdot} =$ $\bar{X}_{2\cdot} =$	493 41.1
June	124 63 83	51 81 106	$T_{31} =$ 508 $\bar{X}_{31} =$ 84.7	20 26 38	44 127 52	$T_{32} =$ 307 $\bar{X}_{32} =$ 51.2	$T_{3\cdot} =$ $\bar{X}_{3\cdot} =$	815 67.9
September	72 100 67	87 68 9	$T_{41} =$ 403 $\bar{X}_{41} =$ 67.2	40 263 129	45 100 115	$T_{42} =$ 692 $\bar{X}_{42} =$ 115.3	$T_{4\cdot} =$ $\bar{X}_{4\cdot} =$	1,095 91.3
			$T_{\cdot 1} = 1,297$ $X_{\cdot 1} = 54.0$			$T_{2\cdot} = 1,277$ $X_{2\cdot} = 53.2$	$T_{\cdot\cdot} =$ $\bar{X}_{\cdot\cdot} =$	2,574 53.6

SUM OF SQUARES

TOTAL: $\sum\limits_{ijk} X_{ijk}^2 - \dfrac{T^2}{nrc} = 7^2 + 19^2 + \cdots 100^2 + 115^2 - \dfrac{2{,}574^2}{(6)(4)(2)}$

$= 245{,}518 - 138{,}030 = 107{,}488$

MONTHS: $\dfrac{1}{nc} \sum\limits_i T_i^2 - \dfrac{T^2}{nrc} = \dfrac{1}{(6)(2)} (171^2 + \cdots + 1{,}095^2) - 138{,}030$

$= 177{,}962 - 138{,}030 = 39{,}932$

EXAMPLE **82**. *continued*

CREEKS: $\dfrac{1}{nr} \sum_{j} T_j^2 - \dfrac{T^2}{nrc} = \dfrac{1}{(6)(4)}(1{,}297^2 + 1{,}277^2) - 138{,}030$

$= 138{,}039 - 138{,}030 = 9$

CREEK–MONTH INTERACTION: $\dfrac{1}{n}\sum_{ij} T_{ij}^2 - \dfrac{1}{nc}\sum_{i} T_i^2 - \dfrac{1}{nr}\sum_{j} T_j^2 + \dfrac{T^2}{nrc}$

$= \dfrac{1}{6}(69^2 + 317^2 + \cdots + 307^2 + 692^2) - 177{,}962 - 138{,}039 + 138{,}030$

$= 190{,}036 - 177{,}962 - 138{,}039 + 138{,}030 = 12{,}065$

DEVIATIONS: $\sum_{ijk} X_{ijk} - \dfrac{1}{n}\sum_{ij} T_{ij}^2 = 245{,}518 - 190{,}036 = 55{,}482$

ANALYSIS OF VARIANCE

SOURCE	SUM OF SQUARES	DEGREES OF FREEDOM	MEAN SQUARE	F
MONTHS	39,932	3	13,311	$\dfrac{13{,}311}{1{,}387} = 9.60$
CREEKS	9	1	9	$\dfrac{9}{1{,}387} = .006$
INTERACTIONS	12,065	3	4,022	$\dfrac{4{,}022}{1{,}387} = 2.90$
DEVIATIONS	55,482	40	1,387	
TOTAL	107,488	47		

The F for months has 3 degrees of freedom in the numerator and 40 in the denominator. The probability is less than .005, so the effect of months must be considered significant. There is obviously no effect of creeks, the F value being nearly zero. The F for interaction with 3 degrees of freedom in the numerator and 40 in the denominator has a probability of between .05 and .025. Whether this agrees with one's criterion of significance is a matter of personal choice. It is significant at the 5 per cent but not the 1 per cent level. Perhaps the best statement which can be made is that there is strong ground for suspecting an interaction, although this interaction is not nearly as important as the seasonal effect as evidenced by their relative mean squares.

The lack of difference between the creeks was surprising to the investigators who had assumed from superficial examination that the creeks did differ. It is this assumption which makes locality a fixed rather than random factor in the analysis, for the observations were designed in part to check this assumption.

The large effect of season is not surprising. The winter months are a time of very low population density, while as the year proceeds the populations of insects increase, reaching a peak in the early fall and then dying out again as the colder weather appears. While it is not apparent from the data presented that there is a yearly cycle, some samples were taken in the months of October 1953 to April 1954, and these observations, which have not been included in the analysis, show the repetition of the density cycle clearly.

The fairly large interaction between location and season deserves close scrutiny because it is unexpected. While it is perfectly possible from a mathematical standpoint to have no main effect of one factor and yet a significant interaction of that factor with another, it is not reasonable from a biological point of view. The lack of main effect of creeks can mean either that the creeks really are important in determining population density but the differential between creeks balances out through the year so that no main effect appears, or else that the creeks are really identical for all practical purposes. If this latter were the case, there ought to be no interaction between creek and month, since the creeks are, in a sense, exact duplicates of each other. The former possibility, although apparently shown by the observations, seems rather unlikely, for it implies that there is a real effect of creek in each month and that these effects are so neatly balanced as to produce absolutely no difference between creeks on the average over the entire year.

This apparent paradox brings out an extremely important consideration in the analysis of variance which, although ignored up to this point, must be discussed especially in connection with two-factor analyses. A basic assumption of the entire technique of analysis of variance is that of *homogeneity of error variance*. It is assumed that although the means for each cell may differ, the true variance of the population within each combination of factors is the same, irrespective of level. The mean square due to deviations is meant to be an estimate of this variance, calculated by pooling the sum of squares for each cell and dividing by the total number of degrees of freedom. If, however, the within-cell variances are different, this pooled calculation is not really an estimate of any parameter but a kind of average with no special meaning. The null hypothesis about interaction states that the variation from one cell mean to another is due entirely to the variation within cells and it is for this reason that the interaction is tested by the F ratio

$$F = \frac{\text{Mean square between cells corrected for main effects}}{\text{Mean square within cells}}$$

But if each cell has a different variance, this F ratio has not the meaning ascribed to it.

A glance at Hassler's data in Example 82 shows a very obvious increase in variance with cells as the mean number of organisms increases. This correlation is far from perfect, but December in both creeks has a much lower variation from sample to sample than does September of the next year. These differences in variance are clearly quite large, and this evidence coupled with the dubious significance of the calculated interaction and the total lack of effect of creeks forces the conclusion that interaction, if present, is not very important and has been overestimated.

There is no general recommendation that can be made for data with highly heterogeneous variances from cell to cell. There will always be some difference, of course, but unless it is obvious, the problem is just as well ignored. Despite protestations to the contrary, many practicing statisticians do just this. If there are very large differences in variability from one cell to the next, the interaction sum of squares should be regarded suspiciously, and it will generally be biased upwards. In cases of real doubt, the zoologist can do no better than to consult with a competent statistician, who will be able to assess the damage and recommend curative measures. Better still, if there is reason to suspect from previous experience that there will be considerable heterogeneity in a projected investigation, a consultation with the statistician *before* the investigation will often produce a constructive plan of sampling, designed to prevent problems in analysis.

Three-factor designs. The most complex analysis with which a zoologist is likely to deal involves three factors. Such models are not uncommon and should be nearly as frequent as two-factor models if proper importance is given sex differences. A very large number of measurements differ between sexes, and a two-factor model with variables like time and space ought generally to include the effect of sex as a third factor.

In the analysis of a three-factor model, there are eight components into which the total sum of squares can be partitioned. These are:

1. Main effect of A
2. Main effect of B
3. Main effect of C
4. AB interaction
5. BC interaction
6. AC interaction
7. ABC interaction ("second order interaction")
8. Deviations within cells

To fix the notation, suppose that factor A is present at r_1, factor B at r_2 levels, and factor C at r_3 levels, and within each cell there are n observations. Any given observation will be denoted by X_{ijkl} so that

X_{ijkl} = lth observation in the kth level of factor C in the jth level of factor B in the ith level of factor A

$\bar{X}_{ijk}$ = mean of the $i, j,$ and kth cell

$\bar{X}_{ij}$ = mean of the ith level of A and jth level of B over all levels of C

$\bar{X}_{ik}$ = mean for the ith level of A and kth level of C over all levels of B

$\bar{X}_{jk}$ = mean for the jth level of B and kth level of C over all levels of A

$\bar{X}_i$ = mean for the ith level of A over all levels of B and C

$\bar{X}_j$ = mean for the jth level of B over all levels of A and C

$\bar{X}_k$ = mean for the kth level of C over all levels of B and A

and $\bar{X}$ = grand mean of all observations

The sums of squares for each component together with the appropriate degrees of freedom are shown in Table 10. Because of lack of space the mean squares are not listed, but as usual these are simply the sum of squares for a component divided by its degrees of freedom.

TABLE 10. Analysis of variance for a three-factor design.

SOURCE	SUM OF SQUARES	DEGREES OF FREEDOM
Main effect of A	$n\, r_2 r_3 \sum_i (\bar{X}_i - \bar{X})^2$	$r_1 - 1$
Main effect of B	$n\, r_1 r_3 \sum_j (\bar{X}_j - \bar{X})^2$	$r_2 - 1$
Main effect of C	$n\, r_1 r_2 \sum_k (\bar{X}_k - \bar{X})^2$	$r_3 - 1$
AB interaction	$n r_3 \sum_{ij} (\bar{X}_{ij} - \bar{X}_i - \bar{X}_j + \bar{X})^2$	$(r_1 - 1)(r_2 - 1)$
BC interaction	$n r_1 \sum_{jk} (\bar{X}_{jk} - \bar{X}_j - \bar{X}_k + \bar{X})^2$	$(r_2 - 1)(r_3 - 1)$
AC interaction	$n r_2 \sum_{ik} (\bar{X}_{ik} - \bar{X}_i - \bar{X}_k + \bar{X})^2$	$(r_1 - 1)(r_3 - 1)$
ABC interaction	$n \sum_{ijk} (\bar{X}_{ijk} - \bar{X}_{ij} - \bar{X}_{ik} - \bar{X}_{jk} + \bar{X}_i + \bar{X}_j + \bar{X}_k - \bar{X})^2$	$(r_1 - 1)(r_2 - 1)(r_3 - 1)$
Deviations	$\sum_{ijkl} (X_{ijkl} - \bar{X}_{ijk})^2$	$(n - 1) r_1 r_2 r_3$
TOTAL	$\sum_{ijkl} (X_{ijkl} - \bar{X})^2$	$n r_1 r_2 r_3 - 1$

Obviously, the three main effects, the three first-order interactions, and the deviation sum of squares are exactly analogous in form to their respective counterparts in the two-factor analysis, and they have the same underlying meaning. The only new factor is the second-order interaction which accounts for the differences among cell means after they are corrected for the differences due to main effects and first-order interactions.

Calculation. Computation is most easily accomplished by formulae exactly analogous to those previously introduced for two-factor analyses.

A effect:
$$\frac{1}{nr_2 r_3} \sum_i T_i^2 - \frac{T^2}{nr_1 r_2 r_3}$$

B effect:
$$\frac{1}{nr_1 r_3} \sum_j T_j^2 - \frac{T^2}{nr_1 r_2 r_3}$$

C effect:
$$\frac{1}{nr_1 r_2} \sum_k T_k^2 - \frac{T^2}{nr_1 r_2 r_3}$$

AB interaction:
$$\frac{1}{nr_3} \sum_{ij} T_{ij}^2 - \frac{1}{nr_2 r_3} \sum_i T_i^2 - \frac{1}{nr_1 r_3} \sum_j T_j^2 + \frac{T^2}{nr_1 r_2 r_3}$$

BC interaction:
$$\frac{1}{nr_1} \sum_{jk} T_{jk}^2 - \frac{1}{nr_1 r_3} \sum_j T_j^2 - \frac{1}{nr_1 r_2} \sum_k T_k^2 + \frac{T^2}{nr_1 r_2 r_3}$$

AC interaction:
$$\frac{1}{nr_2} \sum_{ik} T_{ik}^2 - \frac{1}{nr_2 r_3} \sum_i T_i^2 - \frac{1}{nr_1 r_2} \sum_k T_k^2 + \frac{T^2}{nr_1 r_2 r_3}$$

ABC interaction:
$$\frac{1}{n} \sum_{ijk} T_{ijk}^2 - \frac{1}{nr_3} \sum_{ij} T_{ij}^2 - \frac{1}{nr_1} \sum_{jk} T_{jk}^2 - \frac{1}{nr_2} \sum_{ik} T_{ik}^2 +$$
$$\frac{1}{nr_2 r_3} \sum_i T_i^2 + \frac{1}{nr_1 r_3} \sum_j T_j^2 + \frac{1}{nr_1 r_2} \sum_k T_k^2 - \frac{T^2}{nr_1 r_2 r_3}$$

TOTAL:
$$\sum_{ijkl} X_{ijkl}^2 - \frac{T^2}{nr_1 r_2 r_3}$$

Random Factor Models

The calculations involved in an analysis of variance are the same whether the model is random or fixed, but different meanings are ascribed to the mean squares in the two models. As a result, the difference between random and fixed models lies entirely in the method of testing hypotheses and in the estimation of the relative effects of different factors. We have already pointed out that a mean square is analogous to a sample variance in that it is the sum of squared deviation divided by degrees of freedom. For a fixed model, this is an analogy only. Since the effects of the various levels of a factor A in a fixed model are simply constants, not random samples from a distribution, there is no parameter σ_A^2 which the mean square can properly be said to estimate. In a random model, however, there is some distribution of the effects of the different levels, and a mean square *is* a sample estimate of some true population variance or combination of variances.

For a random model, an extra column containing the *expected mean squares* is added to the analysis of variance table. These expected mean squares are the variances, in symbolic form, that are estimated by each mean square, and an inspection of this column will show precisely how to test various null hypotheses and estimate various effects. No numerical values are entered in the expected mean square column since the true variances are not known. Rather it is a mnemonic device to aid in the analysis.

One-factor designs. In a one-factor design, the total sum of squares was partitioned into two components, one associated with random error, the other associated with differences between levels. These were then converted to mean squares by dividing each sum of squares by its degrees of freedom.

The mean square associated with error,

$$\frac{1}{N-k} \sum_i \left[\sum_j (X_{ij} - \bar{X}_i)^2 \right]$$

is a pooled estimate of the true error variance σ_e^2.

The mean square associated with levels (main effect mean square),

$$\frac{1}{k-1} n \sum_i (\bar{X}_i - \bar{X})^2$$

is an estimate of the sum of two variances. It contains a contribution both from the error variance, σ_e^2, and from the variance due to differences between levels, σ_A^2. It is more precisely an estimate of

$$n\sigma_A^2 + \sigma_e^2$$

The one-factor analysis of variance table for a random model will then look like Table 11, which is identical to that for a fixed model except for the extra column containing the expected mean squares. The sums of squares and the mean squares are identical with those calculated for the fixed model and are abbreviated as S.S.$_1$, S.S.$_2$, M.S.$_1$, and M.S.$_2$, respectively.

TABLE 11. Analysis of variance for a random one-factor design.

SOURCE	SUM OF SQUARES	DEGREES OF FREEDOM	MEAN SQUARE	EXPECTED MEAN SQUARE
Main effect	$n \sum (\bar{X}_i - \bar{X})^2$	$c - 1$	$\dfrac{n}{c-1} \sum (\bar{X}_i - \bar{X})^2$	$n\sigma_A^2 + \sigma_e^2$
Sampling deviations	$\sum_{ij} (X_{ij} - \bar{X}_i)^2$	$N - c$	$\dfrac{1}{N-c} \sum_{ij} (X_{ij} - \bar{X}_i)^2$	σ_e^2
TOTAL	$\sum_{ij} (X_{ij} - \bar{X})^2$	$N - 1$		

As the expected mean square column shows, if σ_A^2 is zero, then the error mean square and the main effect mean square are both simply estimates of σ_e^2. The obvious test of the null hypothesis is then

$$F = \frac{\text{M.S.}_1}{\text{M.S.}_2}$$

with $k - 1$ degrees of freedom in the numerator and $N - k$ in the denominator. If the numerator is not significantly larger than the denominator as judged by the probability of the F test, the null hypothesis is accepted, and it must be assumed the σ_A^2 is zero—that is, that there is no main effect. On the other hand, if F is so large as to correspond to a very small probability, there is evidence that σ_A^2 is different from zero.

The value of σ_A^2 can be estimated directly from the mean squares. The difference between M.S.$_1$ and M.S.$_2$ is obviously an estimate of $n\sigma_A^2$ so that σ_A^2 can be estimated from the quantity

$$\frac{1}{n}(\text{M.S.}_1 - \text{M.S.}_2)$$

In one respect, Table 11 is less general than the one constructed for the fixed model. It is assumed here that the number of observations, n, is the same in all levels. There is no great complication in allowing unequal sample sizes, as far as the test of the null hypothesis is concerned, since this test for a one-factor design is the same for random and fixed models. A great complication, however, occurs in the process of estimation if n is not constant. It is not possible in a book of this kind to cover every detail of the analysis of variance. In all of the discussion of random models and hierarchical models which follows, it will be assumed that the number of observations is equal for all combinations of factors. If this is not the case, recourse must be had either to more complete descriptions of the analysis of variance, such as Snedecor or Cochran and Cox, or else to the services of a professional statistician.

Two-factor designs. By a process of reasoning similar to that for one-factor designs, the analysis of variance for the random two-factor model can be put in the form shown in Table 12.

TABLE 12. Analysis of variance for random two-factor design.

SOURCE	SUM OF SQUARES	DEGREES OF FREEDOM	MEAN SQUARE	EXPECTED MEAN SQUARE
A effect	S.S.$_1$	$r-1$	$\dfrac{\text{S.S.}_1}{r-1} = \text{M.S.}_1$	$\sigma_e^2 + n\sigma_{AB}^2 + nc\sigma_A^2$
B effect	S.S.$_2$	$c-1$	$\dfrac{\text{S.S.}_2}{c-1} = \text{M.S.}_2$	$\sigma_e^2 + n\sigma_{AB}^2 + nr\sigma_B^2$
AB interaction	S.S.$_3$	$(r-1)(c-1)$	$\dfrac{\text{S.S.}_3}{(r-1)(c-1)} = \text{M.S.}_3$	$\sigma_e^2 + n\sigma_{AB}^2$
Deviations	S.S.$_4$	$rc(n-1)$	$\dfrac{\text{S.S.}_4}{rc(n-1)} = \text{M.S.}_4$	σ_e^2
TOTAL	S.S.$_T$	$nrc-1$		

The sums of squares which are abbreviated by S.S. are those calculated for the fixed model, and the analysis of variance proceeds in the usual way to obtain the mean squares. It is in the appropriate F ratio for testing the existence of an effect that this random model differs from the fixed. In the fixed model, the existence of an effect due to factor A was tested by the F ratio

$$F = \frac{\text{M.S.}_1}{\text{M.S.}_4}$$

but this is obviously not satisfactory in a random model, as shown by the expected mean squares of M.S.$_1$ and M.S.$_4$. The mean square due to factor A contains components of variance due to that factor, to the AB interaction and to the random deviations, while the mean square for deviations contains only the random deviation variance. An F ratio between these two mean squares would be unity only if *both* σ_A^2 and σ_{AB}^2 were zero. In the random model, then, the ratio M.S.$_1$/M.S.$_4$ is a test not of the factor A

alone but of both that factor and the interaction. Inspection of the expected mean squares reveals that the correct F ratio is

$$F = \frac{\text{M.S.}_1}{\text{M.S.}_3}$$

—the mean square due to factor A divided by the *interaction* mean square. The mean squares in the numerator and denominator of this F ratio differ only by amount $nc\sigma_A^2$, so that a significant F ratio is evidence for an effect of the factor A, irrespective of the interaction variance. *In general, the rule for testing the existence of an effect in a random model is to construct an* F *ratio whose numerator differs from its denominator only in a single term which contains the variance to be tested.*

Following this rule, the three appropriate F tests for the two-factor case are:

A effect: $\dfrac{\text{M.S.}_1}{\text{M.S.}_3}$ or $\dfrac{A \text{ effect mean square}}{\text{Interaction mean square}}$

B effect: $\dfrac{\text{M.S.}_2}{\text{M.S.}_3}$ or $\dfrac{B \text{ effect mean square}}{\text{Interaction mean square}}$

AB interaction: $\dfrac{\text{M.S.}_3}{\text{M.S.}_4}$ or $\dfrac{\text{Interaction mean square}}{\text{Random deviation mean square}}$

The degrees of freedom for these tests are those associated with the mean squares in numerator and denominator, respectively, in the table.

Estimates of the three variance components σ_A^2, σ_B^2, and σ_{AB}^2 are found by subtracting the denominator from the numerator in each F ratio and then dividing this difference by the appropriate constant. For example,

Expected value of $\text{M.S.}_1 = \sigma_e^2 + n\sigma_{AB}^2 + nc\sigma_A^2$

Expected value of $\text{M.S.}_3 = \sigma_e^2 + n\sigma_{AB}^2$

Therefore, $\dfrac{1}{nc}(\text{M.S.}_1 - \text{M.S.}_3)$ will estimate $\dfrac{1}{nc}(nc\sigma_A^2) = \sigma_A^2$. In the same way $\dfrac{1}{nr}(\text{M.S.}_2 - \text{M.S.}_3)$ will estimate σ_B^2 and $\dfrac{1}{n}(\text{M.S.}_3 - \text{M.S.}_4)$ will estimate σ_{AB}^2. M.S._4 itself is a direct estimate of σ_e^2.

Three-factor designs. The analysis of variance for the three-factor random model is shown in Table 13. All of the columns have not been shown since the analysis is identical with the fixed model up to the point of testing the various factors.

TABLE 13. Expected mean squares for a three-factor analysis.

SOURCE	MEAN SQUARE	EXPECTED MEAN SQUARE
A effect	M.S.$_1$	$\sigma_e^2 + n\sigma_{ABC}^2 + nr_3\sigma_{AB}^2 + nr_2\sigma_{AC}^2 + nr_2r_3\sigma_A^2$
B effect	M.S.$_2$	$\sigma_e^2 + n\sigma_{ABC}^2 + nr_3\sigma_{AB}^2 + nr_1\sigma_{BC}^2 + nr_1r_3\sigma_B^2$
C effect	M.S.$_3$	$\sigma_e^2 + n\sigma_{ABC}^2 + nr_2\sigma_{AC}^2 + nr_1\sigma_{BC}^2 + nr_1r_2\sigma_C^2$
AB interaction	M.S.$_4$	$\sigma_e^2 + n\sigma_{ABC}^2 + nr_3\sigma_{AB}^2$
BC interaction	M.S.$_5$	$\sigma_e^2 + n\sigma_{ABC}^2 + nr_1\sigma_{BC}^2$
AC interaction	M.S.$_6$	$\sigma_e^2 + n\sigma_{ABC}^2 + nr_2\sigma_{AC}^2$
ABC interaction	M.S.$_7$	$\sigma_e^2 + n\sigma_{ABC}^2$
Deviations	M.S.$_8$	σ_e^2

Following the general rule for testing effects in a random model, the following F tests are obvious from the expected mean squares:

$$ABC \text{ interaction}: \frac{\text{M.S.}_7}{\text{M.S.}_8}$$

$$AC \text{ interaction}: \frac{\text{M.S.}_6}{\text{M.S.}_7}$$

$$BC \text{ interaction}: \frac{\text{M.S.}_5}{\text{M.S.}_7}$$

$$AB \text{ interaction}: \frac{\text{M.S.}_4}{\text{M.S.}_7}$$

The second-order interaction is tested with the deviations mean square in the denominator, while the first-order interactions are tested using the second-order interaction mean square in the denominator. What is to be done, however, with the main effects? The general rule appears to break down, for there are no two mean squares which differ only by an amount which is a function of σ_A^2 or σ_B^2 or σ_C^2. Clearly some added manipulation of the mean squares is required to produce a suitable denominator for the F test. A study of the expected mean squares shows that by adding any two first-order interactions and subtracting from that sum the second-order interaction, an expression will result which is suitable for testing one of the main effects. Thus:

Expected value of $(\text{M.S.}_4 + \text{M.S.}_5 - \text{M.S.}_7) = \sigma_e^2 + n\sigma_{ABC}^2 + nr_3\sigma_{AB}^2 + nr_1\sigma_{BC}^2$ which is identical with the expected mean square for the B effect, with the exception of the factor $nr_1 r_3 \sigma_B^2$, so that this combination of mean squares is the proper denominator for the test of $\sigma_B{}^2$. Following this line of reasoning, the three F tests for the three main effects are:

A effect:
$$\frac{\text{M.S.}_{\cdot 1}}{\text{M.S.}_{\cdot 4} + \text{M.S.}_{\cdot 6} - \text{M.S.}_{\cdot 7}}$$

B effect:
$$\frac{\text{M.S.}_{\cdot 2}}{\text{M.S.}_{\cdot 4} + \text{M.S.}_{\cdot 5} - \text{M.S.}_{\cdot 7}}$$

C effect:
$$\frac{\text{M.S.}_{\cdot 3}}{\text{M.S.}_{\cdot 5} + \text{M.S.}_{\cdot 6} - \text{M.S.}_{\cdot 7}}$$

Having found F ratios appropriate for the needed tests, the problem of degrees of freedom for these tests must be faced. In the usual F test, the number of degrees of freedom in the numerator and denominator are simply those associated with the respective mean squares. Unfortunately, the denominators of the F tests for the main effect contain not one but three mean squares, so that the usual rule for finding degrees of freedom fails. The degrees of freedom, M, for the denominator of these F ratios is found from the rather unfortunate expression

$$\frac{(\text{M.S.}_{\cdot\text{I}} + \text{M.S.}_{\cdot\text{II}} - \text{M.S.}_{\cdot\text{III}})^2}{\dfrac{\text{M.S.}_{\cdot\text{I}}{}^2}{\text{d.f.}_{\cdot\text{I}}} + \dfrac{\text{M.S.}_{\cdot\text{II}}{}^2}{\text{d.f.}_{\cdot\text{II}}} + \dfrac{\text{M.S.}_{\cdot\text{III}}{}^2}{\text{d.f.}_{\cdot\text{III}}}}$$

where $\text{M.S.}_{\cdot\text{I}}$, $\text{M.S.}_{\cdot\text{II}}$, and $\text{M.S.}_{\cdot\text{III}}$ are the three mean squares used to calculate the denominator of F, and $\text{d.f.}_{\cdot\text{I}}$, $\text{d.f.}_{\cdot\text{II}}$ and $\text{d.f.}_{\cdot\text{III}}$ are the degrees of freedom associated with these in the analysis of variance table. The degrees of freedom for the numerator of F are simply those associated with the mean square in the numerator. As an example, to test the existence of an effect of factor A, the F ratio

$$F = \frac{\text{M.S.}_{\cdot 1}}{\text{M.S.}_{\cdot 4} + \text{M.S.}_{\cdot 6} - \text{M.S.}_{\cdot 7}}$$

is used and this has $r_1 - 1$ degrees of freedom in the numerator and

$$M = \frac{(\text{M.S.}_{\cdot 4} + \text{M.S.}_{\cdot 6} - \text{M.S.}_{\cdot 7})^2}{\dfrac{(\text{M.S.}_{\cdot 4})^2}{(r_1 - 1)(r_2 - 1)} + \dfrac{(\text{M.S.}_{\cdot 6})^2}{(r_1 - 1)(r_3 - 1)} + \dfrac{(\text{M.S.}_{\cdot 7})^2}{(r_1 - 1)(r_2 - 1)(r_3 - 1)}}$$

degrees of freedom in the denominator.

There is one way in which this complication of testing can be avoided and that is by testing the interactions first, so that a knowledge of their existence may be used in testing main effects. For example, the F ratio

$$F = \frac{\text{M.S.}_1}{\text{M.S.}_4}$$

would be a test of σ_A^2 if it were known that σ_{AC}^2, which also appears in the numerator, were zero. If a test for the AC interaction shows no significance, it might be assumed that σ_{AC}^2 is in fact zero and that $\text{M.S.}_1/\text{M.S.}_4$ is a test of the main effect. Should all of the interactions turn out to be nonsignificant, which is a distinct possibility, then the deviation mean square can be used as the denominator in the F test of main effects. By eliminating one or more interaction variances from consideration, it is possible to make direct tests in the various main effects without resort to the more complex procedure described above. There are objections to this procedure, the chief one being that it is somewhat biased. At times, the interaction component which was assumed to be zero will really exist, and this will result in an overestimate of the main effect. This defect in the method is offset by others in the more complex technique of manipulating the various mean squares to find a suitable denominator. All in all, there is little to choose except on the basis of simplicity of calculation.

Mixed Models

When the factors in an analysis of variance cannot be classified either as all fixed or all random, the model is said to be "mixed." Samples drawn from a number of sympatric species in a series of randomly chosen localities would provide a mixed model, as species is a fixed factor and localities a random one. The analysis of a mixed model is identical in form with that of a random one. Mean squares are calculated in the usual way, and various effects are tested by the ratio of mean squares which differ only in the component being tested. Table 14 shows the expected mean squares for three types of mixed models: the two-factor mixed model, the three-factor model with one factor fixed, and the three-factor model with two factors fixed. A comparision of these expected mean squares with those contained in Tables 12 and 13 shows the essential differences between random and mixed models. First, there is no variance component σ_A^2 corresponding to the main effect of the fixed factor in the mixed model. In its place, there is simply a constant, K, which measures the effect of the factor. Second, certain of the interaction terms found in the random model are missing in the mixed model. The result of these missing terms is that different mean squares are used in the F ratio for testing various effects. For example, in the random two-factor model the test for significance of the B effect would contain the B effect mean square in the numerator and the

AB interaction mean square in the denominator since these differ only by a term containing σ_B^2. In the mixed model, however, to test the B effect where B is the random factor, the correct F ratio should contain the B effect mean square in the numerator and the *deviations* mean square in the denominator. In each case, the proper F ratio can easily be determined by inspection of the expected mean squares, remembering the rule that the numerator and denominator of F must differ only by a quantity proportional to the component being tested.

TABLE 14. Expected mean squares for mixed models. A. Two-factor mixed model. B. Three-factor mixed model with one fixed factor. C. Three-factor mixed model with two fixed factors.

A.

SOURCE	MEAN SQUARE	EXPECTED MEAN SQUARE
A effect (fixed)	M.S.$_1$	$\sigma_e^2 + n\sigma_{AB}^2 + K_A$
B effect (random)	M.S.$_2$	$\sigma_e^2 + nr\sigma_B^2$
AB interaction	M.S.$_3$	$\sigma_e^2 + n\sigma_{AB}^2$
Deviations	M.S.$_4$	σ_e^2

B.

SOURCE	MEAN SQUARE	EXPECTED MEAN SQUARE
A effect (fixed)	M.S.$_1$	$\sigma_e^2 + n\sigma_{ABC}^2 + nr_3\sigma_{AB}^2 + nr_2\sigma_{AC}^2 + K_A$
B effect (random)	M.S.$_2$	$\sigma_e^2 + n\sigma_{ABC}^2 + nr_1\sigma_{BC}^2 + nr_1r_3\sigma_B^2$
C effect (random)	M.S.$_3$	$\sigma_e^2 + n\sigma_{ABC}^2 + nr_1\sigma_{BC}^2 + nr_1r_2\sigma_C^2$
AB interaction	M.S.$_4$	$\sigma_e^2 + n\sigma_{ABC}^2 + nr_3\sigma_{AB}^2$
AC interaction	M.S.$_5$	$\sigma_e^2 + n\sigma_{ABC}^2 + nr_2\sigma_{AC}^2$
BC interaction	M.S.$_6$	$\sigma_e^2 + nr_1\sigma_{BC}^2$
ABC interaction	M.S.$_7$	$\sigma_e^2 + n\sigma_{ABC}^2$
Deviations	M.S.$_8$	σ_e^2

C.

SOURCE	MEAN SQUARE	EXPECTED MEAN SQUARE
A effect (fixed)	M.S.$_1$	$\sigma_e^2 + n\sigma_{ABC}^2 + nr_2\sigma_{AC}^2 + K_A$
B effect (fixed)	M.S.$_2$	$\sigma_e^2 + n\sigma_{ABC}^2 + nr_1\sigma_{BC}^2 + K_B$
C effect (random)	M.S.$_3$	$\sigma_e^2 + nr_1r_2\sigma_C^2$
AB interaction	M.S.$_4$	$\sigma_e^2 + n\sigma_{ABC}^2 + K_{AB}$
AC interaction	M.S.$_5$	$\sigma_e^2 + nr_2\sigma_{AC}^2$
BC interaction	M.S.$_6$	$\sigma_e^2 + nr_1\sigma_{BC}^2$
ABC interaction	M.S.$_7$	$\sigma_e^2 + n\sigma_{ABC}^2$
Deviation	M.S.$_8$	σ_e^2

Hierarchical Models

All the designs that have been discussed to this point were strictly *factorial*, in that every level of each factor appears with every level of all other factors. It is common in zoological investigations to collect data in a very different way. In these investigations, the various levels of factor B may be quite different within each level of factor A. If measurements are made on a number of allopatric species or subspecies each from several different localities, these localities cannot be the same for any two species. Factor A, species, and factor B, locality, are not factorially related and cannot be treated in the usual way. In a sense, the one-factor case which we have discussed is a special case of these *hierarchical*, or *nested*, designs, since the same animals are not measured at each level of the factor. The factor and the individuals are not factorially but hierarchically related.

The hierarchical model can be extended to any number of factors without in any way complicating the analysis. The example given of the allopatric subspecies—each sampled in a number of localities—might be extended to include sampling within each of the localities at a number of substations and at each substation for a number of days, the precise identity of the days being different for each substation. If this sampling scheme is written out symbolically, the reason for the name "hierarchical" is obvious. Thus:

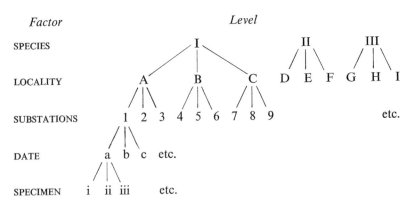

The factors might be described as
1. species
2. localities in species
3. substations in localities in species
4. days in substations in localities in species
5. specimens in days in substations in localities in species

Now, any two adjacent factors really make up a simple one-factor model with replications. Taking only specimens-in-days and days, the specimens are repeated observations in each level of the factor days. Moving one step upward in the hierarchy, days-in-substations are repeated observations in each level of the factor substations. This process may be repeated until the top of the hierarchy is reached, where the last one-factor analysis is that of localities-in-species as repeated observations in each level of the factor species. There are four one-factor comparisons, each one contained within the next one higher up in the hierarchical pattern, and all of these can be combined into a single analysis.

To illustrate the analysis of such a hierarchical (or nested) model, it is sufficient to introduce three factors, say, locality, subsamples in each locality, and specimens in each subsample of each locality.

Let a = number of localities

b = number of subsamples in each locality

c = number of specimens in each subsample

$\bar{X}$ = mean of all localities

$\bar{X}_i$ = mean of the ith locality

$\bar{X}_{ij}$ = mean of jth subsample in the ith locality

X_{ijk} = kth specimen (or observation) in the jth subsample of the ith locality.

TABLE 15. Analysis of variance for a hierarchical model.

SOURCE	SUM OF SQUARES	DEGREES OF FREEDOM	EXPECTED MEAN SQUARE
Locality	$bc \sum_{i}(\bar{X}_i - \bar{X})^2$	$a - 1$	$\sigma_e^2 + c\sigma_s^2 + bc\sigma_l^2$
Subsample	$c \sum_{ij}(\bar{X}_{ij} - \bar{X}_i)^2$	$(b - 1)a$	$\sigma_e^2 + c\sigma_s^2$
Specimens	$\sum_{ijk}(X_{ijk} - \bar{X}_{ij})^2$	$(c - 1)ab$	σ_e^2
TOTAL	$\sum_{ijk}(X_{ijk} - \bar{X})^2$	$abc - 1$	

The analysis of variance table is of the form shown in Table 15.

All of the entries progress in an entirely systematic way so that the design is easily generalized to any number of factors. The pattern of expected mean squares is such that the F tests for any factor are easily seen to be the ratio of any two adjacent mean squares in the table. Thus, a test of the effect of localities would be

$$F = \frac{\text{Mean square for localities}}{\text{Mean square for samples}}$$

and so on.

In addition, estimates of variance components for any factor which is truly random can be obtained by subtracting the denominator from the numerator in the corresponding F test and dividing this difference by the multiplier of the variance component. As an example, the variance due to subsamples would be estimated by

$$\frac{1}{c}(\text{M.S. for subsamples} - \text{M.S. for specimens})$$

Should the factor be a fixed level variable, as, for example, species, it is of course meaningless to calculate such a component, but the F test is valid nevertheless.

Calculation. Beginning at the top of the hierarchy, the sums of squares can be computed as follows:

Factor A: $\quad \dfrac{1}{bc}\sum_i T_i^2 - \dfrac{1}{abc}T^2$

Factor B: $\quad \dfrac{1}{c}\sum_{ij} T_{ij}^2 - \dfrac{1}{bc}\sum_i T_i^2$

Observation: $\quad \sum_{ijk} X_{ijk}^2 - \dfrac{1}{c}\sum_{ij} T_{ij}^2$

TOTAL: $\quad \sum_{ijk} X_{ijk}^2 - \dfrac{1}{abc}T^2$

where T = grand total of all observations

T_i = total in the ith level of factor A

T_{ij} = total in jth level of factor B in the ith level of factor A

X_{ijk} = kth measurement in the jth level of B in the ith level of A

a = numbers of levels of A

b = numbers of levels of B

c = numbers of observations in each level of B

These formulae are easily generalized to any number of factors, since they proceed in a perfectly systematic manner. Thus, if there were five factors as well as individual observations within the lowest factor in the hierarchy, the calculating formulae would be

Factor A: $\quad \dfrac{1}{bcdef} \sum_i T_i^2 - \dfrac{1}{abcdef} T^2$

Factor B: $\quad \dfrac{1}{cdef} \sum_{ij} T_{ij}^2 - \dfrac{1}{bcdef} \sum_i T_i^2$

Factor C: $\quad \dfrac{1}{def} \sum_{ijk} T_{ijk}^2 - \dfrac{1}{cdef} \sum_{ij} T_{ij}^2$

and so on. Finally, the observation sum of squares would be

$$\sum_{ijklmn} T_{ijklmn}^2 - \dfrac{1}{f} \sum_{ijklm} T_{ijklm}^2$$

A problem common in experimental taxonomy is that of distinguishing geographical races. Animals from different geographical populations may differ in any number of morphological characteristics for two reasons. First, there may be average genetic differences among the populations, and, second, environmental differences, acting during morphogenesis, may result in morphological differences even if the various geographical populations are genetically alike. In order to separate these causes of interpopulational variability, it is common practice to collect animals (or plants) from the different populations and raise progeny from these in a controlled environment. In animals where this is possible—frogs, insects, small mammals, and the like—the effect of environment can be separated from real genetic differences among the local populations. If, in the controlled environment, there are significant differences among the populations, then it must be assumed that these are due to real genetic differentiation. If, on the other hand, there is no demonstrable difference under controlled conditions, it must be concluded that the observed differences in nature are the direct result of environmental variation from locality to locality.

Example 83 illustrates the use of a hierarchical analysis of variance in such a problem. The example uses original data on the dipteran fly *Drosophila persimilis* from three localities in western North America. Specimens taken directly from nature were not measured but were allowed to produce offspring in the laboratory. Several samples of offspring from each locality were scored for the total number of bristles on the fourth and fifth sternites of the adult.

EXAMPLE 83. Hierarchical analysis of variance. Data are the number of sternal bristles in samples of *Drosophila persimilis* from three localities in western North America. (Original data)

	SAMPLE 1	SAMPLE 2	SAMPLE 3	SAMPLE 4
Yolla Bolly, California	27	26	28	29
	31	28	31	25
	30	29	31	28
	30	31	28	27
	27	29	33	30
Hope, British Columbia				
	35	33	32	32
	33	33	36	35
	33	31	33	31
	35	33	33	34
	38	37	33	33
Nojogui, California				
	41	41	37	45
	34	40	42	38
	40	43	36	31
	41	37	41	36
	42	41	37	43

a = number of localities = 3
b = number of samples in each locality = 4
c = number of specimens in each subsample = 5

A. Locality sum of squares

$$T_1 = 27 + 31 + 30 + \cdots + 28 + 27 + 30 = 578$$
$$T_2 = 35 + 33 + 33 + \cdots + 31 + 34 + 33 = 673$$
$$T_3 = 41 + 34 + 40 + \cdots + 31 + 36 + 43 = 786$$
$$T = 578 + 673 + 786 = 2{,}037$$

$$\text{Sum of squares} = \frac{\sum_i T_i^2}{bc} - \frac{T^2}{abc}$$

$$= \frac{1}{20}(578^2 + 673^2 + 786^2) - \frac{2{,}037^2}{60}$$

$$= 70{,}240.45 - 69{,}156.15 = 1{,}084.30$$

EXAMPLE **83**. *continued*

B. Sample sum of squares

$$T_{11} = 27 + 31 + 30 + 30 + 27 = 145$$
$$T_{12} = 26 + 28 + 29 + 31 + 29 = 143$$
$$T_{13} = 151$$
$$T_{14} = 139$$
$$T_{21} = 174$$
$$T_{22} = 167$$
$$T_{23} = 167$$
$$T_{24} = 165$$
$$T_{31} = 198$$
$$T_{32} = 202$$
$$T_{33} = 193$$
$$T_{34} = 193$$

Sum of squares

$$\frac{1}{c} \sum_{ij} T_{ij}^2 - \frac{1}{bc} \sum_i T_i^2 = \frac{1}{5}(145^2 + 143^2 + \cdots + 193^2 + 193^2) - 70{,}240.45$$
$$= 70{,}276.20 - 70{,}240.45 = 35.75$$

C. Specimen sum of squares

$$\sum_{ijk} X_{ijk}^2 - \frac{1}{c} \sum_{ij} T_{ij}^2$$
$$= 27^2 + 31^2 + 30^2 + \cdots + 31^2 + 36^2 + 43^2 - 70{,}276.20$$
$$= 70{,}607.00 - 70{,}276.20 = 330.80$$

D. Total sum of squares

$$\sum_{ijk} X_{ijk} - \frac{T^2}{abc} = 70{,}607.00 - 69{,}156.15 = 1{,}450.85$$

E. Degrees of freedom

LOCALITIES:	$a - 1 =$	2
SAMPLES:	$a(b - 1) =$	9
SPECIMENS:	$ab(c - 1) =$	48
TOTAL:	$abc - 1 =$	59

EXAMPLE 83. *continued*

The complete analysis for this example is then:

SOURCE	SUM OF SQUARES	DEGREES OF FREEDOM	MEAN SQUARE	EXPECTED MEAN SQUARE
LOCALITIES	1,084.30	2	542.15	$\sigma_e^2 + 5\sigma_s^2 + 20\sigma_l^2$
SAMPLES	35.75	9	3.97	$\sigma_e^2 + 5\sigma_s^2$
SPECIMENS	330.80	48	6.89	σ_e^2
TOTAL	1,450.85	59		

F-tests

$$\text{Localities: } F = \frac{542.15}{3.97} = 136.56$$

with degrees of freedom = 2 and 9

$$\text{Samples: } F = \frac{3.97}{6.89} = .58$$

with degrees of freedom = 9 and 48

Estimates of variances

Estimate of σ_e^2 = M.S.$_3$ = 6.89
Estimate of $\sigma_s^2 = \frac{1}{5}$ (M.S.$_2$ − M.S.$_3$) = − .58
Estimate of $\sigma_l^2 = \frac{1}{20}$ (M.S.$_1$ − M.S.$_2$) = 26.91

As the analysis of variance shows, there is a marked effect of locality. The value of *F* for localities is 136.56, which for 2 degrees of freedom in the numerator and 9 in the denominator corresponds to a probability between .005 and .01. There is no effect of samples since the *F* ratio testing this effect is less than unity, and this is reflected in the negative estimate of σ_s^2. By the definition of a variance, it can never be negative, but, of course, an estimate of variance can be and often is negative especially when the true variance is close to zero. No particular interpretation can be put on the estimate of σ_e^2, but it is very much smaller than σ_l^2 so that most of the variability in the sample is due to locality differences. The reasonable conclusion from this analysis is that the populations of Nojogui, Hope, and Yolla Bolly are genetically different and constitute distinct geographical races (under one commonly accepted criterion of race).

Complex Analyses

The analysis of variance as a general technique has reached a very high stage of development in the hands of biometricians faced with an almost infinite variety of situations demanding statistical analysis. In large part, the more complex methods have been designed for the use of agricultural scientists whose experiments often reach an unbelievable height of complexity. The zoologist may find, from time to time, that his observations cannot be fitted to one of the simpler schemes of analysis which have been reviewed in this chapter.

Such situations will arise when there are unequal numbers of observations in various levels, or when the error sum of squares varies widely from cell to cell in the analysis, indicating heterogeneity of variance. Another common complexity is the mixture of factorial and hierarchical designs in a single analysis. If sex had been introduced as a factor in the example of the hierarchical analysis just given, a different approach would have been required since sex is clearly a factorial variable, not a hierarchical one.

There are two courses open to the zoologist faced with a complex analysis of variance. First, and best, he may consult a statistician with a good understanding of biological problems. If possible, such consultation should *precede* the actual collection of the data since statistical science can prevent far more ills than it can cure. A common, and justified, complaint of biometricians is that biologists will come to them with a laboriously collected mass of hopeless data. Often the data have been collected in such a way as to make a proper analysis impossible, or else the question of interest to the biologist cannot be answered at all from the information in hand. One channel of consultation open to all biologists is the "Query Department" of *Biometrics*, the journal of the Biometrics Section of the American Statistical Association. While the biometricians who contribute to this service will not undertake a complete numerical analysis of a zoologist's problem, they will outline the proper procedure to follow, together with reasons for their preference.

A second source of information in difficult cases is the excellent book of Snedecor (1956) which, in its latest edition, contains a variety of analyses for especially refractive cases. The book of Cochran and Cox (1957) may also be useful for the analysis of variance although considerable emphasis is placed by them on the design of agricultural experiments.

CHAPTER THIRTEEN

Tests on Frequencies

The various methods of testing the difference between populations and the techniques of correlation and regression discussed in previous chapters apply to continuous variates whose distributions are assumed to be roughly normal. When discrete or discontinuous variates are considered, many of these methods break down and special techniques are required. The present chapter will be devoted to methods for the solution of three common problems which arise from discrete variates. The first is that of the "goodness-of-fit" of the observed frequencies to some hypothetical frequency distribution. While this problem arises in many ways in biology, as for example the fit of the observed frequencies to some Mendelian proportions in genetics, the zoologist is more likely to meet this situation when he desires to fit the observations to some theoretical distribution like a normal or Poisson distribution. The second aspect of tests on frequencies concerns the similarity of distributions of two or more populations. Finally, there is the question, closely related to the second, of correlation, or *association*, between two or more attributes.

The χ^2 Test

All three of these questions can be dealt with by variations in a single testing procedure, the χ^2 test. The χ^2 test has already been introduced for testing the significance of the difference between two binomial proportions, and as such it is only a special case of testing the similarity of two frequency distributions. For the purposes of dealing with frequency data, χ^2 is defined as

$$\chi^2 = \sum \frac{(O-E)^2}{E}$$

where O is the observed frequency in a given class—that is, the number of individuals falling in that class, E is the expected or theoretical number of observations that should fall in that class according to the hypothesis being tested or distribution being compared, and the summation extends over all the classes in the sample.

The χ^2 distribution, which is tabulated in Appendix Table III, depends upon the degrees of freedom, and for a given number of degrees of freedom, the larger the value of χ^2, the smaller the probability. As the formula for χ^2 shows, the greater the deviation of the observed number in a class from its expected value, the greater the value of χ^2 and the lower the probability. This suggests that the probabilities are a measure of how frequently the observations would differ from the expected value by a given amount, if the population from which the sample is taken really did not differ from the theoretical distribution. The table gives the probabilities of observing a value of χ^2 under the null hypothesis. If the observed χ^2 is very large and the probability correspondingly very small, the null hypothesis would be rejected, and one would be forced to assume that there is some real difference between the observations and the theoretical expectation.

The three problems that χ^2 analysis helps to solve differ only in the way in which the expected values are found and the number of degrees of freedom associated with the test.

Goodness-of-Fit

Assume that a sample of animals has been measured with respect to some continuous variate and the normality of the distribution of this variate is to be tested. There is an infinity of normal curves, and the observations could be tested for their fit to any one of these, but the idea of "testing for normality" assumes not that a particular normal curve is in question, but rather whether the distribution will fit any normal curve. Since a normal curve is specified by its mean μ and standard deviation σ, the obvious procedure in testing the normality of an observed distribution is to use $\bar{X}$ and s, the sample mean and standard deviation, to specify a normal curve. Having calculated $\bar{X}$ and s, the observations should be grouped into a convenient number of classes of equal length in the manner discussed in Chapter 3. The class limits are then converted into deviations from $\bar{X}$, and each of these deviations is then divided by s. The class limits are now standardized normal deviates (see page 135), so that the probability of falling within the classes can be read off from the table of the standardized normal distribution. This procedure has already been shown in detail in Example 47, page 137, and is repeated for reference in Example 84. Having computed the expected, or theoretical, numbers for each class under the assumption that the variate is really normally distributed, these may be compared with the observed number in each class and χ^2 for the goodness-of-fit calculated. This has been done in Example 85.

EXAMPLE 84. Fitting a theoretical normal distribution to the observed distribution of Example 29, page 79.

TAIL LENGTH (MM.)	CLASS LIMITS AS DEVIATIONS		$\dfrac{d}{s}$		NORMAL PROBABILITIES	THEORETICAL FREQUENCIES	OBSERVED FREQUENCIES
51.50–53.49...	−8.93–	−6.93	−2.92–	−2.26	.0102	.88	1
53.50–55.49...	−6.93–	−4.93	−2.26–	−1.61	.0418	3.59	3
55.50–57.49...	−4.93–	−2.93	−1.61–	−.96	.1148	9.87	11
57.50–59.49...	−2.93–	.93	−.96–	−.30	.2136	18.37	18
59.50–61.49...	−.93–	1.07	−.30–	.35	.2547	21.90	21
61.50–63.49...	1.07–	3.07	.35–	1.00	.2045	17.59	20
63.50–65.49...	3.07–	5.07	1.00–	1.66	.1102	9.48	9
65.50–67.49...	5.07–	7.07	1.66–	2.31	.0381	3.28	2
67.50–69.49...	7.07–	9.07	2.31–	2.96	.0089	.77	1
				TOTAL	.9968	85.73	86

$\bar{X} = 60.43$

$N = 86$

$s = 3.06$

EXAMPLE 85. Calculation of χ^2 for goodness-of-fit of the observations of Example 29 to a normal distribution. Expected frequencies are calculated in Example 84.

CLASS	OBSERVED O	EXPECTED E	$O-E$	$(O-E)^2$	$\dfrac{(O-E)^2}{E}$
51.50–53.49	1	.88 ⎫	.47	.22	.05
53.50–55.49	3	3.59 ⎭			
55.50–57.49	11	9.87	1.13	1.28	.13
57.50–59.49	18	18.37	.37	.14	.01
59.50–61.49	21	21.90	.90	.81	.04
61.50–63.49	20	17.59	2.41	5.81	.33
63.50–65.49	9	9.48	.48	.23	.02
65.50–67.49	2	3.28 ⎫	1.05	1.10	.27
67.50–69.49	1	.77 ⎭			
TOTAL	86	85.73		$\sum \dfrac{(O-E)^2}{E} = .85 = \chi^2$	

$\chi^2 = .85$, with 4 degrees of freedom

In accordance with the definition of χ^2, each observed value was subtracted from its corresponding expected value, that difference squared, and the result divided by the expected value. The sum of these fractions is equal to χ^2. The first two classes were lumped together as well as the last two in order to make the expected numbers in the resulting enlarged classes somewhat greater. χ^2 as calculated is upwardly biased when the expected value of a class is too small, and in fitting a frequency distribution, very small expected numbers (less than 1) can be avoided by lumping consecutive classes as has been done here.

Before the resulting χ^2 can have a probability attached to it, the degrees of freedom must be determined. There are seven classes actually used in the computation of χ^2 after lumping. The number of degrees of freedom in a χ^2 is equal to the number of classes entering the calculation minus the number of *restrictions* placed on the expected frequencies by the observations themselves, or, to put it another way, the number of constants derived from the observations which were used in calculating the expected values. In finding the theoretical values, the constants N, $\bar{X}$, and s, all taken from the observations, were used. Therefore, the number of degrees of freedom is

$$\text{d.f.} = 7 - 3 = 4$$

Entering the χ^2 table with 4 degrees of freedom, it can be seen that the observed χ^2 of .85 has a very high probability lying between .90 and .95. The difference between the observations and the expectation is not significant, and it may be safely assumed that this variate, tail length in *Peromyscus maniculatus*, is normally distributed.

In order to illustrate the method of finding the degrees of freedom, which is usually the most troublesome matter in tests of goodness-of-fit, Example 86 has been introduced to show a test of observations against a Poisson distribution. In Chapter 8, it was shown that the probability that a Poisson variate takes the value X is

$$\frac{e^{-np}(np)^x}{X!}$$

In this case $\bar{X}$, the sample mean, is the estimate of np, the population mean. The calculations for finding the theoretical or expected values for each class are shown in detail in Example 45.

The absolute frequencies to be expected are found as for the normal distribution by multiplying the relative frequencies by $N = 30$.

In Example 86, χ^2 for the goodness-of-fit is calculated. There are four classes entering into the calculation of χ^2 after lumping, and there are two constants which enter into the calculation of the expectations, $\bar{X}$ and N, so the degrees of freedom are

$$\text{d.f.} = 4 - 2 = 2$$

The table of χ^2 shows that for 2 degrees of freedom a χ^2 of 1.13 has a probability between .75 and .50, so that the agreement between observed and expected distributions is satisfactory.

EXAMPLE 86. Calculation of χ^2 for goodness-of-fit of the data of Example 44 to a Poisson distribution.

NUMBER OF SPECIMENS PER SQUARE (X)	OBSERVED DISTRIBUTION (O)	EXPECTED DISTRIBUTION (E)	$O-E$	$(O-E)^2$	$\dfrac{(O-E)^2}{E}$
0	16	14.4	1.6	2.56	.18
1	9	10.5	−1.5	2.25	.21
2	3	3.9	−.9	.81	.21
3	1 ⎫	.9 ⎫			
4	1 ⎬ 2	.3 ⎬ 1.2	.8	.64	.53
5 and over	0 ⎭	.0 ⎭			
TOTAL	30	30.0		$\sum \dfrac{(O-E)^2}{E} = 1.13 = \chi^2$	

$\chi^2 = 1.13$, with 2 degrees of freedom

If occasion should arise to fit an observed distribution to some predetermined normal, Poisson, or other distribution, the parameters of which are not estimated from the observation but are postulated *a priori*, then the number of degrees of freedom will be one less than the number of classes. This is because the only constant used is N, the sample size, which is always necessary to convert the theoretical relative frequencies to absolute frequencies commensurate with the sample. *A degree of freedom is lost for every constant calculated from the observations.*

The Variance Ratio Test

In fitting an observed distribution to a Poisson or binomial distribution it is usually necessary, as in Example 86, to lump together several of the more extreme classes in order to bring the expected number in the smallest class above 1. In lumping, however, the power of the test to detect a discrepancy between observed and theoretical distributions is seriously curtailed. The loss of 1 degree of freedom in Example 86 virtually assures that no great disagreement between expected and observed distributions will be detected, since it is often in the more extreme classes that the observed distribution differs from the theoretical frequencies. To avoid this loss of power, it is possible to avoid lumping altogether, but such a procedure will often grossly overestimate the discrepancy due to an inflated χ^2 value. For the Poisson and binomial distributions there is an alternative test procedure based upon the fact that the mean and variance of such distributions are related. For a Poisson distribution, the mean is equal to the variance, while for the binomial,

$$\sigma^2 = npq = \frac{\mu(n-\mu)}{n}$$

If an observed distribution fits a binomial or Poisson distribution, the observed sample variance of the distribution should be equal to the theoretical variance of the distribution to which it is being fitted. But the theoretical variance is related to the theoretical mean by the relationships given above, and the theoretical mean, in turn, is arbitrarily made equal to the sample mean when fitting an observed distribution to an expected one. Thus, within the limits of sampling error, if an observed distribution fits a Poisson distribution,

$$\sigma^2 = \bar{X} = s^2$$

while if it fits a binomial distribution,

$$\sigma^2 = \frac{\bar{X}(n-\bar{X})}{n} = s^2$$

These two hypotheses about the equality of a sample variance, s^2, and a theoretical variance, σ^2, can be tested by the ratio

$$\chi^2 = \frac{s^2 N}{\sigma^2}$$

which has the χ^2 distribution with $N - 1$ degrees of freedom ($n - 1$ for the case of the binomial distribution). It must be remembered that the hypothesis being tested is "two sided," while the χ^2 distribution is "one sided." Thus, if s^2 is much *larger* than σ^2, a large value of χ^2 will result with a correspondingly low probability. On the other hand, if s^2 should be much *smaller* than σ^2, the value of χ^2 will be low and the probability as judged from the χ^2 table will be small. Nevertheless, if s^2 is smaller than σ^2, this warrants rejection of the null hypothesis just as much as if s^2 were too large. This difficulty is dealt with simply by using the tabulated probability in the χ^2 table when s^2 is greater than σ^2, but using one minus this probability when s^2 is smaller than σ^2.

Example 86 provides an instructive example of the use of this variance ratio test. The observed variance, s^2, was calculated in Example 45 as 1.034. The sample mean was found to be .73, so that

$$\frac{s^2 N}{\sigma^2} = \frac{s^2 N}{\overline{X}} = \frac{(1.034)(30)}{.73} = 42.49$$

From the χ^2 table, the value 42.49 with 29 degrees of freedom has a probability of .05 so that the difference between the observed and theoretical Poisson distribution would be considered barely significant at the 5 per cent level. In contrast to this, the usual χ^2 test for goodness-of-fit as calculated in Example 86 shows a very good agreement with the theoretical distribution. Calculation of the value for Example 86 without lumping the last two observed classes gives a χ^2 of 2.24 which, for 3 degrees of freedom, corresponds to a probability of .50. For this case, then, lumping of the classes with small expectations had little effect, while both goodness-of-fit tests differ markedly in their results from the variance ratio test.

Here, as before, the problem arises as to how sensitive a test the zoologist wishes to use and how much of a difference between observation and hypothesis is *biologically* significant. The very sensitive variance ratio test says, in effect, that the observations do not really fit a Poisson distribution too well because the observed variance is too high. The χ^2 goodness-of-fit test, on the other hand, says that the observations are not too far from a Poisson distribution or some J-shaped distribution not very different from a Poisson model. If the exactness of fit to the Poisson model is critical to the zoological problem, then the result of the more sensitive test should be used, while if a more general resemblance to the Poisson distribution is sufficient, the less sensitive test will do. The data themselves show a tendency for an excess of observations in both extreme classes (0 specimens and 4 specimens), an excess which is confirmed by the variance ratio test. Such an excess may give rise to an hypothesis on the process of fossilization which will in turn lead to further investigation, or it may be regarded as trivial. This is a zoological rather than a statistical decision.

Association

Correlation is possible only between variates with definitely ascertainable numerical values and when each variate takes a considerable number of different values. The last two chapters have suggested how wide a variety of important problems may be treated by the methods of correlation and regression, but there remain many problems of a similar sort not subject to these methods.

Association is a relationship in which some category of observations tends to occur together with a category of some other given sort of observation more often than can be ascribed to chance alone. It reveals the existence of some kind of connection between two or more sorts of observations. Correlation is a special sort of association in which all the categories are numerical and each set of observations is divided into multiple categories. Other methods are more desirable when categories are not numerical or when numerical categories are too few to give reliable results by correlation methods. The general types of association not efficiently treated by correlation can be summarized and exemplified as follows:

1. Between a variate with multiple categories and a variate with few categories, e.g., between depth of burrow and larger or smaller animals.
2. Between two variates with few categories, e.g., between counts of dorsal and anal fin rays of fish, the distribution of each covering only two or three classes.
3. Between a variate with multiple categories and an attribute, e.g., between weight of fishes of a given species and geographic location.
4. Between a variate with few categories and an attribute, e.g., between number of cuspules on a tooth and stratigraphic occurrence of a fossil mammal.
5. Between two attributes, e.g., between sex and susceptibility to disease.

The same general method can be applied to all these different problems and to any analogous to them. The variety of problems that can be treated by general methods of association is, indeed, much greater than of those that can be dealt with by correlation, and their importance is not less. It should be noted, also, that a variate for which only inadequate or inaccurate data are at hand can often be tested for association even though a correlation coefficient could not be based on it. It is necessary that the data suffice only for a reasonably good division into two or more categories. For instance, association may be tested by merely dividing a sample into smaller and larger observations by rough measurement or without actual measurement. Likewise, a series of observations of a variate with multiple categories can be arbitrarily divided at any point into smaller and larger observations by rough measurement or without actual measurement. Likewise, a series of observations of a variate with multiple categories can be arbitrarily divided at any point into two parts and its association with some other variate or with an attribute tested, a procedure that may greatly simplify problems and reduce the work involved in studying them.

Contingency Classifications

The simplest instances of association are those in which each set of observations has two categories. For the combination of the two sets, there are then four possible categories, and data arranged in this way are said to be placed in a fourfold or 2 × 2 classification. For instance, in studying the association of sex and susceptibility to disease, one set of observations has only the two categories, male and female, and the other only the two, well and diseased. The combination has the four categories:

>Male and well
>Male and diseased
>Female and well
>Female and diseased

This can also be arranged as a dichotomous classification:

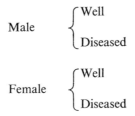

In practice, it is usually most convenient and comprehensible to arrange the data in what is called a "contingency table," a set of rectangular cells with the categories of one set of observations labeled at the top, those of the other at the left side, the corresponding frequencies entered in the cells, and the totals of rows and columns to the right and below the table. Such a table is shown in Example 87.

In order to test whether such data show any association, it is first necessary to establish what the frequencies would be if there were no association, i.e., if the two sorts of things observed were completely independent. Obviously, the numbers of observations in the two samples have nothing to do with association, nor have the total numbers of observations falling into any one category. The marginal totals, in other words, have no direct bearing on association, and in any specific problem they are to be taken as given and immutable. The next step then, is to see what distribution of frequencies within the cells would give the marginal totals actually observed and would show complete theoretical agreement with the hypothesis that the two sets of observations do not influence each other.

EXAMPLE 87. Contingency table of geographic locality and number of serrations on last lower premolar in closely similar members of the fossil mammalian genus *Ptilodus*. (Original data)

NUMBER OF SERRATIONS ON P_4

	LESS THAN 14	MORE THAN 13	TOTALS
Montana	8	21	29
New Mexico	6	0	6
TOTALS	14	21	35

GENERAL FORM OF SUCH 2×2 TABLES

FIRST ATTRIBUTE OR VARIATE

SECOND ATTRIBUTE OR VARIATE		1ST CATEGORY	2ND CATEGORY	TOTALS
	1st category	a	b	$a + b$
	2nd category	c	d	$c + d$
	TOTALS	$a + c$	$b + d$	$a + b + c + d = N$

These conditions would be fulfilled if the total $(a + b)$ were so divided that a and b had the same ratio as $(a + c)$ to $(b + d)$ and if all the other analogous ratios between cells were made equal to those between marginal totals. This would show complete independence, for then the cell frequencies would follow entirely from the totals, which are not affected by any dependence between the two sets of observations. The conditions to be filled are, then, to keep the same totals and also to make

$$\frac{A}{B} = \frac{(a + c)}{(b + d)}$$

$$\frac{C}{D} = \frac{(a + c)}{(b + d)}$$

$$\frac{A}{C} = \frac{(a + b)}{(c + d)}$$

$$\frac{B}{D} = \frac{(a + b)}{(c + d)}$$

in which capital letters are used to indicate theoretical frequencies consistent with independence and lower-case letters to represent the observed totals. All the theoretical frequencies could be calculated by using these as simultaneous equations, but it is more convenient to use formulae by which each theoretical frequency can be calculated separately and directly from the marginal totals. The formulae best used are

$$A = \frac{(a+b)(a+c)}{N}$$

$$B = \frac{(a+b)(b+d)}{N}$$

$$C = \frac{(c+d)(a+c)}{N}$$

$$D = \frac{(c+d)(b+d)}{N}$$

The formulae can easily be remembered by the rule that the theoretical frequency of any cell is the total for the row in which it occurs multiplied by that for the column in which it occurs and divided by the total frequency.

A contingency table can be made with any number of cells, and the rule for finding the theoretical frequencies is the same whatever the size of the table. In practice, there are seldom both many rows and many columns in a table. The simplest and also the most common are 2 × 2 tables; 2 × 3, 2 × 4, and 2 × 5 tables are not uncommon, and 3 × 3 or 3 × 4 tables may also be useful occasionally. Larger tables are cumbersome and are seldom necessary. If a set of observations is on a variate and has many categories, it is usually better to lump these into two or, exceptionally, three. Attributes seldom have many categories and can often also be lumped if they are too finely subdivided for ease of handling.

The work of calculating theoretical frequencies in a simple 2 × 3 table is shown in Example 88.

It is not to be expected that the frequencies actually observed in samples will correspond exactly with the theoretical frequencies or with their nearest integral value, even if the variates and attributes studied are really completely independent in the population. Chance necessarily plays a part, and the chance of complete agreement is always very small. What is needed, then, is to determine the probability that deviations from the theoretical frequencies equal to those observed could have arisen by chance in sampling a population in which the true proportions were those indicated by the theoretical frequencies. If such deviations could have arisen by chance, the data do not prove that the hypothesis of independence is inapplicable. If they could not have arisen by chance, then there is a significant disagreement with the hypothesis of independence, and it follows that there is significant association in the population.

EXAMPLE 88. Contingency table of number of serrations and length of last lower premolars of the fossil mammal *Ptilodus montanus* and calculation of theoretical frequencies on the hypothesis of complete independence. (Original data)

LENGTH	SERRATIONS			TOTALS
	13	14	15	
8.0 mm. or more	4.1 / 0	9.8 / 14	1.0 / 1	15
7.9 mm. or less	3.9 / 8	9.2 / 5	1.0 / 1	14
TOTALS	8	19	2	29

The theoretical frequencies are entered in the upper left corners of the cells. They are calculated as follows:

$$\frac{15 \times 8}{29} = 4.1 \qquad \frac{15 \times 19}{29} = 9.8$$

$$\frac{15 \times 2}{29} = 1.0 \qquad \frac{14 \times 8}{29} = 3.9$$

$$\frac{14 \times 19}{29} = 9.2 \qquad \frac{14 \times 2}{29} = 1.0$$

To test whether there is significant association, the expected values are compared with the observed by χ^2 and the degrees of freedom are arrived at in the following way. If the contingency table has r rows and c columns, there are rc classes or cells. Thus a 2×2 table will have 4 classes, a 3×4 table 12 classes, and so on. In finding the expected values for each cell or class, the r row marginal totals, the c column marginal totals, and the total sample size N were used. The number of constants from the observations is apparently then $r + c + 1$. Actually, N is not an independent constant because if all the marginal totals are known, N follows directly. In addition, all of the $r + c$ marginal totals are not independent because if all but one of them are specified, the last can be calculated. There are then only $r + c - 1$ *independent* constants, which are calculated from the raw observations for use in finding the expected values. The number of degrees of freedom is then:

$$\text{d.f.} = rc - (r + c - 1) = (r - 1)(c - 1)$$

This result is completely general, and in any contingency table of no matter what size, the degrees of freedom are the number of rows minus one, multiplied by the number of columns minus one. Thus a 2×2 table has $1 \times 1 = 1$ degree of freedom, while a 4×3 table will have $3 \times 2 = 6$ degrees of freedom. This rule for finding degrees of freedom applies *only* when all the expected values for the classes are calculated from the marginal totals.

Short-Cut Methods

If the theoretical frequencies are not desired for any reason except the calculation of χ^2, this calculation may be made more simply in certain cases.

For a 2×2 contingency table,

$$\chi^2 = \frac{N(ad - bc)^2}{(a + b)(c + d)(a + c)(b + d)}$$

which is the same formula given in Chapter 10 for testing the difference between two binomial proportions.

The calculation of χ^2, both directly and by calculating the contribution to it of each cell, and its use in testing significance of association are shown in Example 89.

Testing this way is usually a necessary preliminary to a reliable zoological conclusion about association, but it does not give a direct answer to many of the questions legitimately referred to the data. These may usually, however, be answered on a logical basis by reference to the contingency table, and for this essential purpose it is generally advisable or necessary to calculate the theoretical frequencies. Thus, in Example 89 it is plain that the observations show excess frequencies in cells b and c and deficiency in cells a and d and hence that the nature of the association is that fewer dorsal rays are more often associated with fewer anal rays, more dorsal rays more often with more anal rays, fewer dorsal rays less often with more anal rays, and more dorsal rays less often with fewer anal rays than would be expected if the two variates were independent. In other words, the association plainly has the same nature as a positive correlation. The calculations also show that cell b contributes the most to χ^2 and hence departs most from the hypothesis of independence, and that cell c contributes the least and departs the least from the hypothesis.

TESTS ON FREQUENCIES **319**

EXAMPLE **89**. Test of association by χ^2. Dorsal and anal rays of the flying fish *Exocoetus obtusirostris*. (Data from Bruun, 1935)

ANAL RAYS	DORSAL RAYS		TOTALS
	12–13	14–15	
14	8.79 / 5	5.21 / 9	14
13	18.21 / 22	10.79 / 7	29
TOTALS	27	16	43

A. FROM SUMS OF CONTRIBUTIONS OF CELLS

$$\frac{(A-a)^2}{A} = \frac{3.79^2}{8.79} = 1.634$$

$$\frac{(B-b)^2}{B} = \frac{3.79^2}{5.21} = 2.757$$

$$\frac{(C-c)^2}{C} = \frac{3.79^2}{18.21} = .789$$

$$\frac{(D-d)^2}{D} = \frac{3.79^2}{10.79} = 1.331$$

$$\chi^2 = 1.634 + .789 + 2.757 + 1.331 = 6.511$$

B. DIRECT FROM RAW DATA

$$\frac{N(ad-bc)^2}{(a+b)(c+d)(a+c)(b+d)} = \frac{43(5 \times 7 - 9 \times 22)^2}{14 \times 29 \times 27 \times 16} = \frac{1{,}142{,}467}{175{,}392} = 6.514$$

If one classification has two categories and the other has two or more, the theoretical frequencies and the cell contributions to χ^2 can be calculated as just explained and exemplified; but in some cases another method may be more convenient and equally or more enlightening. The data are set up in a contingency table with two rows and two or more columns. For each column, the ratio of the number in the first row to the total is calculated, that is,

$$\frac{a'}{a' + b'}$$

if a' is taken as a value in the first row and b' as the corresponding value in the second row. The analogous ratio is calculated for the row totals, that is,

$$\frac{a}{a + b}$$

where

$a = $ the total for the first row
$b = $ the total for the second row

The value of χ^2 is then

$$\chi^2 = \frac{\sum a' \left(\frac{a'}{a' + b'}\right) - a\left(\frac{a}{a + b}\right)}{\frac{a}{a + b}\left(1 - \frac{a}{a + b}\right)}$$

When there are many columns, this method may be quicker; and it is also advantageous when, as happens, the ratios $a'/(a' + b')$ and $a/(a + b)$ have a logical and pertinent connection with the subject of the investigation. The calculation is shown in Example 90.

For the most general $r \times c$ contingency table, there is no way to avoid calculating the expected values for each cell. There is, however, a short cut formula which uses these expected numbers but avoids entirely the calculation of the deviations of observed from expected frequencies. The usual χ^2 formula

$$\chi^2 = \sum \frac{(O - E)^2}{E}$$

is algebraically equivalent to the formula

$$\chi^2 = \sum \left(\frac{O^2}{E}\right) - N$$

where N is the total number of observations. For any contingency table, this latter formula will be a more efficient calculating device than the defining formula for χ^2, since, as we have pointed out repeatedly, the necessity of computing deviations increases the chance for numerical error.

EXAMPLE 90. Ratio method of calculating χ^2 in a $2 \times c$ table. Mortality of young and observation substations of the tree-swallow *Iridoprocne bicolor*. (Data from Low, 1934)

	SUBSTATIONS						TOTALS
	A	B	C	F	H	M	
Hatched but not fledged (a')	27	11	11	3	5	10	$a = 57$
Fledged (b')	28	4	9	21	14	12	$b = 88$
Totals ($a' + b'$)	55	15	20	24	19	22	$a + b = N = 155$
Mortality ratio $\left(\dfrac{a'}{a' + b'}\right)$	.4909	.7333	.5500	.1250	.2632	.4545	$\dfrac{a}{a + b} = .4323$
$a'\left(\dfrac{a'}{a' + b'}\right)$	13.2543	8.0663	6.0500	.3750	1.3160	4.5450	$\left(\dfrac{a}{a + b}\right)a = 28.9641$

$$\sum \left[a'\left(\frac{a'}{a' + b'}\right)\right] = 33.6066 \qquad 1 - \left(\frac{a}{a+b}\right) = .5677$$

$$\chi^2 = \frac{\sum\left[a'\left(\dfrac{a'}{a'+b'}\right)\right] - a\left(\dfrac{a}{a+b}\right)}{\left(\dfrac{a}{a+b}\right)\left[1 - \left(\dfrac{a}{a+b}\right)\right]} = \frac{33.6066 - 28.9641}{(.4323)(.5677)} = \frac{4.6425}{.24542} = 18.92$$

Degrees of freedom $= (r - 1)(c - 1) = (2 - 1)(6 - 1) = 5$
P (from table) $< .01$

P is much less than .01, the deviation from independence is certainly significant, and mortality and substations are surely associated in some way. The data as here presented do not suggest why they are associated. They do, however, show that for some reason substations B and C had remarkably high mortality while F and H had remarkably low mortality, A and M being not far from average. The calculation in this example is easier by this method than would be calculating theoretical frequencies and contributions to χ^2 for each 12 cells, and the quantities $a'/(a' + b')$ are ratios (mortality ratios) highly pertinent to the problem and desirable in themselves as well as for getting χ^2.

Small Samples

We have already discussed at some length the problem of testing the 2 × 2 contingency table when sample size is small (Chapter 10, page 189). To correct χ^2 for small samples, the absolute value of the deviation in each cell is diminished by .5 before squaring. Thus, if the deviation in a given cell were 1.2, the adjusted deviation would be .7, while if the deviation were −.4, the adjusted deviation would be +.1, and so on. If the short cut formula for a 2 × 2 table is used, this correction cannot be made as such, since the deviations are not calculated. However, an algebraically equivalent correction is to add to the quantity $ad - bc$ an amount equal to $N/2$ when $ad - bc$ is negative and to subtract $N/2$ when $ad - bc$ is positive. An illustration of this correction is given in Example 91.

EXAMPLE 91. Calculation of adjusted χ^2 from a small sample. Weights and depths of burrows of the ground squirrel *Citellus columbianus columbianus*. (Data from Shaw, 1926)

	WEIGHT	DEPTH OF BURROW <29.5	>29.5	TOTALS
RAW DATA	>452	1	10	11
	<452	8	3	11
	TOTALS	9	13	22
SAME DATA, ADJUSTED FOR THE LOW FREQUENCIES	>452	1.5	9.5	11
	<452	7.5	3.5	11
	TOTALS	9	13	22

$$\chi^2 = \frac{(1.5 \times 3.5 - 9.5 \times 7.5)^2 \times 22}{11 \times 11 \times 9 \times 13} = 6.77$$

or $$\chi^2 = \frac{22 \left(1 \times 3 - 10 \times 8 - \frac{22}{2}\right)^2}{11 \times 11 \times 9 \times 13} = 6.77$$

P is slightly less than .01, and the association may be taken as significant. Larger ground squirrels dig deeper burrows. χ^2 calculated without adjustment is 9.6.

To summarize the recommendations for the use of the corrected χ^2 already outlined in Chapter 10:

1. When N is *greater* than 40 and the smallest observed frequency is 10 or *less*, use the adjusted χ^2.
2. When the smallest observed frequency is greater than 10, use the unadjusted χ^2.
3. When N is less than or equal to 40, calculate both the adjusted and unadjusted values of χ^2. If both indicate a significant difference, reject the hypothesis, while if both indicate no significant difference, do not reject. If the unadjusted χ^2 is significant while the adjusted χ^2 is not significant, the hypothesis must be regarded with suspicion although there is no definite evidence for its rejection. An alternative is to use Fisher's exact test (1950).

For $r \times c$ contingency tables in which r and c are not both equal to 2, it is not possible to use Yates' correction to χ^2. For such tables, adjacent rows or adjacent columns should be lumped to raise the cell frequencies. It is not possible to lump only two cells together in a row or column, since the table must always be strictly rectangular with in entry in each cell. Thus, a 5×8 table may be reduced to a 5×7 or 4×8 by lumping two adjacent rows or columns. If lumping can be done either in rows or in columns, it is best to choose the procedure which will maintain the higher number of degrees of freedom. In lumping the 5×8 table, the resultant 5×7 is preferable to the 4×8, since the former has 24 degrees of freedom while the latter has only 21. For a $2 \times c$ table, of course, there is no choice and columns must be lumped. Adding the two rows together will destroy the table and the test.

The general rule to determine when lumping is necessary is that no cell should have an expected frequency less than 1, and not more than 20 per cent of the cells should have expected frequencies less than 5. If it is not possible to reduce the table by lumping to conform to these rules, it is best to abandon any reasonable hope of a statistical test. Samples with such small numbers can give no reliable information.

The Meaning of a Test of Association

The χ^2 test for association is a method for determining whether or not there is a significant association between two variables. The larger the observed value of χ^2 for a given number of degrees of freedom, the more likely it is that the variables are indeed associated. It is tempting to go one step further than this and reason that the actual value of χ^2 is a measure of the intensity of association, larger χ^2 values signifying stronger association.

Unfortunately, *this is quite wrong in general*, although there are circumstances under which a larger χ^2 does mean a closer association. A simple example of a 2 × 2 contingency table will show why χ^2 does not measure the intensity of association. Example 92A is a hypothetical contingency table for a sample of 100 specimens scored for two attributes. Since all the marginal totals are equal, each cell should have 25 per cent of the individuals in it if the variables were independent. Actually, two of the cells have 30 per cent and two have 20 per cent. As the calculation in the example shows, χ^2 with 1 degree of freedom equals 4. Example 92B shows exactly the same situation, but with 200 instead of 100 specimens. Each cell ought to have 25 per cent of the observations in it under the null hypothesis, but again two cells contain 30 per cent and two contain 20 per cent of the observations in the same relationship as Example 92A. The χ^2 calculation now produces a χ^2 with 1 degree of freedom equal to 8. Thus doubling the sample size has doubled the χ^2 value, although the probability of falling in a given cell and thus the intensity of association is the same for both examples. Indeed, both samples have been chosen from the same population. This result is completely general, and for a given set of probabilities in each cell, χ^2 is directly proportional to sample size N. Thus χ^2 is not suitable as measure of the intensity of association.

How, then, can one measure the degree of association between two variables? As R. A. Fisher points out: "To measure the degree of association, it is necessary to have some hypothesis as to the nature of the departure from independence to be measured." Thus, in our hypothetical Example 92, the deviation in each cell of the table from the expected proportion was 5 per cent. In a 2 × 2 contingency table, since all cells have deviations of the same absolute value, this deviation might be used as a measure of association. On the other hand, it might be more desirable to use a relative measure—for example, the deviation in a particular cell divided by its expected proportion. In the hypothetical example, this would be 5 per cent/25 per cent = 20 per cent.

In any case, the measure of association will be some number calculated from the observations on the basis of a postulated relationship, while the *test* of association will be a contingency χ^2. If χ^2 is significant, then it can be said that the measure of association is also significantly different from zero; but this is quite different from saying that χ^2 *is* the measure of association.

Test of Homogeneity

Two or more samples are said to be homogeneous with respect to a given variable if they do not indicate a reasonable difference in the

EXAMPLE 92. Hypothetical example to demonstrate the dependence of χ^2 on sample size.

A.

		CHARACTER A		
		Present	Absent	
CHARACTER B	Present	30	20	50
	Absent	20	30	50
		50	50	100

$$\chi^2 = \frac{N[(ad - bc)^2]}{(a + b)(c + d)(a + c)(b + d)} = \frac{100\,(500^2)}{6{,}250{,}000} = \frac{2{,}500}{625} = 4$$

B.

		CHARACTER A		
		Present	Absent	
CHARACTER B	Present	60	40	100
	Absent	40	60	100
		100	100	200

$$\chi^2 = \frac{200\,(2000^2)}{100{,}000{,}000} = \frac{800{,}000{,}000}{100{,}000{,}000} = 8$$

distributions of that variable in the corresponding populations. To test this relation, exactly the same procedure is used as for association. In a sense, homogeneity can be looked at as a special case of independence of two attributes, except that one of the attributes is now the particular population from which the sample is taken. If the populations have the same distributions, then there will be no association between population and the variate under consideration. A test for homogeneity of two samples is simply a $2 \times c$ contingency table, c being the number of classes into which the measured variate is divided. To compare more than 2 populations, an $r \times c$ contingency table test is used.

Examples 93A and 93B show tests for homogeneity between two and three populations, respectively. The contingency χ^2 has been calculated by the short form for a $2 \times c$ table in Example 93A, while the full formula for χ^2 has been used on the 3×3 table in Example 93B.

EXAMPLE 93. Use of χ^2 to test homogeneity of samples.

A. Two samples of length of P_4, both identified as the fossil mammal *Ptilodus montanus* from approximately the same horizon and locality but collected at different times by different institutions. (Original data)

CALCULATION OF χ^2 BY RATIO METHOD

CLASSES	SAMPLE 1	SAMPLE 2	TOTALS	RATIOS	SAMPLE 1 × RATIOS
7.5 and below	5	3	8	.6250	3.1250
7.6–7.8 ...	9	6	15	.6000	5.4000
7.9–8.1 ...	7	9	16	.4375	3.0625
8.2–8.4 ...	8	14	22	.3636	2.9088
8.5 and above	7	7	14	.5000	3.5000
TOTALS	36	39	75	.4800	17.9963

Total of sample 1 × total of ratios = 17.2800

$$\chi^2 = \frac{17.9963 - 17.2800}{.48 \times .52} = 2.87$$

Degrees of freedom = 4 $P = .50 - .70$

The difference is not significant, and the samples are consistent with the hypothesis that both are from the same population.

More Complex Cases

When more than two variates have been observed in a population, the simple homogeneity and association tests obviously do not apply, since the observation cannot be put in a single $r \times c$ table. It might be that the sex and number of scales are observed on reptiles from a number of populations, or that both the color and the presence or absence of wing spots are noted from several different samples of moths. Such cases require an analysis or partitioning of χ^2 into various components, analogous to the partitioning of variance in similar situations for continuous variates. Unfortunately, χ^2 often does not possess convenient additive properties like the sum of squares in the analysis of variance, and although a considerable amount has been written concerning methods of partitioning χ^2, no full agreement exists among statisticians for many important cases, and the methods presented often have a degree of statistical complexity rarely justified in zoological practice. For these reasons, we will treat complex χ^2 situations in a fairly simplified manner, without attempting to analogize them with analysis of variance techniques and will deal only with such cases as the zoologist is most likely to meet.

EXAMPLE 93. *continued*

B. Three samples from different localities. Hair counts of winter pelage of the deer mouse *Peromyscus maniculatus rubidus*. (Data modified from Heustis, 1931)

CALCULATION BY CONTINGENCY TABLE. (THE DATA HAVE BEEN ARTIFICIALLY SIMPLIFIED FOR CLEARER EXEMPLIFICATION OF METHOD.)

HAIR TYPES	SAMPLES			TOTALS
	Coos Bay	Port Orford	Eugene	
BLACK OVERHAIRS	22.3 / 24	19.1 / 16	28.6 / 30	70
LARGE BANDED HAIRS	12.4 / 9	10.6 / 12	16.0 / 18	39
FUR HAIRS	103.3 / 105	88.3 / 90	132.4 / 129	324
TOTALS	138	118	177	433

CONTRIBUTIONS TO χ^2

$$\frac{(24 - 22.3)^2}{22.3} = .130 \qquad \frac{(18 - 15.9)^2}{15.9} = .250$$

$$\frac{(19.1 - 16)^2}{19.1} = .503 \qquad \frac{(105 - 103)^2}{103} = .028$$

$$\frac{(30 - 28.6)^2}{28.6} = .069 \qquad \frac{(90 - 88.2)^2}{88.2} = .033$$

$$\frac{(12.4 - 9)^2}{12.4} = .932 \qquad \frac{(132 - 129)^2}{132} = .087$$

$$\frac{(12 - 10.6)^2}{10.6} = .185 \qquad \text{TOTAL} \qquad 2.217 = \chi^2$$

Degrees of freedom = 4 $\chi^2 = 2.22$ $P = .7$

The difference is not significant, and the samples are consistent with the hypothesis that all are from the same population.

The simplest approach is to examine some specific examples which illustrate the more important kinds of observations and their appropriate analysis. These illustrations are taken from Lamotte's work (1951) on polymorphism in the land snail *Cepaea nemoralis*. The characters scored are shell color (yellow or rose) and the presence or absence of black bands on the shell. Every specimen can be scored for both of these variates. Collections were made in a large number of European localities, live snails having been picked up at random in each locality. In addition, broken shells were also collected at random, the breakage having been caused by predation. Birds carry the shells to some height and then drop them on rocks in order to break the shell and expose the fleshy contents.

One question of interest to Lamotte was whether the presence or absence of banding had any relation to predation pressure. If such a relationship did exist, it might indicate that one form or the other was more conspicuous and thus more often attacked by the birds. Example 94 shows the result of collections made in six localities, all of a very similar vegetational and physiographic aspect.

EXAMPLE 94. Proportions of the form "unbanded" among broken and intact shells in several colonies of *Cepaea nemoralis* from the Somme valley. (Data from Lamotte, 1951)

COLONY	BROKEN SHELLS			INTACT SHELLS		
	UNBANDED	BANDED	% UNBANDED	UNBANDED	BANDED	% UNBANDED
No. 780	30	68	30.6	195	342	36.7
No. 791	54	48	52.9	276	237	53.4
No. 798	211	280	42.9	886	1,227	41.9
No. 801	65	99	39.6	32	61	34.4
No. 809	37	37	50.0	63	79	44.3
No. 818	1,160	1,025	53.1	3,856	4,031	48.9
TOTAL	1,557	1,557	50.0	5,308	5,977	47.0

The question of an association between presence of bands and predation pressure is answered by a contingency table of the form.

	Banded	Unbanded
Broken	a	b
Intact	c	d

If the proportion of banded shells were the same in both broken and intact collections, the contingency χ^2 would be small, the probability high, and the conclusion would be that banding has no effect on predation. The first and most obvious approach is simply to use the grand totals over all colonies in this 2 × 2 contingency table. The table would then have the form.

	Banded	Unbanded	TOTAL
Broken	1,557	1,557	3,114
Intact	5,977	5,308	11,285
TOTAL	7,534	6,865	14,399

The contingency χ^2 for this table with 1 degree of freedom is

$$\chi^2 = \frac{(1{,}557 \times 5{,}308 - 1{,}557 \times 5{,}977)^2 \times 14{,}399}{(7{,}534)(6{,}865)(3{,}114)(11{,}285)} = 8.596$$

which has a probability of about .001. There are two difficulties in using this χ^2 on the totals which make it unreliable. First, it is affected by differences among colonies in the proportion of banded shells. If there is heterogeneity among samples from which a total χ^2 is calculated, that total is no longer a valid quantity for a χ^2 test. A second objection is that one colony may contribute a disproportionate amount to the test of association. If one colony shows a strong association between banding and predation and in addition it has a disproportionately large sample size, the total χ^2 will be unduly distorted. Both of these objections, in fact, apply to the data of Example 94. It is apparent from inspection of the per cent unbanded in each colony that there is some difference among colonies in this respect. If these differences are not significant, that is, if there is no significant heterogeneity among the colonies in respect to banding, then the total χ^2 is not invalidated. We shall return presently to the test for heterogeneity among the colonies. The second objection also applies because colony No. 818 contains 70 per cent of all the individuals in the total sample. Thus the χ^2 for totals is mainly a reflection of this one colony.

To answer the question of association and yet avoid these problems, it is necessary to go to the individual colony entries. Each colony is independent of the others and a 2 × 2 χ^2 can be calculated for each separately. For colony No. 780 the 2 × 2 table would be of the form

	Banded	Unbanded	TOTAL
Broken	68	30	98
Intact	342	195	537
TOTAL	410	225	635

and

$$\chi^2 = \frac{(68)(195) - (342)(30)}{(410)(225)(98)(537)} \times 635 = 1.18$$

and so on for the remaining five colonies. Example 95 shows these χ^2 results, together with their associated probabilities. Only colony No. 818 shows a significant association between banding and predation. Since this is the colony from which the very large sample was taken it is clear that the χ^2 on totals was a biased figure.

These six χ^2 values may be combined into one because of a fortunate additive property of χ^2. The total of a number of independent χ^2 values is itself distributed as χ^2 and has a number of degrees of freedom equal to the sum of the individual degrees of freedom. Two tests are independent of each other if the data from one do not enter in any way (including calculation of expected values) into the other. This criterion applies to the six snail populations since each was tested by a 2 × 2 contingency table involving only the observations from the particular population in question. Summing the six χ^2 values in Example 95 gives 15.93 as χ^2 with 6 degrees of freedom since each 2 × 2 test had a single degree of freedom. A χ^2 of 15.93 with 6 degrees of freedom has a probability of .015, which although small is considerably larger than that for the χ^2 calculated from the totals added over colonies.

EXAMPLE 95. Results of contingency χ^2 for the relationship between banding and shell condition in each colony of Example 94.

COLONY	% UNBANDED		χ^2	P	x
	BROKEN	INTACT			
No. 780	30.6	36.7	1.18	.30	−1.09
No. 791	52.9	53.4	1.20	.30	−1.10
No. 798	42.9	41.9	.18	.67	+ .42
No. 801	39.6	34.4	.69	.40	+ .83
No. 809	50.0	44.3	.62	.41	+ .79
No. 818	53.1	48.9	12.06	.001	+3.47
TOTAL			15.93		+3.32

Even this method of testing the association is not entirely satisfactory. While it does not suffer from the disproportionate sample size of colony No. 818, it is strongly affected by the disproportionate association in that colony. The other five colonies showed no evidence of association at all, but colony No. 818 had such a large value of χ^2 that it overbalanced the others. Another objection to addition of the six χ^2 values is that no account is taken of the direction of the deviation in each colony, since χ^2 is always positive. Thus colonies Nos. 780 and 791 both have a lower percentage of unbanded shells in the broken sample than in the intact, while the other four colonies have deviations in the opposite direction. Any biological conclusion about the effect of shell banding upon predation must take into account the direction of these deviations. If all the deviations are in the same direction then no matter what the value of χ^2, there would be a strong suspicion that there is some association. On the other hand, the fact that the deviations are not consistent, makes the significant χ^2 of somewhat doubtful meaning. A solution to this problem is shown in the last column of Example 95. The values listed here are the square roots of the χ^2 values for each colony, denoted as χ. To each χ a sign has been given depending upon the direction of the deviation. It is entirely arbitrary whether a higher percentage among broken shells will be considered a positive or negative deviation, but having made some convention, it is adhered to in the rest of the table. Thus the first two colonies have a minus sign, the remaining four a plus sign in accordance with the change in the direction of the association. The total of these χ values, taking sign into account, is given at the bottom of the table. This quantity when divided by the square root of the number of degrees of freedom is a standardized normal deviate. That is,

$$\frac{\Sigma \chi}{\sqrt{\text{d.f.}}} = \frac{3.32}{\sqrt{6}} = 1.36$$

can be looked up directly in the normal tables. The probability that the standardized normal deviate falls within the range ± 1.36 is .826, or the probability of falling outside these limits is $1 - .826 = .174$. This is a much higher probability than that obtained by either of the two previous calculations and, moreover, bears directly on the biological question. The conclusion is that banded and unbanded shells are equally liable to predation by birds although the very high value of χ in colony No. 818 suggests that this locality may be different from the others in this respect.

Having now disposed of the problem of association, it is of some interest to look at the problem of heterogeneity among the colonies. The first approach to this problem would be to make a χ^2 test of homogeneity on the entire 4 × 6 table. There are four classes: broken–banded, broken–unbanded, intact–banded and intact–unbanded, and each of these classes appears in the six populations. Such a homogeneity χ^2 is incorrect, however, because of the nature of the sampling. The 4 × 6 contingency χ^2 will have a large contribution due to the difference in the ratio of broken to intact shells from colony to colony. This difference is not a reflection of any biological situation but simply of the methods of collection. The collector picked up a sample of unbroken shells at random with respect to banding and then a sample from the same locality of broken shells at random with respect to banding, but the number of broken and unbroken shells in each colony are simply functions of how hard he looked for each, when he decided he had enough of each, how extensively the area had been preyed upon, and so on. The ratio of broken to unbroken shells has no particular meaning, and it must not enter into the calculation of homogeneity. The proper way to test the homogeneity from population to population is to calculate two separate 2 × 6 contingency χ^2 values, one for broken shells and one for unbroken. Each has 5 degrees of freedom. When this is done, the results are

Broken	Intact
$\chi^2 = 35.95$	$\chi^2 = 73.54$
$P < .001$	$P < .001$

The colonies differ in their proportions of banded and unbanded shells among both live snails and broken shells.

Another case of a complex χ^2 analysis, also taken from Lamotte's work, is shown in Example 96. Although superficially resembling the previous example, there is one important difference. While broken and unbroken shells were really separate samples from each colony and no biological relationship existed between the frequency of broken and intact shells, the relative numbers of yellow to rose shells in Example 96 is a biological fact, not an artifact of collection. A random sample of shells was collected from each locality and then classified as to color and banding, so that the ratio of yellow to rose is a sample estimate of the true proportion in each colony.

It is important that the zoologist distinguish in his own material between the first case in which only one of the variates has been randomly sampled, and the second case in which both have been so sampled.

EXAMPLE 96. Morphological composition of five colonies of *Cepaea nemoralis* with respect to shell color and banding. (Data from Lamotte, 1951)

COLONY	YELLOW		ROSE	
	UNBANDED	BANDED	UNBANDED	BANDED
No. 683	4	106	20	59
No. 405	12	162	55	71
No. 153	15	64	25	62
No. 152	8	92	5	21
No. 194	3	48	13	39

Part of the analysis of this second case is the same as for the first case. To test for association between shell color and banding, five separate

2 × 2 contingency tables are constructed, one for each population. Thus for colony No. 683, the association between color and banding would be tested by the following tables:

	Yellow	Rose	TOTAL
Unbanded	4	20	24
Banded	106	59	165
TOTAL	110	79	189

$$\chi^2 = \frac{(4 \times 59 - 106 \times 20)^2 \, 189}{(110)(79)(24)(165)} = 19.49$$

and so on, for the remaining four colonies. The results of these contingency tests are shown in Example 97.

EXAMPLE 97. Results of the calculation of χ^2 as a test of association between shell color and banding for the data of Example 96.

COLONY	% UNBANDED YELLOW	% UNBANDED ROSE	χ^2	P	χ
No. 683	3.64	25.32	19.49	.001	4.41
No. 405	6.90	43.65	56.91	.001	7.54
No. 153	18.99	28.74	2.15	.15	1.47
No. 152	8.00	19.23	2.81	.09	1.68
No. 194	5.88	25.00	7.17	.007	2.68
TOTAL			88.53		17.78

Using the 5 per cent level as a criterion of significance, two of the five colonies, No. 152 and No. 153, do not show any significant association. When the individual χ^2 values for the five colonies, each with 1 degree of freedom, are added, however, the result with 5 degrees of freedom is clearly significant. The percentages of unbanded shells in the yellow and rose groups show that this association is consistent, in that there is in every case a lower percentage of unbanded among the yellow shells than among the rose shells. The combined χ^2 is, in the light of this fact, biologically meaningful and shows clearly that an association between banding and color exists. As a check on this, the values of χ have been computed, and since all of the deviations are in the same direction, all the χ have the same sign. The total χ is 17.78 with 5 degrees of freedom, so that

$$\frac{17.78}{\sqrt{5}} = 7.95$$

is a standardized normal deviate. This deviate is so large that it is not even given in the normal tables, and its probability is much less than .001.

Having determined that there is a distinct association between shell color and banding, the next problem is that of the heterogeneity among colonies. The total heterogeneity among the colonies may be calculated from the entire 4 × 5 table. This total heterogeneity is conveniently analysed into two contributing components, one due to color and the other to banding. There is, in addition, a contribution due to the specific interaction of color with banding, but it is not a simple matter to isolate this source of variation and it will be ignored. Heterogeneity due to color is an estimate of how much the colonies differ from each other only because of the differences in the relative numbers of yellow and rose shells. In like manner, heterogeneity due to banding is that part of the difference among colonies due to differences in proportion of unbanded to banded shells, holding color constant. The interaction component, which in the analysis presented here is partially contained in the other two, is that part of the variation not ascribable to the independent effects of color and banding but to the particular association between them. If such a component exists, it means, first, that there is some association between color and banding, which has already been shown, and, second, that this association differs from colony to colony. From the calculation of the association between color and banding, it does not seem likely that there is much heterogeneity in this association. All of the colonies show a higher proportion of unbanded among rose shells.

The analysis of χ^2 may be put into a tabular form as in Example 98A. The rose *vs.* yellow, or color, component of heterogeneity is calculated from a 5 × 2 heterogeneity table in which the numbers of rose and yellow shells for each colony are given, irrespective of their banding (Example 98B). The χ^2 is calculated in the usual manner for an $r \times c$ table with $(r - 1)(c - 1) = 4$ degrees of freedom.

The banded *vs.* unbanded heterogeneity is calculated separately for the two colors. Thus, within yellow shells the homogeneity table is as shown in Example 98C, while for the component within rose shells the appropriate table is given in Example 98D. Finally, the total heterogeneity is calculated from the grand 4 × 5 table of Example 96 in the usual way.

The reason for calculating the heterogeneity χ^2 of banded and unbanded shells separately for each color is to avoid confusing the effect of

EXAMPLE **98**. Homogeneity χ^2 analysis of the data from Example 96.

A. Breakdown of χ^2 components for homogeneity

SOURCE	DEGREES OF FREEDOM	χ^2	P
Rose vs. yellow	4	34.07	<.001
Banded vs. unbanded in yellow	4	16.06	.01−.001
Banded vs. unbanded in rose	4	12.92	.01
Total heterogeneity	12	60.30	

B. Contingency table for calculating the rose vs. yellow χ^2 components

COLONY	YELLOW	ROSE	TOTALS
No. 683	110	79	189
No. 405	174	126	300
No. 153	79	87	166
No. 152	100	26	126
No. 194	51	52	103
TOTAL	514	370	884

C. Contingency table for calculating the χ^2 component of banded vs. unbanded in yellow

COLONY	UNBANDED	BANDED	TOTAL
No. 683	4	106	110
No. 405	12	162	174
No. 153	15	64	79
No. 152	8	92	100
No. 194	3	48	51
TOTAL	42	472	514

D. Contingency table for calculating the χ^2 component of banded vs. unbanded in rose

COLONY	UNBANDED	BANDED	TOTAL
No. 683	20	59	79
No. 405	55	71	126
No. 153	25	62	87
No. 152	5	21	26
No. 194	13	39	52
TOTAL	118	252	370

color with that for banding. Precisely the same technique was followed in the first example to find the heterogenetiy due to banding without the confusing effect of the classification of shells as intact or broken. The difference between the analyses of the first and second example is in the fact that two more components of heterogeneity were calculated in the second. Had they also been calculated in the first, the breakdown of χ^2 would have been:

Source	d. f.
Broken *vs.* intact	5
Banded *vs.* unbanded in broken	5
Banded *vs.* unbanded in intact	5
Total heterogeneity	15

However, broken *vs.* intact heterogeneity is meaningless since, as we pointed out, it is not a reflection of any biological situation. This means in addition that the total heterogeneity is also meaningless because it contains the broken *vs.* intact component. The only important and logical contrasts are the second two in the table, which were in fact calculated for the first example.

The reader should observe that although the degrees of freedom for the three components of χ^2 add to the number for the total χ^2, the χ^2 values themselves do not. The total of the three components of χ^2 in the second example is 63.05, while the total heterogeneity χ^2 is 60.30. This is a small discrepancy but a real one, which always exists in χ^2 analysis with the exception of certain special cases which are not of general interest to the zoologist.

Even more complex cases than these can be handled by the same methods, and one will be illustrated in a schematic way in order to show the application of the method of subdividing the components of χ^2. Suppose that instead of two variables, three were observed in a number of populations. The three variables are denoted by *A*, *B*, and *C*, and the two possible states of each variable by capital and small letters. Thus *A* might be male and *a* female; *B* yellow and *b* rose; *C* banded and *c* unbanded. Then the observations would be put in the form of Table 16A. The analysis of heterogeneity would have the breakdown as shown in Table 16B.

No matter how many variates are observed, an extension of this method will result in a breakdown of χ^2 into components. As in the first example, if any of the contrasts—*A vs. a*, *B vs. b*, or *C vs. c*—do not have any biological meaning due to the nature of the sample, they should be ignored, and the total heterogeneity must also be discarded as meaningless.

TABLE 16.

A. Form for observations when three variables, each with two states, have been observed in six populations.

POPULATION	A				a			
	B		b		B		b	
	C	c	C	c	C	c	C	c
1								
2								
3								
4								
5								
6								

B. The components of χ^2 for a heterogeneity analysis of data in the form of Table 16 A.

SOURCE	TABLE	DEGREES OF FREEDOM
A vs. a	6×2	5
b vs. b in A	6×2	5
B vs. b in a	6×2	5
C vs. c in A in B	6×2	5
C vs. c in A in b	6×2	5
C vs. c in a in B	6×2	5
C vs. c in a in b	6×2	5
Total heterogeneity	6×8	35

Tests of association between any two of the variates must also be fractionated into components in which the third variate is held constant. Thus, to test the association of factor B and C, a 2×2 table

	B	b
C		
c		

must be constructed within A for each of the six populations and then within a for each of the six populations. All twelve χ^2 values, each with 1 degree of freedom, are then added together to get a total χ^2 with 12 degrees of freedom, testing the association of the factors B and C. A similar breakdown must be made for the other two association tests, A with C and A with B. In this way, the association of A and C would be tested by twelve 2 × 2 tables of the form

	A	a
C		
c		

each table holding both population and the factor B constant.

CHAPTER FOURTEEN

Graphic Methods

Almost any numerical data in zoology and many that are not numerical can be represented graphically. A good graph spreads before the eye in a unified and comprehensive way a picture of facts and of relationships that cannot be so clearly grasped, if at all, from any verbal or strictly numerical representation. Sometimes a graph may in itself permit an adequate solution of the problems arising from the data, but more often it does not supplant calculation and direct numerical treatment. Usually the two supplement each other, the graphic method giving an immediate and suggestive résumé of what the written methods reduce to exact values, prove, and interpret.

The most important graphic methods are those concerned with frequency distribution (see Chapters 3-8) and with correlation and regression (Chapter 11). These and a few other graphic methods (e.g., for the comparison of single specimens, Chapter 10) have already been adequately explained and exemplified. There are, however, many other sorts of graphs. The possibilities are, indeed, almost unlimited, but only the more important, with examples of various sorts and some suggestions as to general principles, can be considered here. With so much basic knowledge and some ingenuity, special graphic methods can readily be devised for any particular problem.

Types of Diagrams

Most useful diagrams, although not all, belong to one of the following types:

1. POINT DIAGRAMS. In these the point method of representing frequency distribution (page 49), scatter diagrams of correlation (page 219), and the like are used.
2. LINE DIAGRAMS. These include frequency polygons (page 49), regression and other trend lines (page 224), theoretical curves like the normal curve (page 134), and any other diagram that

relates the original discrete observations to some form of continuous line.

3. BAR DIAGRAMS. In these a line or a rectangle represents each category or variate, and its length is proportionate to the corresponding value. Histograms (page 49) are a special type of bar diagram. Others are mentioned below.

4. AREA DIAGRAMS. In these, a figure of standard shape is subdivided into areas proportional to values to be represented. The most useful type, the pie diagram, uses a circle and subdivides it into sectors by drawing radii.

5. THREE-DIMENSIONAL DIAGRAMS. These include correlation surfaces, etc., discussed below.

6. PICTORIAL DIAGRAMS. This large and miscellaneous group includes maps, diagrammatic pedigrees and phylogenies, graphic representations of numerical properties by actual pictures of animals used in various ways, and many other methods. They include some concepts and methods not primarily numerical, but many are analogous to numerical methods, and some could be reduced to numbers if desired.

Most of these various types of diagrams involve a system of coordinates, a mesh, net, or field of some sort, such that position, linear distance, angle, slope, or the like has a definite numerical value in the diagram. Most important are rectangular coordinates (Figs. 22A, B, and C). All the diagrams given on preceding pages of this book have rectangular coordinates, and their general nature is already sufficiently clear. Arithmetic coordinates (Fig. 22A), those usually employed, represent any two equal differences in values along one axis by equal linear distances. Usually the scales are the same for the X- and Y-axes if these represent analogous variates, and this is preferable when practical. Sometimes, however, the ranges of X and Y are so greatly unequal that an awkward or impossibly large diagram can only be avoided by giving the larger variate a smaller scale. When the two variables are not analogous, for instance, when one is a value of a variate and one a frequency, as in histograms, there is no necessary relationship between the scales, and they are adjusted in each case to produce a convenient and enlightening result.

Rectangular coordinates may also be logarithmic or semilogarithmic (Figs. 22B and 22C). On a logarithmic scale, equal linear distances represent not equal absolute differences but equal ratios. Thus, on arithmetic coordinates the distance between points scaled as 10 and 100 is ten times that between 1 and 10, but on logarithmic coordinates the distances are equal because $10/100 = 1/10$. Logarithmic coordinates are logarithmic on both X- and Y-axes, while semilogarithmic coordinates are arithmetic

GRAPHIC METHODS 341

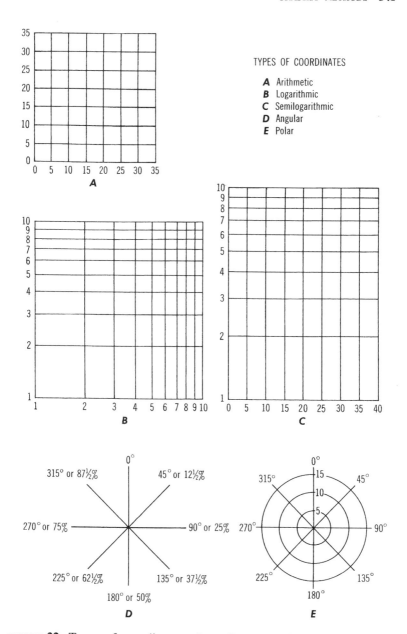

FIGURE 22. Types of coordinates. *A* to *C:* rectangular, (*A:* arithmetic; *B:* logarithmic; *C:* semilogarithmic or arithlog). *D:* angular. *E:* polar.

on the X-axis and logarithmic on the Y-axis. Such coordinates are often used for plotting rates, ratios, geometric progressions, and the like, because on them a geometric progression is plotted as a straight line, equal lines correspond with equal ratios, and equal slopes represent equal rates of change. Semilogarithmic coordinates are most commonly used for time series, plotting time arithmetically on the X-axis. They have the added advantage that if two comparable variates are being plotted in the same field and one is much larger than the other, the smaller is exaggerated and the larger minimized; the comparison is thus clearer and more convenient than on arithmetic coordinates. Paper ruled logarithmically and semilogarithmically can be purchased. If such paper is not readily available, the same result can be obtained (but more laboriously) on arithmetic coordinates by plotting the logarithms of the values appearing

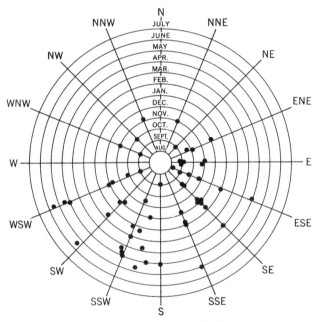

FIGURE 23. A graph on polar coordinates. Bird-banding data on herring gulls banded at Beaver Islands near St. James, Michigan, and recovered during the first year (data from Eaton, 1934). The angular distances, or directions of radii, indicate directions of the compass away from the banding place, and the concentric circles, or distances from the center, represent the dates of recovery and hence elapsed time and age, in months.

in the data. It should be noted that there is no 0 on a logarithmic scale: its base line is 1 since the logarithm of 1 is 0. (The logarithm of 0 is $-\infty$ which, of course, cannot appear on the graph.)

Angular and polar coordinates (Figs. 22D and E, and 23) are also occasionally used but are relatively unimportant. Angular coordinates represent a value by the angle between two lines diverging from a given point. There is thus only one scale, and values must almost necessarily be percentages or other fractions of a total, facts that make angular coordinates of very limited value except in the special form of pie diagrams. Polar coordinates are angular coordinates with another scale added—distance from the central point. They are of considerable value when one of the variables is in fact an angle or falls readily into circular form. For instance, they could be used to plot frequencies of cranial angles, and they are used to plot periodic annual or seasonal phenomena, dividing the angular scale into 30-degree segments, each representing a month.

The most common graphic representations of data are line diagrams on arithmetic rectangular coordinates. These are so widely used that a set of standards has been drawn up for them by a committee representing many fields of study. The essentials of these recommendations are as follows, with some modification and explanation pertinent to the special interests of this book:

1. The general arrangement should be from left to right, that is, with lower values of X to the left and higher to the right, and from bottom to top, with lower values of Y below and higher above.

2. Quantities should, as far as possible, be represented by or proportionate to linear magnitudes. In histograms and curves generally, areas are also important and necessary representations; but in histograms, specifically, these should be kept strictly proportionate to a linear magnitude (that of Y) by keeping the horizontal intervals equal.

3. The zero lines should, if possible, be shown on the diagram, and if this leaves a large blank space, it may be eliminated by a jagged break across the diagram. This recommendation is, however, unnecessary for much zoological work. The absence of the zero line is not misleading to anyone used to such diagrams if the scales are clearly marked.

4. Coordinate lines that are natural limits, such as those for 0 or for 100 per cent or that are otherwise exceptionally important should (or may) be emphasized; and others should not.

5. On logarithmic coordinates, the limiting lines of the diagram should be powers of 10.

6. No more coordinate lines should be drawn than are necessary to guide the eye. It is often sufficiently clear, and is generally neater, simply to give scales at the left and bottom of a diagram and not to draw in any other coordinates.

7. The curve (or other noncoordinate diagram line) should be sharply distinguished from the coordinates, usually by being made heavier.

8. It is often advisable to emphasize individual observations, as distinct from a line based on them, as crosses or distinct dots on the diagram.

9. Scales should be along the axes (seldom applicable to zoological diagrams) or at the left and at the bottom. Other pertinent data, formulae, etc., may, if desired, be arranged along the other two sides of the diagram or written within it.

10. The numerical data on which a diagram is based, if not clearly ascertainable from the diagram, should be given beside it or in the accompanying text.

11. Lettering should be clearly legible either as the diagram appears or after rotating it 90 degrees clockwise.

12. Diagrams should be clearly titled and should, as far as convenient, be self-explanatory without reference to an accompanying text.

Special Types of Graphic Frequency Distributions

The usual frequency polygons and histograms are limited to distributions of the absolute frequencies of a single variate with determinate numerical classes. Other types of graphs are necessary to represent such distributions for

1. relative rather than absolute frequencies;
2. more than one variable; or
3. attributes or variates in which the classes are not numerically determined.

The representation of relative values, of frequencies or any other variables, is discussed on page 63.

The simplest method of representing the frequency distribution of more than one variable on a single diagram is simply to superimpose separate frequency polygons on the same field (see Fig. 6, page 56). They may be distinguished by the nature of the line used—solid, dashed, dotted, etc.— or by shading the enclosed areas differently. If the magnitudes involved are about the same, the same scales may apply to both or all the distributions included, but it may be necessary to give them separate scales. Such diagrams tend to become too involved to follow easily, and they should be avoided unless really simple, clear, and illustrative of an important relationship. Histograms can occasionally, but rarely, be combined in the same way without undue loss of clarity.

A second method particularly useful for histograms is to plot the combined distribution of two samples of the same variate, showing the contribution of the second by marking it off above the first and shading its area (Fig. 24A). Or, what amounts to the same thing, one sample can be plotted first and another then added above it. Three or more samples can be added together and plotted on the same chart in this way, and frequency polygons may be used instead of histograms. For clarity, it is important that the samples really be analogous and the variates homologous. It would, for instance, be valid and useful to plot in this way distributions of the same variates for males and females of one species collected together or for two geographic samples of the same species; but it would usually be merely confusing to combine data on one variate for unrelated species or on two unlike variates for a single species.

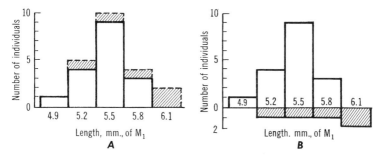

FIGURE 24. Combined histograms of two comparable distributions. Length of first lower molar in the extinct mammal *Anisonchus sectorius* (original data). *A:* samples added to each other; the clear areas represent a sample from New Mexico and the crosshatched areas one from Montana. *B:* the same data, with specimens from New Mexico above the horizontal axis and those from Montana below that axis.

A less common solution of the same problem is to scale values on the Y-axis both above and below its zero point at the intersection of the X-axis, and to plot one distribution in the ordinary way above the X-axis and one below it, as if reflected in a horizontal mirror (Fig. 24B).

Finally, it is possible to show the relationship of two frequency distributions clearly by introducing a third dimension. The method is to prepare a correlation table and then to erect on each cell a column proportionate in height to the cell frequency. If the cells can be made very numerous and very small and still retain frequencies in most of them, the tops of the columns can be blended into a nearly smooth surface with hills and valleys corresponding to regions of greater and less frequencies. This is called the "frequency surface" or "correlation surface," and it gives an almost ideal plastic representation of correlation. The construction of such three-dimensional models and their representation in a figure for publication are, however, too elaborate and too time-consuming for them to be used very extensively. Their reduction to two dimensions for a figure can be by perspective drawing or other oblique projection or by contour mapping like that of topographic maps.

A scatter diagram of the correlation of three variables can be made by laying out the appropriate scales, two in a horizontal plane on a wooden or composition base and one vertical, and representing each triple observation by the head of a pin, its length determined on the vertical scale, inserted at the proper point on the horizontal base. Of several possible methods of representation on paper, perhaps the most practical (if the observations are not too numerous) is to represent each observation by a circle on the field of the horizontal scales, with the third value given as a number in the circle.

In the study of hybridization between species, Anderson (1954) has developed a pictorial scatter diagram for illustrating simultaneously the distribution of a large number of morphological characters. Each specimen is represented on the scatter diagram by a circle or ellipse which may be variously shaded or blacked in to denote from which population or species the particular individual was taken. Thus, if there are only two populations, one may be represented by open circles, the other by circles filled in. If there are more than two, various degrees of blacking in the circles may be used. Two of the many variates which have been measured are chosen for the horizontal and vertical axes of the diagram. Anderson prescribes the following criteria for the choice of these two variates:

1. The variates should have a low measurement error.
2. There should exist many intermediate values for the variate. If one of the many variates is continuous, this would be a suitable choice.
3. The scatter diagram should fairly clearly divide the populations from each other into groups. That is, the two variates should be efficient discriminators of the populations or species.

The remaining variates are then assigned scores—that is, the values for each variate are grouped so that a given specimen will have a value from 0 to 4 or 1 to 5 for each variate. This method of scoring has been explained on page 14 in the discussion of "hybrid" indices. Each variate is then pictured on the scatter diagram as a small "ray," or line, projecting out from the circles. The position of each ray on the circumference of the circle denotes which variate is being pictured, and the length of the ray indicates the score for that variate.

Figure 25 illustrates this method for Sibley's towhee material (1954). There are four populations pictured, each shown by a different type of shading of the circles. Seven variates have been observed, and two of these,

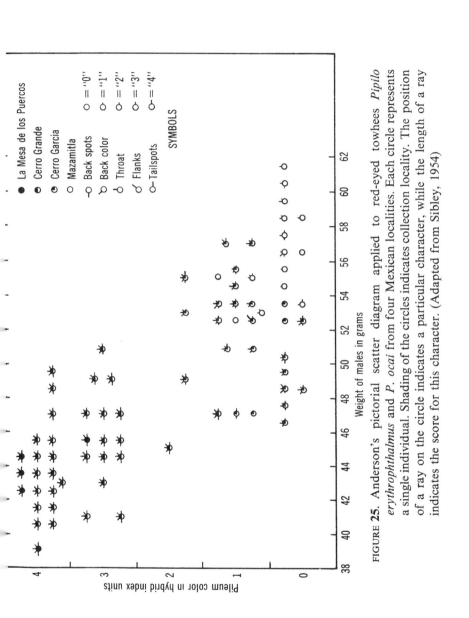

FIGURE 25. Anderson's pictorial scatter diagram applied to red-eyed towhees *Pipilo erythrophthalmus* and *P. ocai* from four Mexican localities. Each circle represents a single individual. Shading of the circles indicates collection locality. The position of a ray on the circle indicates a particular character, while the length of a ray indicates the score for this character. (Adapted from Sibley, 1954)

body weight and pileum color in index units or scores, have been placed on the abscissa and ordinate of the scatter diagram. The remaining five variates are represented by the rays, each with its specific position on the circle and each with a length proportional to the score. The author has chosen his variates judiciously, for two of the populations are confined to the lower right-hand portion of the diagram and two to the upper left. In addition, within the lower right-hand group the two populations are concentrated in different regions. The rays representing the five remaining variates show the same picture. All of the individuals at the upper left have high scores for the five variates, while those at lower right show lower scores as indicated by the shortness of the rays.

In addition to the clarity with which the relation among the seven variates and four populations is shown, this pictorialized scatter diagram has another use. If one of the variates chosen should be a poor discriminator of the different populations or species, there will be no clustering of rays of a given length, but short rays and long rays will appear scattered throughout the diagram. Such a variate can then be dropped from consideration as not pertinent to the problem.

A simple and almost always sufficient solution of the problem of graphic frequency distributions of attributes (and of numerically indeterminate variates) is to use a bar diagram. As in a histogram, each class or category is represented by a rectangle (or it may be, in this case, simply by a vertical line), with its height proportionate to the frequency represented. In bar diagrams, unlike histograms, a short space is generally left between successive categories, and each is separately labeled instead of being scaled continuously along the base of the diagram. The categories of attributes seldom have necessary or logical order, and the usual practice is to arrange them in the order of their frequencies, the highest to the left. It is a great advantage of this helpful and elegant method that almost any number of contiguous bars can be placed in one category, each representing a different sample, so that comparisons are greatly facilitated. It is advisable in such cases to shade the bars differently for the different samples. A single diagram can thus show the distributions of an attribute in males and females, in young and adults, in different years, in samples from different localities, etc. A bar may be given in each category to show an average value for the samples represented. Samples may also be added vertically or their component subsamples represented in the same way as for histograms of variates (see Figs. 26 and 27).

Pyramid diagrams may be used to represent distributions of attributes or, especially, variates. They are constructed by taking the rectangles of bar diagrams and histograms, turning them so that they are horizontal, and piling one on top of the other, centered on a vertical line, so that they look like an edgewise view of a stack of coins of different diameters but the same thickness. They have little advantage over ordinary bar diagrams and histograms and some disadvantages, and they are rarely used. They do, however, have two special applications pertinent to zoology. In ecology, the so-called "pyramid of numbers" and related "pyramids" are well shown in pyramid diagrams. The vertically superposed classes represent size groups, successive steps in nutrition chains, or the like, and the horizontal extent, or the area for each class, represents relative or absolute frequencies or masses. The age composition of a specific population is also well shown in a pyramid diagram. Vertically successive classes are age groups, and horizontal breadth or area is scaled to absolute or relative frequencies. Figures 28A and B are two examples of age pyramid diagrams.

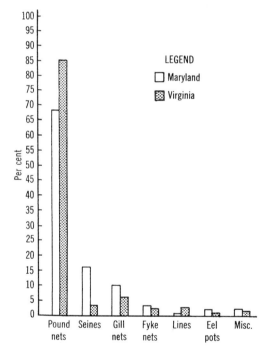

FIGURE 26. Bar diagram comparing categories of two different samples of an attribute. The attribute is method of collecting fishes in Chesapeake Bay during 1920, with the categories shown. Frequencies are given as percentages of total catch. The two samples are the Maryland catch (clear bars) and the Virginia catch (hatched bars). (Data from Hildebrand and Schroeder, 1928)

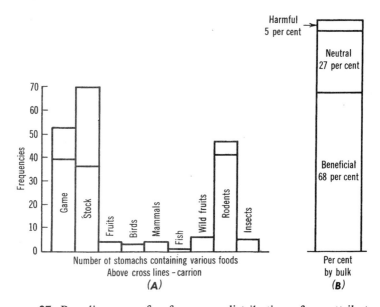

FIGURE 27. Bar diagrams of a frequency distribution of an attribute—food habits of the coyote. *A:* number of stomachs in a sample of 185 containing traces of each of 9 kinds of food; the distance above the crossline in 3 of these represents carrion. *B:* food contents from the same stomachs classified on an economic basis and represented by percentage of bulk. (Adapted from Dixon, 1925)

Comparisons of Ranges and Means

The importance of comparisons of samples of homologous variates is great and has repeatedly been emphasized. The correct graphic representation of such comparisons may convey at a glance all that is most important in a long and complex series of measurements and calculations, and this is one of the most useful of all graphic methods. The simplest form of the best of such methods was early employed by Ruthven (1908). In it a vertical scale is used for the values of the variate, and each sample is represented by a vertical line, the ends of which correspond, on the scale, to the extreme observed values. A crossbar is placed on this line at the arithmetic mean. If the samples have a natural sequence geographically, for instance, a line, preferably broken to distinguish it from other lines in the diagram, connecting the successive means may be added, giving a graphic representation of a sort of trend. It is also helpful to write near the line for each sample a number indicating its total frequency (see Fig. 29).

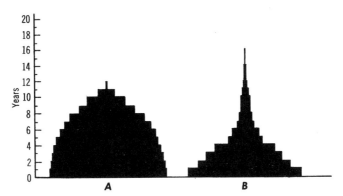

FIGURE 28. Age pyramid diagrams for two recent mammals, *Ovis dalli dalli*, the mountain sheep (*A*) and *Rupicapra rupicapra*, the chamois (*B*). The width of the bottom bar in each pyramid represents the number of individuals of the initial age in the population. Each successive layer is of a width proportional to the percentage of individuals surviving up to that point. The diagram for *Rupicapra rupicapra* shows a high mortality rate at early ages while that for *Ovis d. dalli* shows most mortality to occur in the last 4 to 5 age classes. (Diagrams from Kurtén, 1953, based on computations of Deevey, 1947, and Boulière, 1951)

If the differences between the samples can be expressed numerically, it might be practical and certainly would be useful to lay these values off on a horizontal scale and to make the distances between the vertical lines representing the samples proportionate to the numerical differences between them. This would, for instance, be possible in many cases of samples geographically separated (by miles, by latitude, or by longitude) or samples taken at different times (at different hours, on different days, in different months, etc.), or in environments numerically different (in temperature, in humidity, etc.).

Because the observed range is erratic and tends to vary in a complicated way with sample size, such diagrams may be improved, especially in their indication of probable population overlap, by including a statistical estimate of range (see page 78). If, for instance, the estimate is based on a range of 6σ around the population mean, the sample values ($\bar{X} + 3s$) and ($\bar{X} - 3s$) would be included in the diagram.

An addition to such graphs was devised by Dice and Leraas (1936) and has been used extensively in zoology since that time. Crossbars are added at ($\bar{X} + 2s_{\bar{X}}$) (twice the standard error of the mean, not twice the

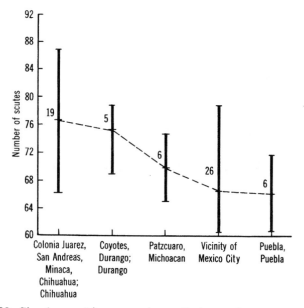

FIGURE 29. Simple graphic comparison of observed range and means. Number of subcaudal scutes in the snake *Thamnophis megalops* from five localities in Mexico. Values of the variate are scaled vertically, not horizontally as in most diagrams. The heavy vertical lines represent observed ranges, and the short cross-lines are the means, connected by a broken line to facilitate comparison. The numbers to the left of the vertical bar are the sizes of the samples. (Adapted from Ruthven, 1908, somewhat modified).

standard deviation of the sample) and $(\bar{X} - 2s_{\bar{x}})$, and these are connected by double vertical lines in order to define clearly the range so laid out. It was thought that, in general, if these ranges overlapped, the difference between the sample means was not significant (at approximately the 5 per cent level), while if they did not overlap, the means were significantly different. There are two difficulties with this method. First, two standard deviations on either side of the sample mean encompasses the 95 per cent confidence interval only in very large samples, so that it would be better to establish the more exact 95 per cent range by using the *t* distribution. The second and far more important objection is that even if exact 95 per cent confidence limits are constructed, there is no general correspondence between overlap of these confidence intervals and the *t*-test for the difference of two means. In fact, the intervals may overlap markedly despite a significant difference between the mean at the 5 per cent level, and, conversely, means which are not significantly different may have nonoverlapping intervals. The correspondence between overlap and significance depends upon the relative sample sizes and standard errors of the two samples, but a few general rules can be given.

1. If the confidence interval for one sample *includes* the *observed mean* for another sample, the two means are certainly not significantly different.

2. If the confidence intervals for two samples are nearly equal in length and these intervals are clearly nonoverlapping, the difference between the samples is significant.

3. As a corollary of (2), if both means are arbitrarily assigned confidence intervals equal in length to the *larger* of the two intervals, and if there is no overlap of these intervals, then the means are significantly different.

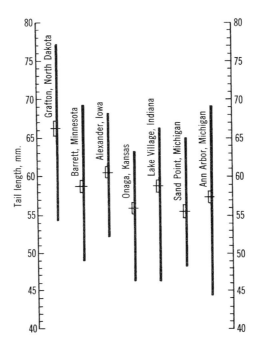

FIGURE 30. Graphic estimation of the significances of the difference between several means. Tail length in the mammal *Peromyscus maniculatus bairdii* from seven localities. Values of the variate are scaled vertically and given on both sides of the diagram to facilitate comparison by means of a straight edge. The vertical lines are observed ranges, and the small rectangles represent the 95 per cent confidence limits for μ, with a cross line at $\bar{X}$.

In any other situation, the overlap or lack of overlap of the graphed intervals is not a reliable measure of significance. Figure 30 shows the Dice-Leraas diagram modified in such a way that the 95 per cent confidence intervals rather than the $2s_{\bar{x}}$ limits are used. For the sample sizes included in Fig. 30 (79 to 107), there is very little difference between the exact 95 per cent limits and the $2s_{\bar{x}}$ limits, but for smaller samples this difference may be considerable. Using rules (2) and (3) above, the diagram shows that the Grafton population is certainly different from all the others, and that both Onaga and Sand Point are significantly different from Alexander, Lake Village, and Barrett. In addition, Ann Arbor and Alexander are clearly nonoverlapping. Rule (1), on the other hand, shows Barrett and Lake Village not to be significantly different from each other, nor are Onaga and Sand Point. All other comparisons are doubtful, however, and t-tests should be made. The advantage of the Dice-Leraas diagram is great in this case, for the number of t-tests necessary has been cut down markedly. There are $n(n-1)/2$ possible tests that can be made among n samples, so that for Fig. 30 there are $7 \times 6/2 = 21$ different t-tests. By inspection, however, 15 of these tests have been eliminated so that only 6 t-tests need to be computed.

The only objection to thus eliminating many of the tests of significance is that exact probability values cannot be assigned to those differences determined by inspection. This is not a serious drawback, however, because the rules are quite conservative, and borderline cases of significance will not be included.

Ruthven's method of connecting the means of samples by a line can conveniently be combined with the corrected Dice-Leraas type of diagram, thus condensing a considerable amount of information into a single picture. An example of this sort of gradient diagram is shown in Fig. 31. The horizontal axis should be in some natural units such as distance, altitude, or even time if the gradient in question is a stratigraphic one. The Dice-Leraas diagram is constructed for each population, the measured variate appearing on the vertical axis. Each population diagram is located at the appropriate point on the horizontal scale, and the sample means are connected by a solid or dotted line. The advantage of this sort of diagram is that it shows not only the difference between samples, but the regression of the measured variate on distance, altitude, time, or whatever the gradient variable happens to be. In Fig. 31 there is a very clear gradient in body weight corresponding to distance from the locality La Mesa de los Puercos in the north to Mazamitla in the south.

Another and equally informative way to show this same relationship is to draw the observed histogram for each population and connect the means of these frequency diagrams by a solid or dotted line. This is particularly useful when the variate is a discontinuous one. Figure 32 illustrates this

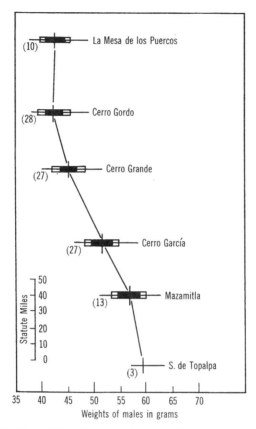

FIGURE 31. Modified Dice-Leraas diagram for demonstrating clines. Weights of hybrid male red-eyed towhees collected in six Mexican localities. Horizontal lines show observed ranges; rectangles mark standard deviation with solid black indicating 95 per cent confidence intervals for the mean. The means are indicated by a vertical line, and the number of specimens for each locality is shown in parentheses. Since the distance on the graph between samples is proportional to the actual distance between localities, the solid line connecting the means is a measure of clinal changes in average weight. (Adapted from Sibley, 1954)

method better than a lengthy description. The advantage of such a method over the Dice-Leraas diagram is that the latter assumes that the variate is roughly normally distributed, while the frequency diagram shows the actual sample distributions.

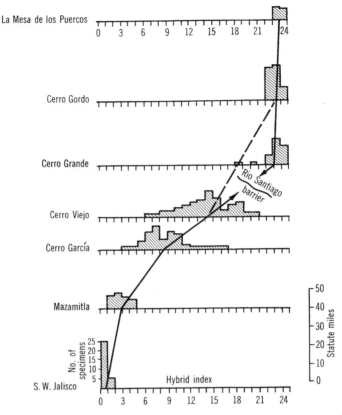

FIGURE 32. Histograms of populations of hybrid red-eyed towhees from Mexico. The horizontal scale is hybrid index, while the vertical scale for each histogram is observed frequency. The vertical distance between histograms, measured at their base lines, is proportional to geographical distance. A solid line joins the means of the distributions to indicate clinal change. The sudden change in mean from Cerro Grande to Cerro Viejo is interpreted by the author as due to a geographic barrier to migration indicated by a wavy line. The dotted line shows the supposed cline in the absence of such a barrier. (Adapted from Sibley, 1954)

Either of these two diagrammatic representations is especially suited for demonstrating altitudinal, geographical, or temporal clines.

It is often desirable to compare graphically the relative rather than the absolute dimensions of a number of animals or groups of animals.

Simpson (1941) introduced the *ratio diagram* for this purpose, and it has since been used widely. Figure 33 is an example of such a diagram for the comparison of five different dimensions of the second metatarsal in various large felines. The first step in constructing the diagram is to convert all the measurements into logarithms (it does not matter to what base). One specimen or group of specimens is then chosen as the standard of comparison. For each dimension the difference between the logarithmic value of the standard and each of the other specimens or groups is calculated. Each dimension X in each group is now represented by a new number:

$$d = \log X - \log \text{standard}$$

Dimensions larger than the standard are thus represented by positive values of d, while those smaller than the standard have negative values. The values of d for the standard group itself are, of course, all zero. The horizontal scale in the diagram is marked off in units of d, while the different variates are placed in some arbitrary order along the vertical scale, as in Fig. 33. The values of d for each group or specimen are now represented by points on the diagram at appropriate distances along the horizontal and vertical scales, and a line is drawn to connect the points for each group. The lines are simply an aid to the eye in interpreting the picture and are not meant to represent any sort of mathematical function. The horizontal distance between any two points in the diagram is proportional to the ratio of the dimensions of the two animals. This is because

$$\log X - \log Y = \log \frac{X}{Y}$$

For instance, if a given dimension in three specimens a, b, and c takes the values $a = 1$, $b = 2$, and $c = 4$, then $a/b = b/c$ and the distance between a and b on the diagram will be equal to the distance between b and c. Any specimen that has the same body proportions as the standard will be represented by a series of points falling on a straight line, even though that specimen is larger or smaller than the standard. Moreover, any two specimens or groups which are proportioned alike will be represented by lines whose forms are alike even if neither is a straight line parallel to the standard. Thus, in Fig. 33 the fossil specimen chosen as a standard is virtually identical with the recent jaguar in proportions of the second metatarsal but very different from the living puma.

Another use of the ratio diagram is in estimating roughly the size of missing parts of specimens. If the available dimensions of a specimen should show a vertical alignment in the ratio diagram indicating a close similarity to the standard, then it is a reasonable inference that missing parts will also fall on this vertical line.

358 QUANTITATIVE ZOOLOGY

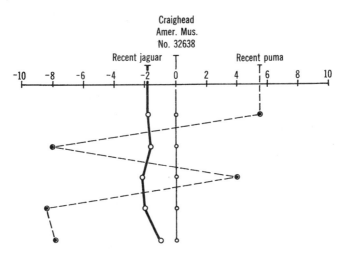

FIGURE 33. Ratio diagram comparing a number of dimensions of the second metatarsal in 3 large felines. The horizontal scale represents the deviation from the standard specimen of the logarithm of each dimension. No vertical scale is used. The recent jaguar is seen to be virtually identical in proportion to the fossil specimen chosen as standard, since the points for the jaguar lie nearly on a straight line parallel to the standard line. (Adapted from Simpson, 1941).

Proportions and Percentages

In some cases—for instance, in the most practical methods of colorimetry—observations are made and recorded as relative rather than absolute values. In other instances, it is helpful to reduce absolute data to relative values, usually to percentages, in order to facilitate comparisons. Samples of different sizes but with variations present in the same proportions may appear very different in a graph if absolute values are used. The plotting of percentages at once reveals their similarity. Such diagrams have, however, the one serious drawback that they do not take into account the varying reliabilities of the observed proportions or their probable significance. They should, therefore, be accompanied by numerical data, in the text, in the legend, or lettered on the diagram itself, by which these probabilities may be judged. It is usually advisable to state the absolute values represented by the percentages or the total frequency included. Further analysis by various of the methods explained in previous chapters may also be necessary.

Any graphic method representing absolute values may also be used to represent relative values. Frequency polygons, histograms, trend lines, and the like all lend themselves to this use. It is, however, often desired to compare relative values from two or more samples but also to retain the graphic record of absolute values. This problem can be neatly solved by plotting the samples on a single diagram and equating critical values involved in the problem—for instance, making the graphic representation of total frequency the same for all samples or making the initial and terminal points of growth curves coincide. The values for the various samples are thus made graphically comparable and relative; the absolute values can be retained by calculating and showing a different scale for each sample. The method becomes too elaborate with many samples but is excellent for the comparison of a few. Figure 34 gives a typical example.

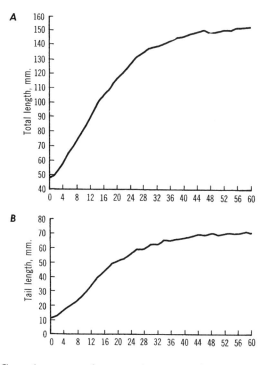

FIGURE 34. Growth curves of two variates recorded by absolute values. Mean measurements on identical samples of the mammal *Peromyscus maniculatus artemisiae*. *A:* growth curve of mean total length. *B:* growth curve of mean tail length. Because of the much smaller absolute values of *B*, the curves appear to be of different types. (Adapted from Svihla, 1934)

As originally published, the growth curves were separate and were given by absolute values on the same scale for both. The two curves then appeared to be very different, and the author concluded that the growth of these two variates was quite distinctive. Figure 35 shows that the growth curves are really almost identical. The apparent difference was not caused by kind or by relative rate of growth but simply by the different initial and terminal absolute magnitudes.

For representing relative frequencies of the different categories of an attribute (or, more rarely, the different classes of a variate), bar diagrams are useful. The samples are represented by rectangles all of the same height, and each is subdivided by horizontal lines into areas proportionate to the frequencies of the various categories. Connecting the homologous

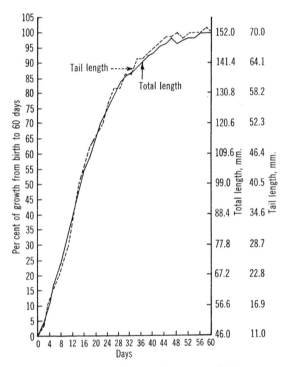

FIGURE 35. Growth curves of two variates recorded as percentages of total growth attained in the period of observation (same raw data as Fig. 34). The continuous line represents growth in total length, the broken line growth in tail length. Reduced to a proper basis for comparison, the two curves are seen to be almost identical in character.

division lines by lines (often broken) from one rectangle to the next generally makes the visual comparison still simpler and more obvious (see Fig. 36).

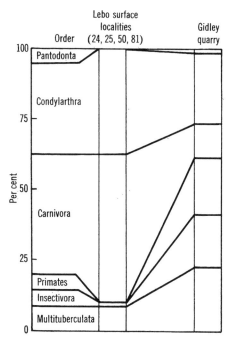

FIGURE 36. Graphic comparison of compositions of two samples by bar diagram. Surface and quarry collections of fossil mammals from the Paleocene of Montana (original data), subdivided by percentages of the various taxonomic groups (orders) represented.

In a variation of this method, each sample is represented by a vertical line clear across the field of the diagram (for instance, from 0 to 100 per cent) and the proportions are recorded by oblique or horizontal lines drawn between the proper points on the various lines, these points being determined in the same way as the division lines of the rectangles when these are used. In drawing the diagram, it is convenient to make the lines or rectangles representing the sample 10 cm. (100 mm.) in length or some integral multiple of this. Then the data in percentage can be directly transferred with no further calculation; using 100 mm., 1 mm. on the line represents 1 per cent in the data. Following Sumner (1927), this method is now generally used for graphic representations of results of tint photometry. The samples in this case are observations through different color screens, with one observation, a relative value as read, in each sample. The length of the vertical line for each sample represents the reading for pure white, and the point graphed is at the relative value of the reading for the object being observed (see Fig. 37).

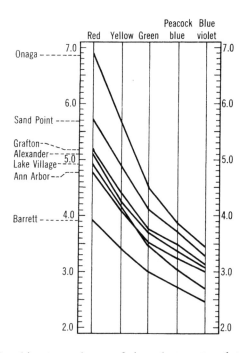

FIGURE 37. Graphic comparisons of tint photometry data. Skins of the mammal *Peromyscus maniculatus bairdii* from seven different localities. The scales on the two sides are tint photometer readings, and each of the five vertical lines between them represents the use of a filter of the designated standard color. Each oblique line passes through the mean values for the sample from the given locality, thus representing the color data for the corresponding sample. (Adapted from Dice, 1937)

Still another variation of this form of graph, one that is often used in showing changes in population composition, is illustrated in Fig. 38. Time is represented on the horizontal axis, and a line is drawn parallel to this axis at some convenient height above it. The area under this line represents the total population. The area is then broken into oblique strips in such a way that the vertical span of each strip at any instant of time represents the proportion of the total population represented by a particular class at that time. Such strips always begin at the lower left-hand end of the diagram and ascend toward the right, becoming at first broader and then narrower as the strip reaches the upper limit. At any moment of time, a vertical line drawn through such a diagram will be broken into segments by the strips, and the lengths of these segments are proportional to the relative frequencies of the classes.

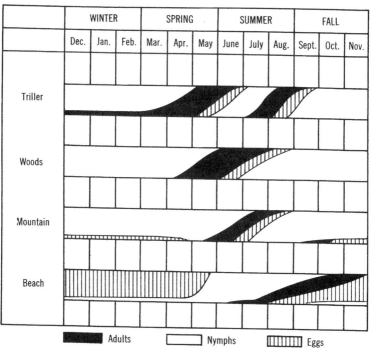

FIGURE 38. Seasonal population composition of four races of the field cricket *Acheta assimilis*. The vertical span of each band at any point on the horizontal scale is proportional to the percentage of the total population which is represented by that developmental stage. (From Fulton, 1952)

If absolute rather than relative frequencies are to be shown in such a diagram, the horizontal line above the time axis can be replaced by an undulating curve whose height above the axis is proportional to absolute numbers. The vertical span of the strips will then be proportional to absolute rather than relative numbers.

Pie diagrams are circles divided into sectors proportionate in area to the relative values of the categories represented. They do not lend themselves well to the representation of anything but proportions or percentages but are useful for this purpose. The procedure is to change the data into percentages and then to convert these into angles, measured in degrees, by multiplying by 3.6. The zero or first radius is drawn vertically upward from the center (to twelve o'clock on a clock dial). If there is no natural order of the categories represented, it is customary to place them in the order of diminishing size clockwise (see Fig. 39). If there is a natural order, the first sector lies immediately to the right of the zero radius, and the others follow clockwise. If there are many very small sectors, the radii may become too crowded or tend to form a black blot near the center; in this case, it is advisable to draw small and large concentric circles and draw the radii only between these, not continuing them to the center. It is also possible to draw one or more still larger circles and to represent on them the proportions of the same data arranged in different and broader categories; for instance, if the categories are taxonomic, an outer ring may show relative abundances by classes or orders and an inner ring or circle the same by suborders or families (see Fig. 40).

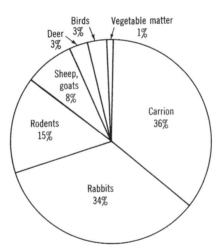

FIGURE 39. Pie diagram used for graphic representation of food habits. Winter stomach contents of coyotes, in percentages. (Adapted from Sperry, 1934)

Pie diagrams convey certain simple ideas more directly to the eye than almost any other sort of graph, but they are not ideal for scientific comparisons. Their most extensive use has been to impress the layman and to popularize the numerical results of some types of investigation. For this purpose, their appeal and their self-explanatory nature are often enhanced by putting in each segment, if it is large enough, an actual picture of the thing that it represents. For instance, if the categories are families of animals, a typical member of the family may be portrayed in each sector. In all cases the sectors should also be labeled in words; and usually it is advisable also to give in each a number indicating the percentage or, rarely, the absolute value that it represents.

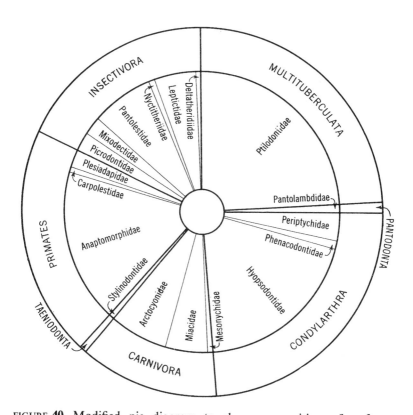

FIGURE 40. Modified pie diagram to show composition of a fauna. Percentages of specimens belonging to various taxonomic groups in a collection of Paleocene mammals from Montana (original data). The outer ring shows the orders represented and the inner ring their constituent families in this fauna. The center is left clear to avoid the confusion of many lines converging to a point.

Time Series and Periodic Phenomena

Most time series really involve at least three variables: time, frequencies, and values of a variate or categories of an attribute. Graphic representation usually eliminates or omits one of these, although all three can be shown if necessary. In series like growth curves, frequencies usually do not appear. The curve is based either on a single individual or on means for a group. In either case, frequencies need not be represented, for they are the same throughout. This is not invariably true of curves based on groups, but as far as possible these should not only have the same frequency throughout, but also be based on precisely the same individuals. If this cannot be done, the deficiencies of the data should be very plainly stated, for the curve will then be relatively unreliable and may be misleading.

Time series showing means may often be improved to convey much additional and important information by plotting a line, distinguished in some way such as by being broken, through the points representing the upper and lower confidence limits in each time class. Changes in dispersion over periods of time are thus represented, and these may be even more enlightening than changes in means.

In other cases there are only one or a few categories of the variable being observed, and the time series is primarily concerned with frequencies, these then being plotted on the vertical axis against time on the horizontal axis. The frequencies may, of course, be represented either in relative or in absolute form. If there is more than one category, lines for each, distinguished in some way, may be plotted on the same diagram (see Fig. 41). The same method may be used in a bar diagram, with one bar for each time class, its height proportionate to the frequency. Additional categories may be represented by subdividing the bars or by including in each time class a separate bar for each category. Times, frequencies, and observed values may also all be shown at once in a diagram by various modifications of scatter or point diagrams (see Fig. 42).

Phenomena that are periodic or cyclical with a definite period can also be represented in any of these ways on rectangular coordinates, but often they are particularly adapted to the use of polar coordinates. In this case, the angular scale represents time, 360 degrees being equated to the length of the cycle, and frequencies or values of a variate are represented by distance from the center.

Pictorial Diagrams

The name "pictorial diagrams" is here used broadly to mean most sorts of diagrams not including a definite numerical scale or not drawn on a system of coordinates, not only those that involve pictures in the literal sense. Diagrams not essentially pictorial may be dressed up with or expressed in terms of pictures to make them more sprightly or to give them popular appeal. This is often done with pie diagrams, as has been mentioned. Another method is to represent classes (e.g., time groups) by separate pictures of the thing being studied and to make the sizes of the

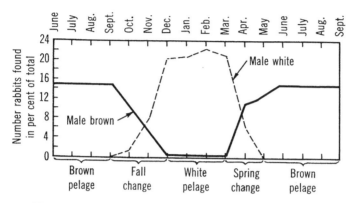

FIGURE 41. Graphic representation of periodic phenomena. Percentages of males in two color phases in total collections of snowshoe rabbits taken throughout the year, showing seasonal shift in dominant color phase. (Adapted from Aldous, 1937)

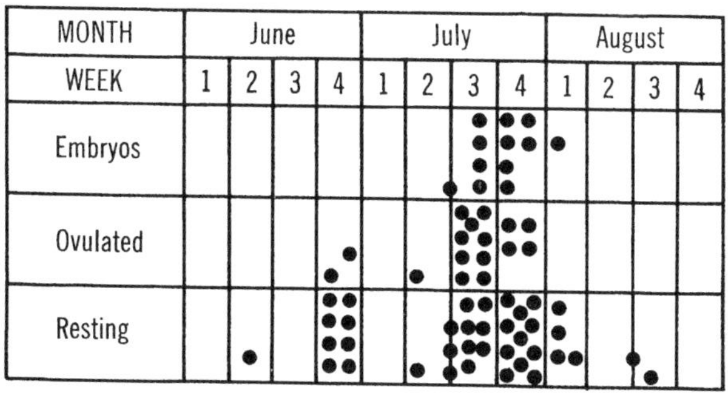

FIGURE 42. Point diagram of a periodic phenomenon. Samples of female Eastern American chipmunks collected during three summer months. Each dot represents an individual, classed according to three physiological (reproductive) conditions. The diagram shows a marked breeding season beginning late in June and culminating in July. (Adapted from Schooley, 1934)

pictures proportionate to frequencies or to values of a variate,[1] or to give each class a number of repeated pictures of equal size in proportion to these frequencies or values.

Many sorts of pictorial diagrams useful in zoology are essentially nonnumerical and so fall outside the scope of this book, but a few may be mentioned. Graphic representations of animal phylogenies are of this type. Some of these are highly ingenious and manage to convey

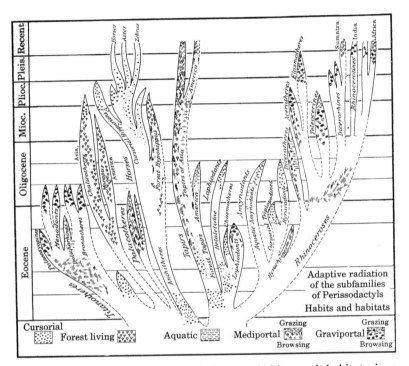

FIGURE 43. Pictorial diagram of phylogeny, habits, and habitats in a group of animals. Fossil and recent perissodactyls, as interpreted by Osborn. The vertical scale represents time, and the horizontal dimension (not scaled) is used to show the spread of the various phyla. Habits and habitats are shown by special patterns. (From Osborn, 1929)

[1] These diagrams, widely used on posters and the like but not in scientific work, are always misleading. If, as is usually done, the height of the pictured person, animal, or thing is made proportionate to a frequency or other value, the differences are exaggerated in apparent value. If the presumed volume of the thing pictured is used for scaling, the differences are strongly minimized to the eye. What the eye actually judges in a printed picture is area, but the use of area for scaling in such diagrams is not so apparently logical as either height or volume and is almost never encountered.

much information in addition to the supposed lines of descent—for instance, about geographic distribution, environment, habits, or morphological changes (see Fig. 43). Usually, time is represented on a vertical scale, in such charts, with the oldest forms at the bottom. Still more striking and convenient in many instances is a circular arrangement with phyla radiating from a central point. Charts of genetic descent and relationship of individual animals also belong to this general group; and they, too, manage to express much more than descent alone, by using symbols of different shape, shading, and subdivision for the various individuals (see Fig. 44). Data dealing with geographic distribution of any sort are also frequently and well represented pictorially on maps, bathymetric charts, cross sections of parts of the earth, pictures of landscapes, etc.

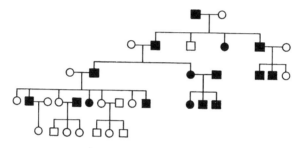

FIGURE 44. Chart representing genetic descent and inheritance. The chart represents a human family descending from the couple represented at the top. Squares represent males and circles females. Those in solid black represent bald individuals. (Adapted from Snyder and Yingling, 1935)

Among pictorial zoological diagrams, two of particular importance deal with essentially numerical idea, even though usually omitting actual numbers, and so merit special comment here. The first of these compares proportions or absolute dimensions of various sorts in ancestral and descendent or otherwise related animals by presenting drawings of the anatomical elements concerned with homologous points connected by lines, usually broken (see Fig. 45). Numerical values of the proportions or absolute dimensions may also be inserted on the figure.

A more complex but very interesting way of making similar comparisons in greater detail and taking into account areas and angles or directions as well as linear dimensions was devised by D'Arcy Thompson (1917) and has since been used chiefly by paleontologists. On a drawing of all or any part of an animal a regular system of rectangular coordinates is superposed.

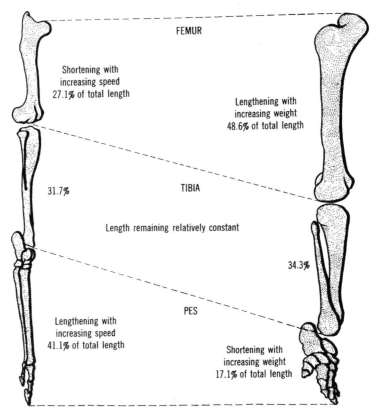

FIGURE 45. Pictorial diagram of contrasting anatomical proportions. Limbs of horse (left) and elephant (right), reduced to the same length, to show the marked difference in proportions of segments correlated with different habits and bulk. (From Osborn, 1929)

The same parts of an allied animal are then drawn, and on it are superposed lines, some or all of which usually must be curved, in such a way that each passes as nearly as is practicable through points homologous with those touched by a corresponding line on the first drawing, and each space between the lines covers approximately the same anatomical features as a corresponding square (or rectangle) on the first drawing. There is no way of making all the points and areas correspond absolutely if the animals differ markedly, but a close approximation is usually possible for forms that are visibly related. An effort is made to make the second system—

the deformed coordinates—as simple as possible, with its lines smooth curves and not strongly irregular (see Fig. 46).

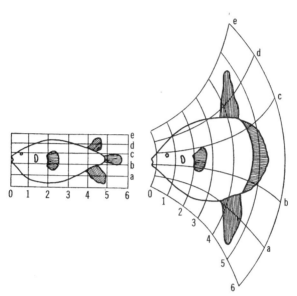

FIGURE 46. Pictorial comparison of proportions by rectangular and deformed coordinates. *Diodon*, left, has approximately normal fish proportions and is laid out on rectangular coordinates. A corresponding deformed coordinate system laid out on *Orthagoristus*, right, shows an almost perfectly regular expansion of the posterior part. (From D'Arcy Thompson, 1917)

The method is most usefully applied by taking an ancestral or a presumably primitive animal as a standard of comparison, laying the rectangular coordinates on it, and then developing from it systems of deformed coordinates for more specialized descendants or relatives. The deformed coordinates then provide a vivid visualization of the differential expansion, contraction, and skewing that have taken place in the course of evolution (see Fig. 47). The method has also been used, not always with complete success but at least always with results more reliable than the usual guesswork, to effect harmonious restoration of missing parts in fossil animals and to reconstruct hypothetical intermediate stages between less and more specialized animals.

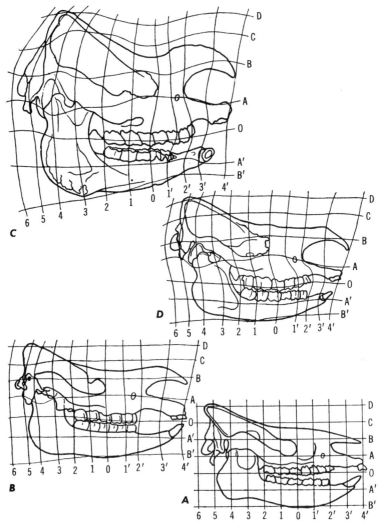

FIGURE 47. Changing proportions in phylogeny shown by deformed coordinates. *C* and *D* are skulls of modern rhinoceroses, *D* more primitive, to which *B* is approximately ancestral. *A* represents the primitive structure of the whole group and is laid out on rectangular coordinates, and the deformation of corresponding coordinates on the other skulls shows progressive irregular changes in proportions. *A: Subhyracodon occidentalis. B: Gaindatherium browni. C: Rhinoceros unicornis. D: Dicerorhinus sumatrensis.* (Adapted from Colbert, 1935)

CHAPTER FIFTEEN

Growth

It would be impossible, in a work devoted to numerical methods in zoology generally, to consider every sort of data or of analysis involved in special zoological studies. There is, however, one group of problems that are of remarkable importance, that demand some numerical methods in addition to those already discussed, and that also illustrate in a clear, useful way the application and adaptation of general methods in approaching a specific problem. These problems—those arising from the numerical phenomena of growth—are therefore given special consideration in the present chapter. In addition to supplementing what has gone before, this chapter is in a sense a review and a large scale example of the whole subject of numerical treatment of zoological data, for most of the principles and many of the methods already expounded here find application. It is also an introduction, elementary but adequate for any but extreme specialists, to what promises to be one of the most important and fruitful fields of strictly zoological and paleontological research, and one that has only recently begun to be appreciated. It will be found also that many of the methods and results based on growth have far wider applications and implications.

The Problems of Growth

There are three quantitative aspects of growth in which the zoologist is interested. The first and basic form of the growth problem is concerned with the change over time in some dimension of an animal. As an animal ages, the weight, surface area, and length of its various organs change in a fairly regular manner, this change with time comprising growth in its strictest sense. It should not be supposed that growth is always positive, for in the process of aging some organs may after a time begin to decrease in size and, in some cases, entirely disappear. Such is the case with the thymus in mammals, which reaches a maximum size at sexual maturity and then slowly atrophies. The change in a single dimension of an animal over time is the basic measurable aspect of growth from which the others are derived.

The second problem of growth is that of the relative sizes of two dimensions of a single animal. If there is some functional relationship between the magnitude of each of two dimensions and time, there will also be some function relating the two dimensions with each other, with the factor of time held constant. For example, suppose the length of some organ were related to the age of the animal by the function

$$\text{Length} = 3 \times \text{age}$$

while the width–age relationship were

$$\text{Width} = 2 \times \text{age}$$

Then obviously, for any given animal measured at any age, one should expect the relation

$$\text{Length} = \frac{3}{2} \text{width}$$

Growth relations are never so simple as this, but the point is best illustrated by such a naïve case. Time could be eliminated from this relation, only because it was assumed to be the same for both length and width measurements. This is only possible if both dimensions are measured simultaneously on the same individual or on two individuals known to be of the same age.

The third aspect of growth, and by the far the most difficult to attack, is concerned with changes of shape. Shape cannot be defined by a single number except in some very unusual cases, and for this reason its quantitative treatment is quite difficult. If one attempts to make the subjective notion of shapes amenable to objective measurement it becomes clear that only in the simplest cases can form be put in numerical terms. The change from a circular to an ellipsoidal form can be numerically described by specifying the relative lengths of the major and minor axes of the ellipse. When these two dimensions are equal, the form is circular; as they become progressively more unequal, the ellipse becomes flatter and flatter. The change from a circle to a square is not so simply described, however. In Fig. 47, D'Arcy Thompson's method of coordinate transformation was illustrated to show how the shape of one skull could be transformed to that of another by deforming coordinates. To the eye, it is obvious that the necessary deformation is greater for D than for B and still greater for C than for either B or D. How can this "degree of deformation" be quantitatively scaled? With sufficient ingenuity, it would be possible to devise a measure of how much the deformed coordinates of C differ from the rectangular one of A, but even so this would describe only the quantity of the deformation and not its quality. What changes in shape can be adequately described are confined to those describable by changes in relative dimensions over time, and it is to such cases that we will confine our attention. Whether or not shape changes are *describable* in these terms, they are certainly the result of differential growth of an organ or organism in different places or along different axes. Shape changes are, then, in theory if not in practice, analyzable into the growth over time of the various dimensions of the organism.

Growth in Time

The simplest sort of growth which can be imagined is additive change or growth by accretion. In an additive growth process, a constant amount is added to the organ in each unit of time, irrespective of the size of the organ.

The relationship between the sizes of the organ and time is given by the function

$$Y_t = Y_0 + Kt$$

where

Y_0 = the initial size of the organ
K = additive growth rate
t = elapsed time
Y_t = the size of the organ at the end of this time.

The assumption underlying an additive growth process is that the new material which has been laid down does not itself grow but is simply added as dead weight, so to speak, at a constant rate. This suggests that additive growth will be found among vertebrates in such organs as scales, feathers, hair, claws, teeth, and some types of horn. It is also found in concretionary structures like otoliths or calculi. Among invertebrates, additive growth is more important because their hard parts, notably the shells of mollusks, tend to grow in this way.

Even among such organs as are made up of precipitated, nongrowing materials, there are two considerations which disturb the simple additive picture. First, animals do not continue to grow through their entire lives but have more or less determinate growth. While bivalve Mollusca may indeed continue to increase the size of their valves more or less continuously throughout life, almost all animals show a definite slowing up of growth in later life, and in many growth may come to a complete end long before death. Because the growth rate is not constant throughout life, the simple additive equation is not a suitable description of the growth process except for relatively short periods even for those structures whose growth is simply additive in nature. The second objection to this additive scheme is a dimensional one. If the linear dimensions of an organ grow by a simple addition process, the area and mass cannot and vice versa. Mass varies as the cube of linear dimensions and area as the square, so that if growth in a linear dimension is represented by

$$L_t = L_0 + Kt$$

growth in area must be given by a function of the form

$$A_t = (\sqrt{A_0} + Kt)^2$$

and growth in mass by

$$M_t = (\sqrt[3]{M_0} + Kt)^3$$

These relations do not always hold true, however. For example, in the growth of hair, the increment is always in length only and not in the other linear dimensions. In such a case, mass and area increase in the same way as does length. For a spherical object, on the other hand, like an otolith, all linear dimensions are increased, so that the growth laws of mass, area, and length are not the same. It is possible that in some cases it is mass, rather than length, which is subject to a constant increment. In such a case,

$$M_t = M_0 + Kt$$

but

$$L_t^3 = L_0^3 + Kt$$

Each situation must be regarded in the light of its peculiar biology of growth, and no general rule can be given. Even *a priori* knowledge about the underlying phenomena of growth is not always reliable, and the elucidation of the growth law for a particular dimension of a particular organ or organism remains a matter for empirical confirmation.

Aside from the cases of deposition of essentially inert material, most growth is of a nonadditive nature. In the discussion of frequency distributions, it was pointed out that most distributions are moderately skewed to the right and it was suggested that the reason for this skew is a correlation between absolute size and variability of size. Looked at in another way, this hypothesis assumes that changes in size are functions of the size itself, larger size being subject to larger changes. Such a relationship would result if, for example, the living material added by growth began to grow immediately. Then the increase (or increment) at any moment would depend upon the amount of material already present and the size of the increments would increase continuously from instant to instant. If this pattern of growth continued in a regular way, the size of an organ at any time would be most easily represented by the expression

$$\log_e X_t = \log_e X_0 + K_g t$$

in which $\log_e$ is the natural logarithm (log to the base e), K_g is the *instantaneous* or *geometric* growth rate, X_0 is the value of the variate at the beginning of the growth period in question, and X_t is the value of the variate at the end of the period. This relationship can also be put in the equivalent form

$$X_t = X_0 \, e^{K_g t}$$

in which case the constant K_g may be referred to as the *exponential* growth rate. The logarithmic form is the more useful because it is essentially a line of the type

$$Y = b + at$$

where $\log_e X_t$ plays the role of Y, $\log_e X_0$ is b, and K_g is a. That is, by converting the variate X to its logarithm, the growth curve becomes rectilinear and the methods outlined in Chapter 11 may be applied to it for statistical analysis.

As in the case of arithmetic growth, it is not to be expected *a priori* that a geometric growth rate will remain constant throughout the entire life of the organism, or that the geometric growth curve will apply in any specific instance. Justification for the use of the geometric growth operation is, again, empirical: the observation that it is, in fact, a reasonable description of a set of observations.

It is appropriate here to consider the role of mathematical models like the geometric growth curve in biology.

The process of formation of a mathematical model is the same as for any hypothesis. Some preliminary observations suggest a *biological* hypothesis which is then translated into mathematical terms. If a number of observations clearly contradict the mathematical model, the hypothesis is inadequate and requires some alteration or outright rejection. On the other hand, if the observations are in accord with the model, it is accepted as a useful descriptive expression, but *the biological hypothesis upon which the model is based is not thereby proved.* Many biological processes may lead to indistinguishable mathematical expressions so that an investigation of the biological system itself is necessary to sort out the correct from the incorrect hypotheses. The successful model will suggest further lines of investigation, but this is the most that it can do.

The establishment of a mathematical "law" for biological observations is then a trial and error process by which models are successively tried and altered until a suitable description expression is found. The great danger of this method lies in the extraordinary power of a mathematical function to fit observations. If enough parameters are introduced into the equation of a curve, that curve will, by suitable choice of parameter values, fit any set of observations perfectly. For example, if the size of an organ is observed at 4 different ages, the curve

$$X = at^3 + bt^2 + ct + d$$

will pass exactly through the observed points if the proper values of the four parameters a, b, c, and d, are used.

Thus, the establishment of a satisfactory mathematical model should not be confused with an elucidation of the underlying biological process. To say that a certain organ grows by the geometric law is to say that the observations fit this curve and nothing more. Simple additive growth can mimic geometric growth if the growth rate changes over time in the proper way, even though the newly formed material is not itself growing.

It is necessary to distinguish between geometric increment and geometric rate of growth. The two are often confused but are quite distinct.

The organ that grows from 50 to 60 grams in 10 days has increased in weight by 20 per cent, its geometric increment, but it has not been growing at 20 per cent per 10 days nor at 2 per cent per day. Supposing, for the moment, that the increment is added instantaneously at the end of each day, the result of a geometric, compound-interest rate of 2 per cent per day is shown by the following series:

Days	Weight	Increment (2% of weight at beginning of day)
0	50	1.00
1	51	1.02
2	52.02	1.0404
3	53.0604	1.061208
4	54.121608	1.08243216
5	55.20404016	1.1040808032
6	56.3081209632	1.126162419264
7	57.434283382464	1.14868566764928
8	58.58296905011328	1.1716593810022656
9	59.7546284311155456	1.195092568622310912
10	60.949720999737856512	

Discontinuous geometric growth at 2 per cent per day would thus cause 50 grams to increase to nearly 61, not to 60 in 10 days. This series is analogous to interest of 2 per cent per day compounded daily. In fact, in such growth the compounding is not daily or at any other fixed discontinuous points but is continuous. The interest, so to speak, is continuously due, is paid instantaneously, and immediately becomes capital and starts paying interest itself.

The most obvious biological model for such geometric growth is the number of cells in a bacterial colony growing on an unlimited food supply. A single cell divides giving rise to two cells each of which divides into two, the resulting four cells then dividing and so on. Every new cell that is added to the population itself adds two new cells in a kind of biological chain reaction. Figure 48 shows this sort of growth in a culture of *Escherichia coli* beginning with a few cells. The logarithm of the number of cells is plotted against time and the figure shows this relationship to be rectilinear as is expected from the geometric growth function.

$$\log X_t = \log X_0 + K_g t$$

The slope of the line in Fig. 48 is the value of K_g and the point of intersection of the line with the ordinate is the value of $\log X_0$.

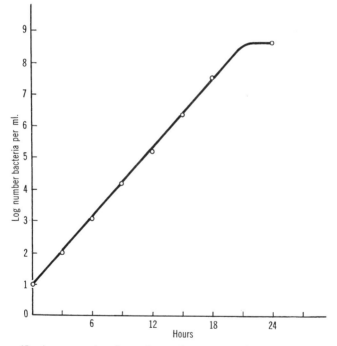

FIGURE 48. An example of nearly perfect geometric growth. Number of cells in a colony of the bacterium *Escherichia coli*. Time is measured on the abscissa and logarithm of number of cells per milliliter on the ordinate. (Adapted from Atwood, Schneider, and Ryan, 1951)

As a complete representation of growth, this model is as inadequate as the additive one, for the same objections apply. The last point in Fig. 48 shows one of these objections coming into play, for the number of living cells in the culture has stopped increasing due to limitation of food supply In higher organisms as well as in bacteria, the instantaneous rate of increase, K_g, is not constant.

K_g is approximately constant over short periods and often for longer periods in embryonic or earliest postnatal life, but it is seldom even roughly constant throughout life. It almost invariably drops as the organ in question nears its definitive size; and even if growth is continuous throughout life, the rate of growth becomes less. Constant multiplicative growth leads eventually to such enormous increments that the slowing down of growth is almost inevitable. For instance, if an animal weighed 10 pounds at birth and grew steadily at 1 per cent per day, in 10 years it would weigh more than 70,000,000,000,000,000 pounds! Yet most animals grow faster than 1 per cent per day in early stages of growth.

K_g may even become negative, signaling a decrease in the size of the organ, as, for example, the case of the thymus cited above. If the geometric growth rate were a constant over the entire life time of the animal, a growth curve would appear as in Fig. 48 (excluding the last point). Because of the lack of constancy of K_g, however, the logarithmic plot of size against time will not be strictly rectilinear but will level off as, in fact, the last part of the curve in Fig. 48 does. An arithmetic plot of size against time is convex upward, rising very steeply at first and gradually leveling off until it becomes horizontal, or nearly so, as the adult size is reached. If, however, the whole curve is available, from the fertilized egg through the embryonic as well as the juvenile growth, the early part is seen to be concave upward, beginning nearly horizontal and curving upward until it becomes steep and runs into the postnatal curve. The point of inflection, where the concavity changes from upward (and to the left) to downward (and to the right) is that at which the increment per unit time—the arithmetic growth rate—is greatest. Birth usually occurs at about this time, so that the earlier part of the curve, with increments increasing more and more rapidly, is mainly or wholly embryonic, while the later part, with increments decreasing more and more slowly, is mainly or wholly postnatal. Most growth observations cover only the postnatal period, and especially its earlier part.

The arithmetic growth rate begins essentially at zero, rises to a maximum at about the period of birth, then falls off and eventually reaches zero when the animal stops growing. This curve is usually strongly skewed because the period of accelerating arithmetic growth is generally much shorter than the period of decelerating growth, so that the peak of the curve is far to the left of its midpoint. The curve of the geometric growth rate usually follows a very different course. This rate is often very high in the earliest embryonic stages, then falls very rapidly and may level out so that the rate is relatively constant, or fluctuates about a nearly horizontal line or trend through much of the embryonic period. Near the time of maximum arithmetic growth, the geometric growth rate normally curves downward again and falls off, at first rapidly, then more and more slowly, until it finally reaches zero when maximum size is attained (see Fig. 49).

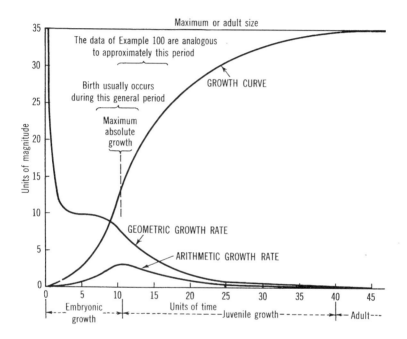

FIGURE 49. Idealized curves of growth and growth rates. Based on many sets of data but generalized and idealized. The relatively longer periods of maturity and senility are not included.

These relationships are subject to innumerable variations and modifications. Growth may proceed by cycles instead of in a single continuous curve. The rates are commonly highly irregular over short periods and show an even trend only over longer intervals. Some organs (like the thymus) may have a very different sort of growth pattern. These modifications cannot be considered here, where the concern is only with the general methods of study and not with its results. It is also often possible to fit a mathematical curve to growth data over longer or shorter periods, but relatively exact methods of doing this may be very complex and laborious and belong in the domain of the specialist and the skilled mathematician. Moreover, their limited zoological significance seldom if ever could really justify the labor.

Sampling for Growth

One of the most important and difficult elements in studies of growth is that of sampling. The technique of sampling for such studies can involve the following procedures, in their approximate order of accuracy.

1. Measuring parts affected by growth in each member of a group of live animals of known ages successively as each reaches certain fixed and preferably equidistant ages. The group should include the same individuals throughout, and the individual records should be kept distinct. This is the only method that gives adequate and complete information on growth. It gives a series of individual records and also a series of means from which population values can be estimated. As far as possible, factors other than age should be held constant, and all that might influence growth (e.g., food, temperature, humidity) should be recorded and their bearing on the matter tested by partial correlation, association, or some analogous process. Generally, this ideal technique can be applied only to laboratory animals, and then it has the disadvantage that it cannot be directly related to what actually happens in a wild population. It is occasionally possible to sample wild animals in this way with groups that can be tagged or otherwise surely identified, that tolerate handling, and that remain available over considerable periods of time. It can, for instance, often be applied to very young or to relatively sedentary birds.

2. Measuring a single individual in the same way.

3. Measuring successive random samples from a population of known age. Individual records have no value as such (unless, as in method 1, the individuals are repeatedly observed and exactly identified); but the means and the dispersion are still useful, although the estimates are less exact and less comparable than for method 1. It may, however, be impractical to take live samples of wild animals, and in any case the animals must be killed if the growth of some element not measurable on live animals is being studied. In this case, every sample after the first is inevitably biased. For instance, if the mean of the first sample happened to be above the population mean, the next sample will tend to be below the population mean, etc. Effects of this bias must be closely checked. For instance, if there were really no growth at all, but if the mean of the first 5 samples of 10 happened to be below the mean for all 10, then the mean for the second 5 samples will certainly be above that for the first 5, and an incautious worker will

conclude that growth has occurred. This inevitable bias is minimized by making the samples as small as possible in relation to the population and yet large enough to reveal significant changes if they occur. All the samples together should be only a small fraction of the available population, preferably not over one-tenth. Thus, if growth in laboratory animals is to be studied from 10 successive dead samples of 10 specimens each, about as small a series of observations as will give really reliable results, the samples should be drawn at random from a homogeneous laboratory stock of at least 1,000 animals. Sampling of wild populations should also be most carefully planned and scrutinized to be sure that the sampling itself is biasing the population as little as possible. Dead samples of single individuals for growth studies are relatively unreliable but are sometimes the best that can be done. Finally, animals that belong to the population but that die while the sampling is in progress should be measured, if possible, and compared with the samples, to determine whether their removal has significantly biased the population. If the mortality is high and the population small, this effect may be very appreciable, especially as the animals that die often average distinctly below the mean size of the population, so that if they are not taken into account growth will appear to be more rapid or more significant than it really is.

4. Successive samples of approximately known mean age but inexactly known individual age can be taken and studied as in method 3. The means obtained in this way for good samples will be about as reliable as those from method 3, but the indicated variation will have less significance, for it is here affected by growth as well as by variation, strictly speaking. This is usually the only or the best method available for wild animals, to which it can be applied when they have a fairly limited and known hatching or breeding period.

5. Successive age groups in a large sample collected over a short period of time can be compared. With well-defined age groups, the method is reasonably reliable. Its disadvantage is that the number of such groups, and hence the number of growth stages available, is almost always small, seldom more than two or three, which severely limits the information as to growth as a continuous process. Caution must also be used in comparing the younger groups with the combined adult groups. The former, if recognizable at all, are usually well defined, often representing animals born in one year, and their mean size and mean age have real and reliable meanings. The adult group generally covers animals with much greater age differences, and the mean size is not a real and reliable measure of animals of any particular single mean age.

6. Age groups separated physiologically or anatomically from samples covering any period of time or, as with fossils, an indefinite period of time,

can be compared. This and, to much less extent, the preceding method are the only ones so far mentioned that are applicable to fossils. This method is inaccurate because the age criteria used are themselves variable as to age of appearance, because numerical values can seldom be given to the mean ages of the groups, because each group generally does not tend to have its values clustered around a mean, and because the successive groups usually do not cover equal or measurably related periods of time. Rate of growth can only exceptionally be studied from such data, and the results are generally somewhat indefinite, but some valuable information can be gathered.

7. It is occasionally possible to get some idea of growth trends from a sample that cannot be or is not sharply subdivided in any way correlated with age. Thus, in a large sample of animals of unknown age and without sharply defined (discontinuous) physiological age criteria, it may yet be valid to assume that total length or some analogous measurement is closely correlated with age. If other measurements are then plotted against this, an idea, necessarily inaccurate but often suggestive, of their growth trend can be obtained. This method also does not give any measure of actual rates of growth, for that of the measurement used as a basis of comparison is not known. It may also give fallacious results because it starts out by assuming the truth of something that it seeks to prove, that is, that a measurement is correlated with age. As will be shown in detail, such data are invaluable in studies of relative growth, but they are of relatively little use in the study of absolute growth.

In order that methods 5 and 6 be valid, the division of a heterogeneous population or series of populations must be accomplished either by means of obvious physiological or morphological changes associated with age, or else by internal evidence from the distribution of measurements in the population.

Age Groups

Since most animals have a more or less sharply defined breeding season, the young are normally born during a definite season or, at least, more commonly during some months of the year than during others. It thus often happens that in collections made over a short period of time the specimens will fall into distinct age groups rather than being evenly distributed as to age. This result is of great zoological significance. It is important in itself, as showing rates of growth, breeding season, and age of maturity. It also is important for taxonomy; for the different age groups produce an effect on distributions almost identical with that of different species, and competent taxonomy must thus bear this possibility in mind. The study of variation has two aspects: variation within a single age group and variation correlated with age. The two must be separated for adequate analysis, and the methods of studying them are not the same.

The degree of distinctness of age groups depends on these factors:

1. Limitation of birth season. The shorter this is in relation to the life span of the animal, the sharper will be the age group. If there is no definite birth season (young are about as likely to appear at one time as another), there will be no age groups. Man well exemplifies a species in which age groups are practically indistinguishable for this reason.

2. Rapidity of growth, or of any changes correlated with age. Obviously if these changes are rapid, the discontinuity between age groups will be correspondingly great; and if they are slow, the discontinuity will become obscure or disappear. In almost all animals, growth is more rapid in early life than in maturity; and consequently the younger age groups are more clearly distinguished than the later. In mature animals, age groups usually are inseparable for this reason. The example given below for *Hemidactylium scutatum* clearly illustrates this (Fig. 50).

3. Variability within any one age group. Highly variable groups will tend to overlap and less variable groups to be better distinguished.

4. Uniformity of growth within the species. This is a special case of the last factor. If some individuals grow much more rapidly than other, they may well overtake the slower members of the next older group; and hence the groups will soon tend to merge.

5. Period of sampling. Age groups can be clear-cut only if the whole sample is taken over a period of time that is short relative to the life span of the animal or, more strictly, relative to the time between hatchings and seasons of parturition. If the sample is evenly distributed or taken at random over a long period of time, the concentration of the sample at definite parts of the life cycle will not be likely to occur, and hence age groups will not be distinguishable. Sampling of recent animals can be controlled to any degree of accuracy in this respect. Even among fossil animals, it will be shown in a later paragraph that seasonal sampling does occur.

Given a sample that is known or may reasonably be assumed to be homogeneous except in age, age groups may be recognized by biological or by statistical methods or by a combination of the two. The appropriate statistical method is obtaining frequency distributions for characters that change with age and then proceeding to observe or to test the normality, modality, and other characters of these distributions that show them to include different groups. The most obvious and generally the most useful characters are linear dimensions involved in growth; and if age groups are recognizable at all, distributions of these dimensions will generally show

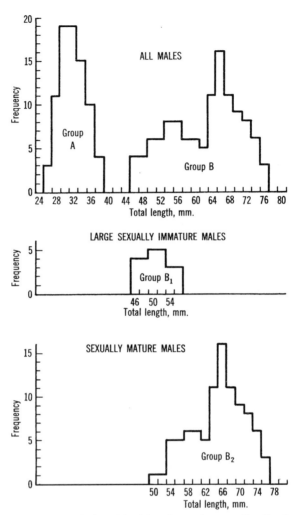

FIGURE 50. Age groups distinguished by discontinuous distribution of size and by physiology. Males of the salamander *Hemidactylium scutatum*. Group A includes individuals in their first year. Group B has been split into group B_1, aged 1 to 2 years, and group B_2, older than 2 years.

distinct bi- or polymodality. The fact that growth is usually differential may also be used by obtaining frequency distributions of appropriate ratios. Thus, in salamanders the tail is often smaller relative to the body in the young than in adults, so that the tail:body ratio may give clear age groups. Similarly, in mammals the head generally grows more slowly than does the body, the legs may elongate more slowly than they become stouter, etc. If age groups exist, testing of all the possible important characters of this sort will almost inevitably reveal and define them.

Biological age grouping depends on the recording for each specimen of a definite character that appears at approximately the same time in the life of any individual of the species. Such groups are not absolutely valid temporally, for there are no such characters that appear at exactly the same epoch in every individual; but many do distinguish years or even shorter periods. There is also the distinction that they do not depend, as do recognizable statistical age groups, on an actual discontinuity in the appearance of young individuals, the criterion not being such a discontinuity in group appearance but a milestone that every individual passes, at whatever season. When clear-cut, however, the biological groups do usually serve to distinguish groups that are also statistically valid and often permit the valid separation of the components of a polymodal frequency distribution.

Most obvious of biological age criteria, for recent animals, is the attainment of sexual maturity. More exact but less often available are periodic phenomena that leave records of their occurrence, such as fluctuations in rate of growth of fish scales or of elephant tusks. Frequently, these are annual but cannot be assumed *a priori* to be so, each case demanding investigation of this point. With the period of fluctuation determined, the age of the animal can be closely determined by counting the alternations of slowly and rapidly deposited tissue, just as is done in determining the age of a tree by the rings in its wood. Among mammals, the loss of deciduous and eruption of permanent teeth are also periodic phenomena that occur at approximately the same rate for all individuals of a species. The fusion of epiphyses and sutures in mammalian skeletons also serves to distinguish young and mature individuals. In mammals and some other groups, the wear of teeth and sometimes consequent changes in coronal patterns may be closely correlated with age and then permit fairly exact and multiple age grouping. An example is given on page 389.

An excellent example of both statistical and biological methods and of their combination in determining age groups is given by Blanchard and Blanchard (1931) for the salamander *Hemidactylium scutatum*. Their data for total length of males are reproduced in Fig. 50. In the histogram of all their observations, the distribution falls into two sharply distinguished groups, here lettered A and B. A, including the smaller individuals, is well defined and seems evidently to comprise a homogeneous group. B, on the other hand, is clearly heterogeneous and has at least two modes. This group includes individuals both sexually mature and immature. On this biological criterion, a group B_1 can be dissected out that appears to be itself homogeneous. The remaining group, B_2, including all the sexually mature males, gives a frequency distribution that is no longer clearly bimodal although it is skewed and tails off to the left in a way that suggests some heterogeneity.

The authors show that the characteristics of the whole distribution can be quite precisely related to biological facts. A, the numerous and sharply defined group of small sexually immature males, consists of individuals in their first year. The gap between this and B represents the growth made by the next preceding generation during the winter months before the appearance of the generation seen in A, a time when no salamanders were being born. Group B_1, large sexually immature males, includes individuals in their second year. In a quantitative sample, B_1 includes fewer individuals than A because of high mortality before reaching the second year and fewer than B_2 because the latter includes slower growth thereafter, so that large variants in the second year are as long as small variants in the third year. B_2 includes the sexually mature males, in their third year and older. It is broad, skewed, and very obscurely polymodal because animals of more than one year are included; but growth is now so slow that yearly age groups can no longer be clearly separated.

The essential statistics of these groups are as follows:

Age group	N	O.R.	$\bar{X}$	s	V
A	81	25.8–38.3	31.93	2.98	9.3
B	117	45.3–76.6	62.15	8.10	13.0
B_1	24	45.3–56.8	49.34	3.20	6.5
B_2	93	49.8–76.6	62.12	6.12	9.4

The coefficient of variation of first-year individuals (A) is unexpectedly high, and the lower figure for the second year (B_1) suggests the advisability of investigating the significance and cause of this apparent change. The variability of B_2 over B_1 agrees with the evidence that the former is still heterogeneous.

If sexual maturity is not determinable or is reached in the period from one breeding period to the next, such samples will generally be divisible only into two groups, a relatively well-defined immature group, born at the last hatching or bearing epoch, and a large heterogeneous but practically indissectible group of older individuals. This is, in fact, the usual condition in collections of many species of vertebrates.

In dealing with fossils, even though individual age can often be approximately determined, it is relatively rare to find natural aggregation into age groups. The season of death (corresponding to date of collection of recent animals) cannot be controlled and can only exceptionally be determined. Many, perhaps most, fossil deposits include animals killed at various different seasons, so that age groups are not so sampled as to be kept distinct. Nevertheless, it does commonly occur that in a given deposit animals are entombed more frequently or exclusively at one time of the year, for instance, because of winter mortality, spring flood, summer drought, or seasonal migrations to pastures and watering places. There are also many instances of mass mortality in which all the fossils at a given horizon represent animals that died at the same time. In such cases, definite age grouping occurs among fossils and numerous examples are appearing now that paleontologists have become more inclined to look for and able to recognize such groupings.

A number of excellent examples of age classification in fossils can be found in the work of Kurtén (1953). The most clear-cut is the case of the fossil carnivore *Ictitherium wongii*. The age classification made was on the basis of eruption and wear of the teeth, and all the specimens fell into one of seven groups clearly differentiated on the basis of upper premolar wear. The three oldest age groups were somewhat less clearly differentiated than the younger specimens, and this intergradation in older specimens is typical when tooth wear is used as a criterion of aging. The youngest group was assumed to be about 8 or 9 months old, as the permanent premolars were newly erupted with no sign of wear. The differentiation of age groups in this material was somewhat more clear-cut than for other forms, due, apparently, to the fact that the teeth wear quite slowly until the enamel layer is pierced, after which wear is much more rapid. A corollary of this, however, is that the age groups are undoubtedly not uniform in time span, the earlier stages requiring a longer time for a given amount of wear.

A second illustration of the use of dentition for age classification is shown in Fig. 51 from Kurtén's work on *Plesiaddax depereti*. Figure 51A is the frequency distribution of the height of the metacone in M^2 and Fig. 51B is the same information for M^3. The distributions are multimodal and have been broken into small unimodal distributions which Kurtén interpreted as age classes. There are two juvenile classes, one of which is not shown and the other of which appears only in Fig. 51A. These were qualitatively differentiated on the basis of milk dentition, and the purpose of the frequency distribution is to differentiate the adult specimens. A given individual did not always fall in the same group on the basis of M^2 and M^3 but the choice always lay between adjacent classes. As for *Ictitherium*, the older individuals were less easily classified than the younger, and there is undoubtedly some subjectivity in the procedure. In finally grouping the specimens, attention was given to the entire dentition, but most weight was placed on the second and third upper premolars, with the resulting estimation of age distribution shown in Example 99.

Kurtén considered in some detail the possible causes of the seasonal

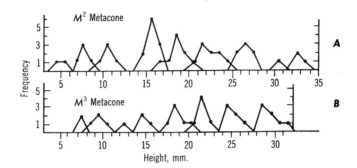

FIGURE 51. Frequency distributions of metacone height in the fossil mammal *Plesiaddax depereti*. A: distribution of metacone heights of M^2. B: distribution of metacone heights of M^3. The polymodal distribution is taken to be evidence of a series of age groups, and the population is broken into age classes on the basis of the modes. (Adapted from Kurtén, 1953)

deposition of these two genera and concluded, albeit somewhat tentatively, that seasonal droughts of regular occurrence were responsible for the wholesale death and fossilization of these more or less distinct age classes. It should be pointed out that seasonal entombment alone is not sufficient to produce distinct fossil age classes but must be coupled with seasonal reproduction. Small mammals like *Mus* or *Rattus*, for example, which produce a number of litters during the year, would not give clear-cut age groups in fossil deposits, no matter what the periodicity of fossilization.

EXAMPLE 99. Estimated age distribution of 76 specimens of the fossil ovibovine *Plesiaddax depereti*, from the lower Pliocene of China. (Data from Kurtén, 1953)

AGE IN YEARS (MIDPOINTS)	FREQUENCY
$\frac{1}{2}$	11
$1\frac{1}{2}$	5
$2\frac{1}{2}$	13
$3\frac{1}{2}$	12
$4\frac{1}{2}$	12
$5\frac{1}{2}$	10
$6\frac{1}{2}$	7
$7\frac{1}{2}$	4
$8\frac{1}{2}$	2

Fitting Growth Curves

Having obtained the values for size at a number of points of time, one should place them on a scatter diagram with time on the abscissa and size on the ordinate. This original plotting of the data is extremely important, for the pattern of the points will suggest the sort of growth process which underlies the observations. Since *a priori* assumptions about the nature of the growth function may be extremely misleading in any particular case, an additive or logarithmic growth function should never be fitted blindly to the data without first inspecting the pattern which the points show on a rectangular coordinate scale. In any event, neither the additive nor exponential curves are likely to be completely adequate representations of growth over considerable periods of time, no matter how much the biology of the situation may seem to suggest one or the other.

If the points do not show any strong curvature, the regression of the form

$$Y = Y_0 + Kt$$

may be fitted to the points, where K is the regression coefficient of Y, the size, on time, t, and Y_0 is the Y intercept (the value of Y when $t = 0$). Although a functional relationship and not a prediction equation is desired here, the regression line of Y on t is best for this case, since time has been deliberately chosen at stated intervals rather than having been allowed to vary randomly. If for some reason the sampling scheme is such as to make time random as well as size, then Bartlett's method for finding the "best fit" line should be used.

If the growth curve is really rectilinear, the value of K will not show any trend with time. That is, the growth rate should be constant for all periods of growth. Obviously, there will be some sampling error for K if all the measurements are not taken on the same individual, but, at most, these will be random fluctuations around the average K fitted to the points. If there is some real curvature in the growth curve, on the other hand, K will differ for each growth period and show an upward or downward trend. Since this change in K may not be obvious from the plot of the raw observations, the best procedure is to estimate K for each period independently by the relation

$$K_i = \frac{Y_i - Y_{i-1}}{t_i - t_{i-1}}$$

Thus K_1, the constant for the first period, would be

$$K_1 = \frac{Y_1 - Y_0}{t_1 - t_0}$$

These values of K are then tabulated and should be plotted against time. If there is a trend in K, it will show up both in the tabulation and the graph, and it will be clear that an additive growth function is not an adequate description of the data. Example 100 shows the pertinent information for such a study in rainbow trout. Figure 52 is the graph of absolute size against time showing, in addition, the plot of the arithmetic and geometric growth rates. The plot of absolute size against time shows a curvature which is reflected in the distinct downward trend of K with time.

EXAMPLE 100. Tabular record of growth and growth rate. Length of head in rainbow trout. (Data from Jenkinson, 1912)

STAGE	AGE IN DAYS FROM SPAWNING	ELAPSED DAYS	NUMBER OF INDI- VIDUALS	MEAN LENGTH OF HEAD IN MM.	INCRE- MENT IN MM.	K IN MM. PER DAY	K_g IN % PER DAY
I	49	0	200	36.1	—	—	—
II	63	14	209	53.4	17.3	1.24	2.80
III	77	28	198	68.1	14.7	1.05	1.74
IV	92	43	200	79.3	11.2	.75	1.02
V	106	57	161	87.3	8.0	.57	.69

The data of Example 100 clearly fit neither an additive scheme nor a logarithmic scheme, because the curve of growth is concave downward whereas the logarithmic function assumes an ever increasing rate of growth. Should the growth curve be concave upward as evidenced by an increasing trend in the arithmetic growth rate K, an attempt can be made to fit the geometric growth equation

$$Y_t = Y_0 e^{K_g t}$$

where K_g is the geometric rate of increase. The only techniques for curve fitting that we have developed are those for rectilinear relations, but as previously noted the geometric growth curve is easily put in a rectilinear form by using its logarithmic equivalent

$$\log_e Y = \log_e Y_0 + K_g t$$

This is of the general form

$$Z = a + bX$$

where Z is $\log_e Y$
a is $\log_e Y_0$
b is K_g
and X is t

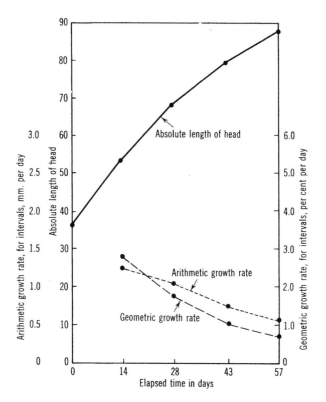

FIGURE 52. Growth curve and growth rate curves. Length of head in rainbow trout (data of Example 100). (Adapted from Jenkinson, 1912)

To fit the points to this curve, all that is necessary is to convert the measurements Y into their natural logarithms. If a table of natural logarithms is unavailable, use may be made of the relation

$$\log_e Y = 2.30259 \log_{10} Y$$

The regression of these logarithms on t is then calculated, and the slope of that regression line is K_g, the geometric rate of increase.

Just as for the rectilinear relationship, the descriptive applicability of the equation holds good only as far as K_g is approximately constant. It is even more difficult to observe a systematic change in the geometric constant K_g by simple inspection of the plot of raw observations against time. Again the values of K_g should be estimated separately for each growth period and plotted against time using the relation

$$K_{g1} = \frac{\log_e Y_1 - \log_e Y_0}{t_1 - t_0}$$

and so on, for all growth periods.

Example 101 illustrates this sort of treatment for measurements made on the aquatic hemipteran *Notonecta*.

EXAMPLE 101. Body length for six developmental instars in *Notonecta undulata* from Minaki, Canada. Means of 37 females followed through development. (Data from Clark and Hersh, 1939)

INSTAR (t)	BODY LENGTH IN MM. (Y)	$\log_e Y$	K_g
1	2.138	.760	—
2	3.188	1.159	.399
3	4.792	1.567	.408
4	6.245	1.832	.265
5	8.648	2.157	.325
6	11.561	2.448	.291

These data have been plotted in Fig. 53 to show the trends. Figure 53A, a simple arithmetic plot of length against time, shows the typical curve, concave upwards, which is expected from a geometric growth function. In Fig. 53B, these same lengths have been converted to logarithms and plotted against time, the resulting line being very nearly rectilinear. The least-squares fit (the usual regression line) for these points results in the equation

$$\log_e Y = .473 + .336\, t$$

or, in its equivalent exponential form,

$$Y = 1.605\, e^{.336t}$$

The solid curves in Figs. 53A and B are these least-squares curves, and it is obvious that the observations fit them very well indeed. There is, nevertheless, a slight suspicion of a curvilinear relation in Fig. 53B, and to check this the values of K_g for each growth period have been calculated, tabulated in Example 101 and graphed in Fig. 53C. This last diagram confirms the suspicion that the growth function is not quite exponential, for despite the fluctuation in K_g with time, there is an evident decreasing trend from the first to the last instar.

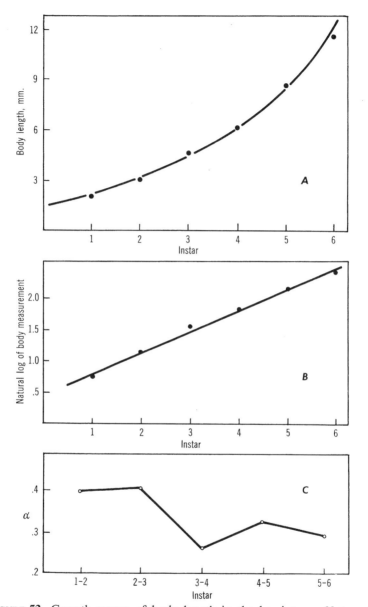

FIGURE 53. Growth curves of body length in the hemipteran *Notonecta undulata*. *A:* arithmetic plot of length. *B:* logarithmic plot of length. *C:* plot of geometric growth constant against time. (Data of Example 101)

It is a safe assumption that no growth curve in nature can possibly have a constant value either for K, the arithmetic growth rate, or K_g, the logarithmic rate, if observations are made over a sufficiently long period of development. Under such circumstances, the only adequate representation of growth by a single function would be one in which K decreased with time in such a way to make the curve reach some maximum point or plateau. A number of such curves have been suggested, but they have no particular biological appeal except insofar as their form is suggestive of the observed facts. They are, moreover, extremely difficult to fit by least-squares methods, and all involve some sort of trial-and-error solutions. Since no biological insight is to be gained by their use, we do not recommend them. Instead, the best treatment of growth data which can be suggested is the procedure followed in this section. If neither the arithmetic nor geometric growth functions are satisfactory in a particular case, a diagram showing the change in K_g or K in time will convey as much information as any more complex function.

Proportions and Size

Even more important than changes of absolute dimensions with growth are changes in proportions. It is a well-known fact that such changes do occur—for instance, that young mammals have larger heads relative to their bodies than do adults of the same species. It is less well known but nevertheless has been clearly shown that the same sorts of changes may appear not only with age but with animals of the same age but of different sizes. That is, smaller adults, for instance, may tend to have relatively larger heads than do larger individuals of the same age and species. It has even been demonstrated in numerous instances that the same sort of relationship may appear between larger and smaller species of a genus or, in rarer cases, even between larger and smaller genera of a family. The relationship of proportions and size thus has far wider implications than are involved in growth alone, although what appears to be a basic law of these relationships was discovered from growth studies and has hitherto found its most important applications in that field.

Allometric Growth

Two dimensions of organs or an organism may, and frequently do, grow in such a way that the ratio between their geometric growth rates remains approximately constant over considerable periods of time. This relationship may be symbolized by

$$\frac{K_{g_Y}}{K_{g_X}} = \alpha$$

in which K_{g_Y} and K_{g_X} are the two geometric growth rates and α is their ratio, which by hypothesis is a constant for a given instance. If α is indeed constant over the time considered, the relationship between the two dimensions, X and Y, can be described by the function

$$Y = b X^\alpha$$

called the "allometric equation." Y and X are the dimensions in question, α is the ratio of geometric rates, and b is another constant specific to the dimensions under consideration. This latter constant has a clear mathematical meaning: it is the value of Y when X^α is unity, or alternatively, the ratio of Y to X when α is unity. The biological interpretation of the constant is somewhat hazy, however. It is related in some way but only obscurely to the basic size difference between X and Y and on this account is sometimes referred to as the "initial growth constant."

This relationship between Y and X may apply either to dimensions of one organism at any given *time* or to the same dimension of two organisms both of the same *age*. It is, as a matter of observation, widely applicable to growth relationships, more so than the geometric growth function itself. This is presumably because the allometric equation is applicable when α, the ratio between the two geometric rates, is relatively constant, while the geometric growth curves require that the rates themselves be constant. K_{g_X} and K_{g_Y} may change considerably during growth, invalidating the geometric growth curve as a descriptive function, but if K_{g_X}/K_{g_Y} remains fairly constant, the allometric equation applies. Thus, the allometric growth equation adequately describes all cases where the geometric function applies to both variates. In addition, for many cases where the geometric or arithmetic growth curves do not apply because of changing values of K, the allometric equation can be applied.

It is entirely an empirical matter and neither mathematically nor biologically to be expected *a priori* that the ratio of the rates, α, will remain approximately constant. The applicability of the allometric equation, like other growth equations, is purely descriptive, and whether the description is useful and adequate in a given case depends on how well it fits the particular data.

The constant α is generally referred to as the *constant of allometry*, although many other names (e.g., "growth ratio") and symbols (e.g., "K") have been proposed for it and are in common use. If this constant is equal to unity, the two dimensions have the same geometric growth rates, and there will be a constant ratio between them no matter what their sizes. That is, when $\alpha = 1$

$$Y = bX$$

so that

$$\frac{Y}{X} = b$$

Two dimensions which maintain a constant ratio of sizes are variously said to grow "isometrically," "isomorphically," or "isogonically." In contrast to this simple situation, which is seldom if ever observed, dimensions whose ratio is constantly changing are said to grow "allometrically," "allomorphically," or "heterogonically." Note that in allometric growth, the ratio of *dimensions*, X/Y, is constantly changing, while the ratio of rates K_{gx}/K_{gy} is constant. We will use the terminology which seems at the present time to be generally agreed upon and say that two dimensions whose ratio changes constantly in accordance with the law

$$Y = bX^\alpha$$

are showing *allometry*, or *allometric growth*, with a *constant of allometry* α. If this constant is less than unity, the relation will be described as *negative allometry;* if greater than unity, *positive allometry*. The relation between Y and X will show positive allometry, if the geometric rate of increase is greater for Y than for X, while it will be negative in the reverse case. There is no biological difference, but only a formal one, between positive and negative allometry, since if

$$Y = bX^\alpha$$

then

$$X = \frac{1}{b^{1/\alpha}} Y^{1/\alpha}$$

It is simply a matter of which variable is taken as the dependent and which as the independent one.

The deviations from this simple allometric relation may be of two sorts. First, α may change continuously during growth, so that at first one dimension is growing with a much greater geometric rate than another, but as time goes on the rates become more equal and may even reverse their relations. This is the case when one organ or dimension ceases to grow before the other. Second, there may be a sharp change in the constant of allometry at a critical stage in the life cycle. This is a common occurrence in arthropods, where the crises marked by moulting may signal a distinct change in the relative growth rate of different structures. Another unfortunate difficulty of the simple allometric formula is that if two dimensions each show simple allometry with a third, their sum cannot conform to this relation. For example, if two segments of a limb each show a simple allometric relation with body size, the limb as a whole will not show this relation unless the value of α happens to be the same for both segments when compared with the body. This objection will not be too serious if the two values of α are very similar or if the absolute sizes of the parts are nearly equal, but in some cases it may present a real difficulty. Despite the reservations and difficulties of regarding the growth of two dimensions as being related by the simple allometric equation, enough empirical evidence exists to render it a practically useful description of this relation in many instances.

Fitting the Allometric Equation

The first step in analyzing relative growth patterns is to make a scatter diagram of the observations on rectangular coordinate graph paper, the ordinate and abscissa representing the two dimensions compared. There is no substitute for an inspection of the raw observations in this manner, and the tendency to plunge head first into a curve-fitting procedure may only introduce excessive complications. A case in point is Example 102 on the dimensions of a sample of the gar pike, *Lepisosteus platyrhincus*.

EXAMPLE 102. Dimensions of a sample of 225 individuals of the gar pike *Lepisosteus platyrhincus* from Englewood, Florida. Arranged by 10 cm. class range increments of standard length. (Data from Hammett and Hammett, 1939)

CLASS RANGE OF STANDARD LENGTH	NUMBER	LENGTH IN CM.			
		STANDARD	TOTAL	BODY	HEAD
30– 39.5	13	37.6	43.8	20.4	10.7
40– 49.5	69	44.8	52.0	24.1	12.9
50– 59.5	50	54.1	62.6	29.0	15.4
60– 69.5	36	64.5	74.3	34.8	18.2
70– 79.5	29	74.1	84.3	40.7	20.7
80– 89.5	15	84.0	95.9	46.5	23.2
90– 99.5	9	93.1	105.1	51.5	24.9
100–109.5	2	106.0	121.0	59.5	28.0
110–119.5	2	116.0	132.5	64.5	31.5

Rather than plotting all of the 225 raw observations for each comparison, the specimens were divided into nine classes based upon their standard length, and the entries in the table represent the mean of the specimens of that class for each dimension.

In Fig. 54, the comparisons of total, head, and body length with standard length have been plotted on a simple arithmetic scale. The abscissa represents standard length in centimeters, while the ordinate, also in centimeters, represents the other dimension to which standard length is being compared. It is obvious from the figure that the relations of standard length to the other three variates are perfectly linear and are adequately represented by the simple expression

$$Y = B + AX$$

Thus, the relative growth of these dimensions is strictly isometric with the constant of allometry, α, equal to unity. Assigning the variable X to the standard length, then the three lines in Fig. 54 have the equations (regressions fitted by least squares)

Total length: $Y = 1.49 + 1.12\,X$
Head length: $Y = 1.42 + .26\,X$
Body length: $Y = -1.55 + .57\,X$

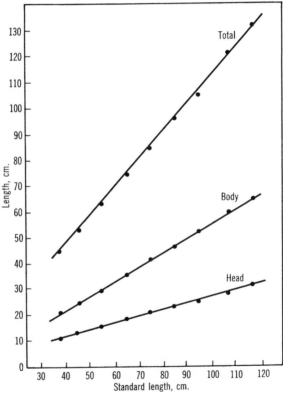

FIGURE 54. Growth curves of total length, body length, and head length, all relative to standard length, in the gar pike *Lepisosteus platyrhincus*. (Data of Example 102, from Hammett and Hammett, 1939)

The three Y intercepts, represented by B in the linear equation, are really spurious since the biology of growth must assume that when $X = 0$, $Y = 0$ in every case. For example, the Y intercept for body length is -1.55 which means mathematically that body length is -1.55 centimeters when standard length is zero, clearly an absurd result. The assumption inherent in the allometric equation of growth is that the line relating the two dimensions passes through the origin. This is entirely reasonable biologically, except in those cases where an organ is entirely missing for part of the growth of the animal and then suddenly forms and begins to grow. Even for such organs, there is usually a small *Anlage* from the beginning of growth of the organism, so that, strictly speaking, all dimensions and organs begin their growth at the same time. The fact that the regression lines have a Y intercept significantly different from zero is simply another demonstration of the point that a growth curve is only a description of the data on hand. The form of the regression undoubtedly changes from embryological differentiation to adulthood, so that the single regression line fitted to the data cannot be used as a basis for extrapolation back to $X = 0$. On the assumption that the Y intercepts are essentially zero, the allometric equation for the comparisons are of the general form

$$Y = bX$$

which is a special case of the more general relation

$$Y = bX^\alpha$$

with α equal to unity. The purpose of this discussion is to show that the linear growth relation and the more general allometric relation are not mutually exclusive: rather that one is a special case of the other. To demonstrate the truth of this statement in practical terms, the data of Example 102 have been fitted to the more general allometric equation. The method is essentially the same as for geometric growth in time. By taking logarithms of both sides, the allometric equation takes the form

$$\log Y = \log b + \alpha \log X$$

This is of the standard linear form in two variables, $\log X$ and $\log Y$, with $\log b$ playing the role of the Y intercept and α, the slope. The logarithms in this equation may be to any base, since a change of base only multiplies both sides of the equation by a constant. In general, the sampling for allometric growth is such that neither dimension can be considered the dependent variable, and the fitting problem is not one of regression, but rather of finding the "best fit" line. Bartlett's method is the best for this purpose, although several other systems have been suggested in connection with growth functions.

Two methods of curve fitting discussed by Teissier (1948), the "major axis" and "reduced major axis" lines, are used from time to time in fitting growth curves. We do not recommend either of these methods, but as they are in the literature they deserve some attention. As we discussed in Chapter 11, the problem of regression is that the squared deviations of the observations from the calculated line may be minimized either in the Y direction or in the X direction, but not in both simultaneously. The methods of "major axis" and "reduced major axis" minimize neither of these deviations but are attempts to find a compromise between them. In the "major axis" method the deviations minimized are the perpendicular distances from the observed points to the calculated line.

The great fault of this method is that the slope of the fitted line is not convertible from one unit of measurement to another. Thus, if one group of animals is measured in centimeters and another in inches, the two regressions can never be compared, whereas least squares regression coefficients, or Bartlett's "best fit" slope can be converted from centimeter to inch units by using the simple relationship

$$1 \text{ inch} = 2.54 \text{ centimeters}$$

The "reduced major axis" method minimizes the products of the X and Y deviations of each point from the estimated line. That is,

$$\sum (dX_i)(dY_i)$$

is made as small as possible. This produces a line that is very close to Bartlett's "best fit" line and does not have the convertibility problem of the "major axis" method.

Representing the line by the equation

$$Y = a + bX$$

where a is the Y intercept and b the slope, then the estimate of b is simply

$$\hat{b} = \frac{s_Y}{s_X}$$

where s_Y and s_X are the familiar sample standard deviations of Y and X, respectively. The estimate of a is

$$\hat{a} = \overline{Y} - \hat{b}\overline{X}$$

where $\overline{X}$ and $\overline{Y}$ are the sample means of Y and X. The "reduced major axis" method is then quite simple, but the great difficulty of it is that no exact confidence intervals can be calculated for b.[1] Since an estimate is of very little use unless a confidence interval can be calculated, this otherwise desirable system of curve fitting cannot be recommended.

---

[1] The standard error for b calculated by the reduced major axis method is given by Kermack and Haldane (1950). A knowledge of the standard error alone, however, is not sufficient to establish confidence intervals or make tests of significance.

In Example 103 the logarithms of standard length (Y) and the other three dimensions in Example 102 are shown, and these logarithms are plotted in Fig. 55, together with the "best fit" lines. The equations for these lines are:

$$\text{Total length:} \quad \log Y = .104 + .976 \log X$$
$$\text{Body length:} \quad \log Y = -.333 + 1.039 \log X$$
$$\text{Head length:} \quad \log Y = -.414 + .923 \log X$$

or, in their equivalent exponential form,

$$\text{Total length:} \quad Y = 1.271 \, X^{.976}$$
$$\text{Body length:} \quad Y = .463 \, X^{1.039}$$
$$\text{Head length:} \quad Y = .385 \, X^{.923}$$

While the value of α is very close to unity for total length and body length, that for head length is suspiciously low and, as a matter of fact, the scatter diagram for head length against standard length does show a detectable downward curvature indicating negative allometric growth. The conclusion of the author was that the dimensions measured showed essentially isometric growth with respect to standard length, and while they did not attempt to fit the allometric equation, the values of .976, 1.039, and .923 for α support this contention in the main. Stated in another way, the geometric growth rates for head length, total length, and body length are very nearly equal to that for standard length.

EXAMPLE 103. Logarithms to the base 10 of dimensions for the data of Example 102.

STANDARD	TOTAL	BODY	HEAD
1.5752	1.6415	1.3096	1.0294
1.6513	1.7160	1.3820	1.1106
1.7332	1.7966	1.4624	1.1875
1.8096	1.8710	1.5416	1.2601
1.8698	1.9258	1.6096	1.3160
1.9243	1.9818	1.6675	1.3655
1.9690	2.0216	1.7118	1.3962
2.0253	2.0828	1.7745	1.4472
2.0645	2.1222	1.8096	1.4983

While the various axial dimensions of *Lepisosteus* seem to be growing roughly isometrically, this is by no means the general rule for all animals. The usual situation is one in which the coefficients of allometry vary considerably from dimension to dimension, and these variations often

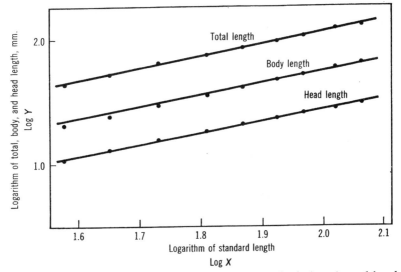

FIGURE 55. Relative growth curves of total length, body length, and head length, plotted logarithmically against the logarithm of standard length, in *Lepisosteus platyrhincus*. (Same data as Fig. 54)

reveal the existence of distinct growth patterns or gradients. The complete data on *Notonecta*, some of which were used in Example 101, demonstrate this phenomenon in a striking way. Example 104, which is reproduced from the work of Clark and Hersh (1939), lists the values of b and α in the allometric equation estimated by the authors for the relation of fifteen dimensions to total body length. Figure 56 is the logarithmic graph of a few of these measurements.

In both males and females, the coefficients of allometry are close to unity, with the exception of those for body width, head width, head length (distance of synthlipsis to vertex), and length of the third tarsus. These latter dimensions are growing allometrically with respect to body length, and especially striking is the value of α for the distance from the synthlipsis to the vertex, which is negative. A negative coefficient of allometry, means that the organ in question is growing *smaller* as the other dimensions grow larger. This must be distinguished from the general case of negative allometry, in which α is less than unity. When α is less than unity but greater than 0, the *relative* size of the organ is becoming smaller with an increase in the other variable, while an α less than zero means a decrease in *absolute* size of Y as X increases.

EXAMPLE **104.** Constants of the allometric growth formula $Y = bx^\alpha$ for the average values of 37 females and 35 males of *Notonecta undulata*. In each case, X is total body length and Y is the measurement listed in the table. (Data from Clark and Hersh, 1939)

	FEMALES		MALES	
	b	α	b	α
Body width	0.616	0.810	0.656	0.765
Head width	0.445	0.790	0.482	0.742
Synthlipsis to vertex	1.039	−0.271	1.020	−0.262
First leg	0.422	1.074	0.470	1.013
Second leg	0.549	1.069	0.605	1.012
Third leg	1.100	0.948	1.178	0.908
Femur 1	0.153	1.066	0.188	0.971
Femur 2	0.190	1.114	0.209	1.060
Femur 3	0.268	1.141	0.286	1.104
Tibia 1	0.140	1.144	0.152	1.101
Tibia 2	0.209	1.059	0.223	1.013
Tibia 3	0.348	0.964	0.373	0.924
Tarsus 1	0.119	1.028	0.138	0.945
Tarsus 2	0.143	1.056	0.150	1.010
Tarsus 3	0.517	0.742	0.503	0.743

Growth Gradients

In addition to the considerable variation in α from one comparison to another, a second interesting feature of these data is the pattern of growth relations which appears. In both males and females, α decreases from the first to the third leg. This decrease is not uniform for the three components of leg length, however. Thus the value of α for the femur increases from anterior to posterior, that for the tibia decreases, while that for the tarsus has its highest value at the second segment and decreases anteriorly and posteriorly from this point.

The appearance of such growth gradients is a fairly general phenomenon in animals. In addition to anterior–posterior gradients, proximal–distal patterns are not unusual. It has been found that where any region of the body shows pronounced allometry, positive or negative, there are usually also allometric relationships within the smaller segments of this region, and that the allometry coefficients for these segments are not capricious and irregular in distribution but follow definite gradients; for instance, in dealing with limbs, the allometry is often most intense in the proximal parts. This is true, for example, of the third segment in *Notonecta*, for both in males and females α is highest and close to unity for the femur, smaller for the tibia, and smallest and most deviant from unity for the tarsus.

Huxley (1932) has given the following allometry coefficients for designated limb segments in young sheep, expressed as the coefficient of allometry relating the weight of skeletal parts in the segments to the weight of the vertebral column.

Part	α
Scapula	1.25
Humerus	1.01
Radius + ulna	.80
Carpals	.67
Metacarpals	.60
Pelvis	1.74
Femur	.95
Tibia + fibula	.84
Tarsals	.70
Metatarsals	.62

These data are represented graphically in Fig. 57.

From such studies, it appears that the body may be considered as covered by a general field of growth potential, and that changes in body proportions are the result of the varying intensities of this potential—intensities distributed in orderly growth patterns.

Allometry of Growth and Size

The simple allometry equation $Y = bX^\alpha$ was derived from a consideration of the relative growth of two dimensions over time. This relation between two dimensions over time may be called *allometry of growth*, and it requires a very special sort of sampling procedure for its measurement. The two measurements must be taken on a single specimen at a series of ages or periods in its growth. This may be done for a large number of animals but the important point is that each individual is followed through its growth, and an allometric growth function is plotted for each.

The alternative sampling scheme, and the one which is most common, consists in sampling a number of individuals whose ages are either known or unknown, making the two measurements on each, and plotting the values of X and Y. There is no guarantee in this case that a larger individual is older than a smaller one, and since the growth rates for different individuals will differ, some of the larger specimens will, in fact, be younger than some of the smaller ones. It is a curious fact, and one that

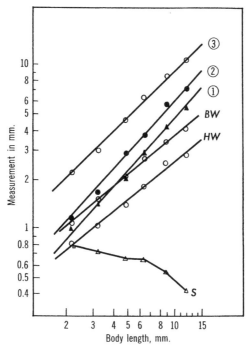

FIGURE 56. Relative growth curves of several measurements in *Notonecta undulata*, plotted logarithmically against the logarithm of body length. *S*. head length; *HW*: head width; *BW*: body width; *1*, *2*, and *3*: lengths of first, second, and third legs. (Data of Example 105, from Clark and Hersh, 1939)

has been explained in many ways, that the simple allometric relation applies to such data as frequently as it does to growth records. Such a relation may be called *allometry of size*, and two dimensions will be related by a simple size allometry provided

1. They are related by simple growth allometry.
2. The allometric constant, α, in the growth function, is the same for all individuals.

It does not have to be assumed that all individuals grow at the same rate but only that the relative growth rates of the two dimensions is much the same for all the animals measured. This is not an unreasonable assumption, since the relative growth rate of two organs or dimensions is a fairly basic characteristic of a species. It is sufficiently basic so that a sample made up of closely related species in a genus, and sometimes even closely related genera, may conform to a simple allometric relation. This last

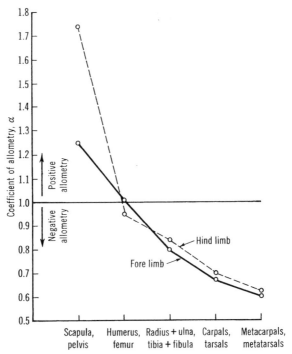

FIGURE 57. Growth gradients in limb segments. Weights of limb bones relative to vertebral column in young Suffolk sheep. Both fore and hind limbs show a strong regular growth gradient, with a distal center of negative allometry.

fact makes the allometric relation a very useful one in paleontology.

Example 102 on *Lepisosteus* is a case of allometry of size. No attempt was made by the authors to follow single fishes through their development, but rather a large random sample of fish was measured irrespective of age. Such a sampling scheme obscures the actual pattern of the relative growth of two dimensions, and it should not be mistaken for observations on allometric growth. It is entirely possible that the various individuals grew with very different allometric relations, or that simple allometry was not followed throughout the development of the animals, yet a random sample of individuals might give measurements mimicking a simple allometric equation. This would depend on the pattern of growth of the individuals and the age distribution in the sample. We do not mean to imply that such data are useless or incorrect but rather that they reflect a different phenomenon than do growth records themselves. Allometry of size is not allometry of growth, and the two should not be confused.

Even some records of growth allometry may obscure individual patterns. The constants in Example 104 were calculated from the average measurements of 37 females and 35 males at a number of stages of growth. These constants then represent the growth allometry of an "average" *Notonecta undulata*. The authors did keep separate records of the curve of allometric growth for each of the individuals, and these separate curves show patterns not apparent for the mythical "average" individual. Thus the constants in Example 104 show a decrease in α progressing from the first to the third leg, but when the individual curves of allometry were examined about half of the females showed the highest value of α at the second segment rather than the first. The same was true for males. It would be erroneous to conclude that the typical individual of this species has an anterior–posterior gradient in relative growth rate. There are two distinct patterns of allometry which only appear from the individual records.

Deviations from Simple Allometry

If an organism is to show simple allometry of growth for two measurements, the ratio of the two geometric growth rates must be constant over the entire period of growth. If a group of organisms is to follow the curve of simple size allometry, either they must all grow allometrically with the same value of α, or else the individual deviations from simple allometry of growth must cancel out in such a way as to mimic a simple allometric relation. In nature neither of these situations is likely to exist in a perfect state, and curves of relative growth or relative size usually show some deviation from simple allometry. As for the more complex curves of growth over time, mathematical expressions have been suggested for the description of these cases, but again these formulae have nothing to recommend them except their empirical success in individual cases.

The method of detecting deviations from simple allometry is different for allometry of growth than for allometry of size because of the way in which the values are distributed. In allometry of growth, there will be a series of points clearly spaced from each other, each representing the relation of one dimension to another at a specific stage of development. If the two dimensions have a simple allometric relation over the entire growth period, α estimated from successive growth periods should show no trend, although there may be random fluctuations due to measurement errors or slight variations in the true value of α. For each growth period, α is estimated by the relation

$$\alpha_1 = \frac{\log_e Y_2 - \log_e Y_1}{\log_e X_2 - \log_e X_1}$$

and so on.

Example 105 is taken from Clark and Hersh's observations on *Notonecta*.

EXAMPLE 105. Average dimensions in millimeters for 37 females of *Notonecta undulata* at six successive instars. (Data from Clark and Hersh, 1939)

INSTAR	BODY LENGTH (X)	HEAD LENGTH (Y)	$\log_e X$	$\log_e Y$
1	2.138	0.818	0.760	−.201
2	3.188	0.747	1.159	−.292
3	4.792	0.683	1.567	−.381
4	6.245	0.681	1.832	−.384
5	8.648	0.572	2.157	−.559
6	11.561	0.447	2.448	−.805

For the entire growth period, the "best fit" curve of simple allometric growth is

$$Y = 1.039 \ X^{-0.271}$$

For each period, α is estimated from

$$\alpha_1 = \frac{\log_e Y_2 - \log_e Y_1}{\log_e X_2 - \log_e X_1}$$

and so on.
For example,

$$\frac{-.292 - (-.201)}{1.159 - .760} = \frac{-.091}{.399} = -.288$$

The results are:

Instar	α
1–2	−.288
2–3	−.219
3–4	−.011
4–5	−.537
5–6	−.845

Figure 58 is a plot of α against instar number to show the changes in the relative growth pattern over time of head length relative to body length. While it was clear that the two dimensions were not related by a simple allometric relation from the scatter diagram of the logarithm of the observations themselves, the pattern of the deviation from simple allometric growth is best seen by the graph of α over time.

The method outlined above is seldom useful for investigating deviations from simple allometry of size since the points on the scatter diagram do not have a logical sequence. Depending upon the age distribution in the sample, the points may be more or less clustered, and older individuals may be smaller than younger ones. The best system for observing deviations from a simple size allometry is that suggested by Richards and Kavanaugh (1945). The "best fit" line for simple allometry is found, and for each value on the abscissa of the scatter diagram the theoretical value for the ordinate is calculated from the "best fit" line. The deviations of the actual value on the ordinate from the theoretical ones are then tabulated, and a new scatter diagram is made of the deviations plotted against the corresponding values on the abscissa. If there is a systematic deviation of the observations from a simple allometric relation, the graph of deviations will show a distinct trend or curvature. Example 106 illustrates this calculation for the relation of body width to the length of the second antenna in a subterranean isopod *Asellus californicus*.

The theoretical value of $\log_e Y$ is found by substituting the observed values of $\log_e X$ into the "best fit" line on the assumption of simple allometric growth. This equation is

$$\log_e Y = -1.00 + .70 \log_e X$$

by Bartlett's method.

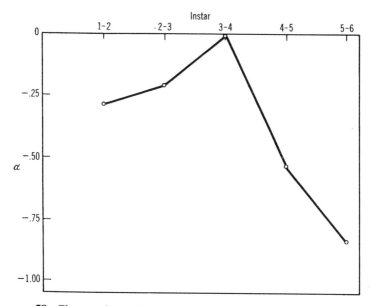

FIGURE 58. Changes in α, the coefficient of allometry, from one instar to another: growth of head length relative to body length in *Notonecta undulata*. (Data from Clark and Hersh, 1939)

EXAMPLE 106. Body width and second antenna length in millimeters for a sample of 18 individuals of *Asellus californicus*. (Data from Miller and Hoy, 1939)

BODY WIDTH (Y)	ANTENNA LENGTH (X)	$\log_e Y$	$\log_e X$	CALCULATED $\log_e Y$	DEVIATION obs.−exp·
.3	.9	−1.20	−.11	−1.08	−.12
.4	1.3	− .92	.26	− .82	−.10
.3	1.1	−1.20	.10	− .93	−.27
.5	1.6	− .69	.47	− .67	−.02
.6	1.7	− .51	.53	− .63	.12
.6	1.9	− .51	.64	− .55	.04
.6	2.1	− .51	.74	− .48	−.03
.7	2.1	− .36	.74	− .48	.12
.8	2.6	− .22	.96	− .33	.11
.8	2.6	− .22	.96	− .33	.11
1.3	6.6	.26	1.89	.32	−.06
1.1	4.3	.10	1.46	.02	.08
1.3	4.3	.26	1.46	.02	.24
1.3	6.0	.26	1.79	.25	.01
1.3	6.0	.26	1.79	.25	.01
1.6	8.0	.47	2.08	.46	.01
1.7	11.1	.53	2.40	.68	−.15
1.7	11.9	.53	2.48	.74	−.21

In Fig. 59, the deviations of $\log_e Y$ have been plotted against $\log_e X$ and the curvilinearity of the relation is obvious. For large and small values of X, Y tends to be smaller than would be predicted by simple allometry, while for intermediate values of X, Y is too large.

It may happen sometimes that the deviation from a simple allometric relation between two variables can be attributed to some sort of natural heterogeneity of the observations. For allometry of growth, it may be that there are distinct crises in development at which time the constant of allometry takes on a new value which is then maintained until yet another developmental crisis. For allometry of size, the sample may consist of two or more distinct groups, differing in size and allometric relations. An example of this latter heterogeneity would be a sample taken without regard to sex. It very often happens that there is a distinct difference both in the size of males and females and in the allometric relationship of two dimensions between the two sexes. If there are such causes of heterogeneity, the scatter diagram relating log Y to log X can be dissected into two or more groups, each showing a good fit to the simple allometry equation, although the values of α and b vary from group to group. Even if there is no obvious biological heterogeneity like sex, the scatter diagram often decomposes itself into two or more groups or clusters with a distinct

GROWTH 413

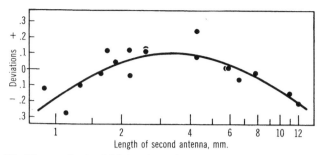

FIGURE 59. The method of Richards and Kavanaugh (1945) for detecting deviations from simple allometry in the relation of body width to second antenna length of the subterranean isopod *Asellus californicus*. On the abscissa are values of second antenna length. On the ordinate are the deviations of the observed body widths from the values predicted by the "best fit" line relating logarithm of body width to logarithm of second antenna length. (Data of Example 106, from Miller and Hoy, 1939)

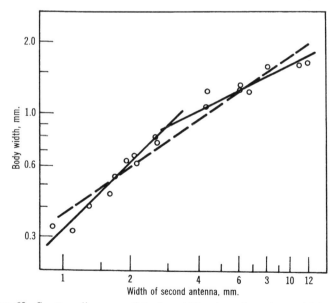

FIGURE 60. Scatter diagram of body width plotted against width of the second antenna in *Asellus californicus*. Both variates plotted logarithmically. The broken line is the "best fit" line to all the points. One solid line is the best fit to the lower points, while the other is the best fit to the upper points. (Data of Example 106, from Miller and Hoy, 1939)

hiatus between them. This is certainly the case of *Asellus* as shown in Fig. 60, the ten smaller individuals forming a cluster separate from the eight larger ones. Fitting two separate allometry functions to these two groups by Bartlett's method results in the two solid lines in Fig. 61, with the equations

Smaller group: $\log_e Y = -1.19 + 1.05 \log_e X$

Larger group: $\log_e Y = - .47 + .42 \log_e X$

Moreover, simple inspection shows that there are no systematic deviations of the observed points within each group from the new allometry functions. What the biological factor is which separates these groups is not clear, although it is suggestive that the smaller group consists of 3 males, 6

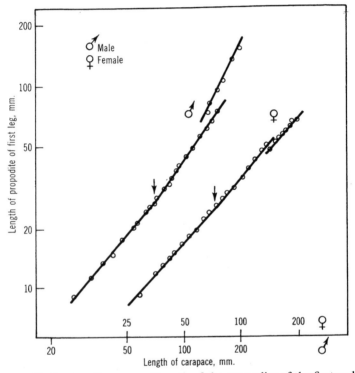

FIGURE 61. Scatter diagrams of length of the propodite of the first walking leg plotted against carapace length in males and females of the crab *Maia squinado*. Both variates plotted logarithmically. In each sex these separate lines have been fitted to three groups of points. The upper group in each sex is separated from the others by a clear hiatus in values. The lower two groups in each sex are separated by a critical moult indicated by an arrow. (Adapted from Teissier, 1948)

females, and 1 of unknown sex, while the larger group is entirely made up of males.

Figure 61 demonstrates the way in which allometry of growth can be dissected into a series of simple allometric curves on the basis of critical stages in development. These observations were made on the relative lengths of the carapace (X) and the propodite of the first walking leg (Y) in the crab *Maia squinado*. Males and females are separately plotted on the logarithmic scale, the arrows representing critical moults in the growth of this crustacean. Teissier, who made these measurements, describes the first moult as the "prepuberty" and the second as the "puberty" moult. Following the prepuberty moult there is a distinct change in the shape of the line for males and a less obvious one, if any, for females. After the puberty moult, however, there is a distinct discontinuity in the line both for males and females. Since the points shown are averages for large numbers of individuals of known growth stage, and since the "puberty" moult is the last one in the development of this crab, the third stage of the allometry line probably represents allometry of size rather than that of growth. This accounts for the slight overlap of the points in the second and third stages.

Any relation which deviates from the simple allometry equation can be dissected into a series of rectilinear relations, but judgment as to whether this is biologically sensible or not rests with the zoologist. It is not clear whether the allometric relation of body width to antenna length in *Notonecta* is really made up of two simple relations or whether α is continually changing over the entire growth of the animals. The dissection of these data into two groups certainly provides a satisfactory description of the growth relationship, but the possibility that there is one relationship with a constantly changing α cannot be ruled out.

Changes in Shape

There is no conventional system for analysis of changes in shape as there is for allometric growth. For one thing, it is not entirely clear what is meant by shape, a point which we have already discussed. In the simplest cases, change in shape is simply change in relative sizes of two dimensions, and as such, shape can be treated by the method of allometric growth. Such a method, however, describes the relation between two dimensions only, while a shape change is often best described in terms of a number of dimensions considered simultaneously. It is in such cases that the use of deformed or transformed coordinates, introduced by D'Arcy Thompson and illustrated in Fig. 47, is necessary. One of the limitations of this method has already been pointed out—that it may be very difficult to reduce the warping of the coordinate system to quantitative terms. This is not to say that such a reduction is impossible, but the important question for the zoologist is to what degree the quantitative description of the qualitative results is biologically interpretable. As a case in point, it is quite interesting to review a rather simple case of the quantitative treatment of changes in shape discussed by Medawar (1945). Figure 62 is taken directly from Medawar's article. It shows the well-known fact that the linear dimensions of the growing human form show considerable allometry, the head becoming relatively much smaller, the appendages and trunk relatively larger. The six figures a–f represent outline drawings of a developing man at various stages in his growth, the age listed below each figure. The horizontal lines are drawn at corresponding points along the axis of each figure, so that the same morphological landmarks are intersected by a given line irrespective of age of development. These landmarks are chosen arbitrarily at the first stage so as to make the lines equidistant from each other and having been chosen must be adhered to rigorously in succeeding figures. Thus, the fifth horizontal line in Fig. 62A is chosen to pass just under the nipples, and it passes through these points in all succeeding figures, although other possible landmarks no longer remain constant with respect to it. The line passing through the nipples in the youngest figure also happens to pass through the elbow joint at this stage, but at later ages this is not true. Precisely what landmarks are chosen is a matter of decision for the zoologist, and a different set will give a very different picture of shape changes. This arbitrariness of choice is the first difficulty of the method.

Having drawn the horizontal lines in each figure, the distance of each line from the base line is measured and this distance is plotted against the distance of the same line from the base in the initial figure. Figure 63 taken from Medawar is the plot of the distance from the base in Fig. 62B against the same distance in Fig. 62A, the reference figure. This process is repeated for each succeeding age, always plotting the distances at any age against those in the initial figure. To each of these scatter diagrams a curve is fitted which represents the transformation of points from the reference figure. To make the method successful, each curve must be of the *same mathematical form*. That is, if the curve relating the distances in the reference figure to those in the next figure is rectilinear,

$$X_1 = a + bX_0$$

then all of the other curves must also be rectilinear, although the values of the estimated parameters a and b will be different for each transformation. For the specific case in question, Medawar has fitted a curve of the form

$$X_1 = bX_0 + cX_0^2$$

We have not discussed methods for fitting such second degree functions, and the reader is referred to Snedecor for a discussion of this technique.

GROWTH   417

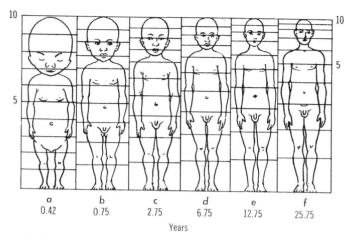

FIGURE 62. Idealized figures of human body proportions at five ages. All figures made to the same total height. Horizontal line chosen at equal intervals in *a* pass through the same morphological landmarks in *b–f*. (Adapted from Medawar, 1945)

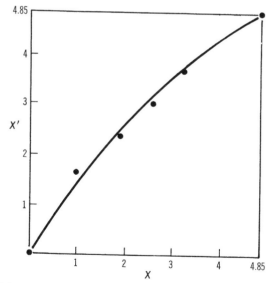

FIGURE 63. Plot of the heights of landmarks in Fig. 62B (on the ordinate) against the heights of these same landmarks in Fig. 62A (on the abscissa). See text (p. 416) for explanation. (Adapted from Medawar, 1945)

The relationships derived by the author for the six figures are:

2*a* and 2*a*: $X_0 = 1.000\ X_0 - 0.0000\ X_0^2$ $t = 0.42$ yrs.
2*a* and 2*b*: $X_1 = 1.428\ X_0 - 0.0882\ X_0^2$ $t = 0.75$
2*a* and 2*c*: $X_2 = 1.616\ X_0 - 0.1270\ X_0^2$ $t = 2.75$
2*a* and 2*d*: $X_3 = 1.780\ X_0 - 0.1699\ X_0^2$ $t = 6.75$
2*a* and 2*e*: $X_4 = 1.885\ X_0 - 0.1824\ X_0^2$ $t = 12.75$
2*a* and 2*f*: $X_5 = 1.990\ X_0 - 0.2042\ X_0^2$ $t = 25.75$

Now all of the figures have been drawn to the same total height which, in this case, was chosen as 4.85 units. This means that when X_0 is 4.85, so are X_1, X_2, X_3, X_4, and X_5. This fact enables one to eliminate one of the parameters in the fitted functions. Substituting $X_0 = 4.85$ and $X_1 = 4.85$ in the general form of the equation

$$X_1 = bX_0 + cX_0^2$$

results in the relation

$$c = \frac{1-b}{4.85}$$

That this is so can easily be checked from the actual fitted values of b and c. In each case, b and c turn out to have this relation. There is only one parameter, b, left in each relation, and this parameter has some value for each stage of growth, which may now be plotted against time. Figure 64

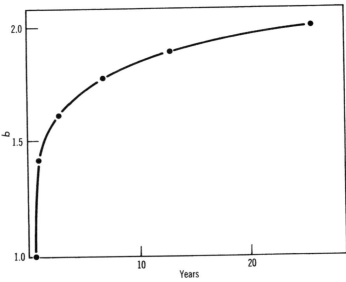

FIGURE 64. Plot of the shape parameter b against age, using the same data as Figs. 62 and 63. See text for explanation. (Adapted from Medawar, 1945)

is a plot of the estimated b at each time, and the estimates are connected by a smooth line.

The result of all this manipulation, then, is to estimate a parameter, b, which changes over time and whose change with time can be plotted. What is this parameter? Has it any detectable biological significance and does the form of its change over time reveal anything about growth? It would seem that there is no particular biological meaning to b. It is a quantity which permits the transformation of one set of coordinates into another, and the change of b with time is some measure of the change of shape with time. If b does not change, shape does not change; if b changes rapidly so does shape. While this shape parameter itself may have no intuitive appeal, its change with time certainly conveys the sense of change in shape itself. It is not clear how the value of b may be compared from one organism to another nor even how the rate of change of b with time can be related from one animal to another. Within a given form, however, changes in shape during growth may be approximately described by it.

The Use of Growth Curves in Systematics

All differences between animals, whether they be morphological, physiological, or behavioral, are grist for the taxonomists mill. Just as a difference in color or structure may be used as a basis for taxonomic decision, so differences in patterns of growth are valid characters upon which systematic conclusions may be based. Moreover, since dimensions of organs or organisms are the chief characters of use to the taxonomist, he is forced to consider the relationship between age and size in order to make proper biological inferences from his statistical results. If the mean length of the femur, say, differs significantly between two groups of animals, this may be evidence of specific difference, but it may as easily be a result of a difference in nutrition, or sex, or *age*. Wherever possible, such factors must be eliminated, either by assuming that the populations compared are homogeneous and alike with respect to these factors, or else that the age distribution, for example, is essentially the same in all the samples studied. If the animals studied have a clear-cut adult stage during which no growth occurs, then the inclusion of adults only in the samples will eliminate the age factor. Mammals, birds, and most insects have an adult stage that is clearly recognizable on qualitative grounds For most other animals, however, growth continues indefinitely, age classes are more or less difficult to recognize qualitatively, and samples are bound to be heterogeneous to an unknown degree. In such samples the only valid method of comparison is one based on the age–size relationship itself, that is, the growth curve. Thus, a comparison of growth relationships between samples from different populations may serve two purposes. First, the growth curve is itself a physiological character useful as a basis for taxonomic inferences, and, second, the growth curve of a given dimension may provide the only valid means of comparing the dimension from sample to sample because of unavoidable heterogeneity of the samples. Even if the animals can be classified as to age, the number of individuals of any given age class in each sample may be so small as to make a comparison of the samples, age class by age class, impractical.

The methods of comparing samples by means of their growth curves are quite simple. In the case that the age of each individual in the sample is known or can be closely estimated, then the first step is to find the most suitable form of relationship between age and the dimension, X, in question. If it should be a geometric growth function, for example, then the regression of $\log_e X$ on $\log_e t$ (age) is found for each sample. The slope of the line for each sample is an estimate of K_g, the geometric growth constant, and the sample values of K_g can be compared by the method suggested in Chapter 11 for measuring the significance of a difference between two regression coefficients. If the regression coefficients are significantly different, this may be evidence of a specific or subspecific difference between the population, while if the regression coefficients do not differ, this may be considered evidence to the contrary. Again, it must be emphasized that the demonstration of a difference is not *per se* proof of a systematic difference. It is only a datum to be used as a basis for a biological decision.

The difference between the Y intercepts of the different growth curves can also be tested by the methods of Chapter 11. In a geometric growth curve written in terms of logarithms, the Y intercept is the logarithm of the initial value of the variate before geometric growth began. This is its mathematical significance, but whether this can be translated into biological terms is doubtful. Since growth is never perfectly in accord with a simple geometric (or arithmetic) scheme, the Y intercept is a reflection of a change in growth pattern from earlier to later stages of growth. If two growth curves differ in their Y intercepts, even if they have identical slopes, this must be taken as evidence of different patterns of growth at some time in development.

In the case where ages cannot be specified for the animals in the sample, recourse must be had to a comparison of allometric curves relating two dimensions in each sample. If the allometric curve is transformed into its logarithmic form

$$\log_e Y = \log_e b + \alpha \log_e X$$

the problem is again one of linear regression with α, the slope, and $\log_e b$ the Y intercept. Bartlett's "best fit" line should be calculated for each sample, and a comparison can then be made of the values of α and $\log_e b$ in each sample by the approximate t-test discussed on page 237.

Appendix Tables

APPENDIX TABLE I. Areas and ordinates of the standardized normal curve. Areas shown are those between $\tau = 0$ and the tabulated value of τ.

τ	AREA	ORDINATE	τ	AREA	ORDINATE
.00	.0000	.3989	1.40	.4192	.1497
.05	.0199	.3984	1.45	.4265	.1394
.10	.0398	.3970	1.50	.4332	.1295
.15	.0596	.3945	1.55	.4394	.1200
.20	.0793	.3910	1.60	.4452	.1109
.25	.0987	.3867	1.65	.4505	.1023
.30	.1179	.3814	1.70	.4554	.0941
.35	.1368	.3752	1.75	.4599	.0863
.40	.1554	.3683	1.80	.4641	.0790
.45	.1736	.3605	1.85	.4678	.0721
.50	.1915	.3521	1.90	.4713	.0656
.55	.2088	.3429	1.95	.4744	.0596
.60	.2258	.3332	2.00	.4773	.0540
.65	.2422	.3230	2.10	.4821	.0440
.70	.2580	.3123	2.20	.4861	.0355
.75	.2734	.3011	2.30	.4893	.0283
.80	.2881	.2897	2.40	.4918	.0224
.85	.3023	.2780	2.50	.4938	.0175
.90	.3159	.2661	2.60	.4953	.0136
.95	.3289	.2541	2.70	.4965	.0104
1.00	.3413	.2420	2.80	.4974	.0079
1.05	.3531	.2299	2.90	.4981	.0060
1.10	.3643	.2179	3.00	.4987	.0044
1.15	.3749	.2059	3.20	.4993	.0024
1.20	.3849	.1942	3.40	.4997	.0012
1.25	.3944	.1827	3.60	.4998	.0006
1.30	.4032	.1714	3.80	.4999	.0003
1.35	.4115	.1604	4.00	.5000	.0001

APPENDIX TABLE **II**. Cumulative Student's t distribution. The body of the table contains values of Student's t; n is the number of degrees of freedom.

Probabilities for confidence intervals

CUMULATIVE PROBABILITY BETWEEN $+t$ AND $-t$

n	.1	.2	.3	.4	.5	.6	.7	.8	.9	.95	.98	.99	.999
1	.158	.325	.510	.727	1.000	1.376	1.963	3.078	6.314	12.706	31.821	63.657	636.619
2	.142	.289	.445	.617	.816	1.061	1.386	1.886	2.920	4.303	6.965	9.925	31.598
3	.137	.277	.424	.584	.765	.978	1.250	1.638	2.353	3.182	4.541	5.841	12.941
4	.134	.271	.414	.569	.741	.941	1.190	1.533	2.132	2.776	3.747	4.604	8.610
5	.132	.267	.408	.559	.727	.920	1.156	1.476	2.015	2.571	3.365	4.032	6.859
6	.131	.265	.404	.553	.718	.906	1.134	1.440	1.943	2.447	3.143	3.707	5.959
7	.130	.263	.402	.549	.711	.896	1.119	1.415	1.895	2.365	2.998	3.499	5.405
8	.130	.262	.399	.546	.706	.889	1.108	1.397	1.860	2.306	2.896	3.355	5.041
9	.129	.261	.398	.543	.703	.883	1.100	1.383	1.833	2.262	2.821	3.250	4.781
10	.129	.260	.397	.542	.700	.879	1.093	1.372	1.812	2.228	2.764	3.169	4.587
11	.129	.260	.396	.540	.697	.876	1.088	1.363	1.796	2.201	2.718	3.106	4.437
12	.128	.259	.395	.539	.695	.873	1.083	1.356	1.782	2.179	2.681	3.055	4.318
13	.128	.259	.394	.538	.694	.870	1.079	1.350	1.771	2.160	2.650	3.012	4.221
14	.128	.258	.393	.537	.692	.868	1.076	1.345	1.761	2.145	2.624	2.977	4.140
15	.128	.258	.393	.536	.691	.866	1.074	1.341	1.753	2.131	2.602	2.947	4.073
16	.128	.258	.392	.535	.690	.865	1.071	1.337	1.746	2.120	2.583	2.921	4.015
17	.128	.257	.392	.534	.689	.863	1.069	1.333	1.740	2.110	2.567	2.898	3.965
18	.127	.257	.392	.534	.688	.862	1.067	1.330	1.734	2.101	2.552	2.878	3.922
19	.127	.257	.391	.533	.688	.861	1.066	1.328	1.729	2.093	2.539	2.861	3.883
20	.127	.257	.391	.533	.687	.860	1.064	1.325	1.725	2.086	2.528	2.845	3.850
21	.127	.257	.391	.532	.686	.859	1.063	1.323	1.721	2.080	2.518	2.831	3.819
22	.127	.256	.390	.532	.686	.858	1.061	1.321	1.717	2.074	2.508	2.819	3.792
23	.127	.256	.390	.532	.685	.858	1.060	1.319	1.714	2.069	2.500	2.807	3.767
24	.127	.256	.390	.531	.685	.857	1.059	1.318	1.711	2.064	2.492	2.797	3.745
25	.127	.256	.390	.531	.684	.856	1.058	1.316	1.708	2.060	2.485	2.787	3.725
26	.127	.256	.390	.531	.684	.856	1.058	1.315	1.706	2.056	2.479	2.779	3.707
27	.127	.256	.389	.531	.684	.855	1.057	1.314	1.703	2.052	2.473	2.771	3.690
28	.127	.256	.389	.530	.683	.855	1.056	1.313	1.701	2.048	2.467	2.763	3.674
29	.127	.256	.389	.530	.683	.854	1.055	1.311	1.699	2.045	2.462	2.756	3.659
30	.127	.256	.389	.530	.683	.854	1.055	1.310	1.697	2.042	2.457	2.750	3.646
40	.126	.255	.388	.529	.681	.851	1.050	1.303	1.684	2.021	2.423	2.704	3.551
60	.126	.254	.387	.527	.679	.848	1.046	1.296	1.671	2.000	2.390	2.660	3.460
120	.126	.254	.386	.526	.677	.845	1.041	1.289	1.658	1.980	2.358	2.617	3.373
∞	.126	.253	.385	.524	.674	.842	1.036	1.282	1.645	1.960	2.326	2.576	3.291
"2-sided"	.9	.8	.7	.6	.5	.4	.3	.2	.1	.05	.02	.01	.001
"1-sided"	.45	.40	.35	.30	.25	.20	.15	.10	.05	.025	.010	.005	.0005

Significance probabilities for t-test

Appendix Table II is adapted from Table III of Fisher and Yates, *Statistical Tables for Biological, Agricultural, and Medical Research,* published by Oliver & Boyd, Ltd., Edinburgh, by permission of the authors and publishers.

APPENDIX TABLE **III**. Cumulative χ^2 distribution. The body of table contains value of χ^2; n is the number of degrees of freedom.

Probabilities for confidence intervals

CUMULATIVE PROBABILITY BETWEEN 0 AND THE TABULATED VALUE OF χ^2

n	.005	.010	.025	.050	.100	.250	.500	.750	.900	.950	.975	.990	.995
1	$.0^4 393$	$.0^3 157$	$.0^3 982$	$.0^2 393$	.0158	.102	.455	1.32	2.71	3.84	5.02	6.63	7.88
2	.0100	.0201	.0506	.103	.211	.575	1.39	2.77	4.61	5.99	7.38	9.21	10.6
3	.0717	.115	.216	.352	.584	1.21	2.37	4.11	6.25	7.81	9.35	11.3	12.8
4	.207	.297	.484	.711	1.06	1.92	3.36	5.39	7.78	9.49	11.1	13.3	14.9
5	.412	.554	.831	1.15	1.61	2.67	4.35	6.63	9.24	11.1	12.8	15.1	16.7
6	.676	.872	1.24	1.64	2.20	3.45	5.35	7.84	10.6	12.6	14.4	16.8	18.5
7	.989	1.24	1.69	2.17	2.83	4.25	6.35	9.04	12.0	14.1	16.0	18.5	20.3
8	1.34	1.65	2.18	2.73	3.49	5.07	7.34	10.2	13.4	15.5	17.5	20.1	22.0
9	1.73	2.09	2.70	3.33	4.17	5.90	8.34	11.4	14.7	16.9	19.0	21.7	23.6
10	2.16	2.56	3.25	3.94	4.87	6.74	9.34	12.5	16.0	18.3	20.5	23.2	25.2
11	2.60	3.05	3.82	4.57	5.58	7.58	10.3	13.7	17.3	19.7	21.9	24.7	26.8
12	3.07	3.57	4.40	5.23	6.30	8.44	11.3	14.8	18.5	21.0	23.3	26.2	28.3
13	3.57	4.11	5.01	5.89	7.04	9.30	12.3	16.0	19.8	22.4	24.7	27.7	29.8
14	4.07	4.66	5.63	6.57	7.79	10.2	13.3	17.1	21.1	23.7	26.1	29.1	31.3
15	4.60	5.23	6.26	7.26	8.55	11.0	14.3	18.2	22.3	25.0	27.5	30.6	32.8
16	5.14	5.81	6.91	7.96	9.31	11.9	15.3	19.4	23.5	26.3	28.8	32.0	34.3
17	5.70	6.41	7.56	8.67	10.1	12.8	16.3	20.5	24.8	27.6	30.2	33.4	35.7
18	6.26	7.01	8.23	9.39	10.9	13.7	17.3	21.6	26.0	28.9	31.5	34.8	37.2
19	6.84	7.63	8.91	10.1	11.7	14.6	18.3	22.7	27.2	30.1	32.9	26.2	38.6
20	7.43	8.26	9.59	10.9	12.4	15.5	19.3	23.8	28.4	31.4	34.2	37.6	40.0
21	8.03	8.90	10.3	11.6	13.2	16.3	20.3	24.9	29.6	32.7	35.5	38.9	41.4
22	8.64	9.54	11.0	12.3	14.0	17.2	21.3	26.0	30.8	33.9	36.8	40.3	42.8
23	9.26	10.2	11.7	13.1	14.8	18.1	22.3	27.1	32.0	35.2	38.1	41.6	44.2
24	9.89	10.9	12.4	13.8	15.7	19.0	23.3	28.2	33.2	36.4	39.4	43.0	45.6
25	10.5	11.5	13.1	14.6	16.5	19.9	24.3	29.3	34.4	37.7	40.6	44.3	46.9
26	11.2	12.2	13.8	15.4	17.3	20.8	25.3	30.4	35.6	38.9	41.9	45.6	48.3
27	11.8	12.9	14.6	16.2	18.1	21.7	26.3	31.5	36.7	40.1	43.2	47.0	49.6
28	12.5	13.6	15.3	16.9	18.9	22.7	27.3	32.6	37.9	41.3	44.5	48.3	51.0
29	13.1	14.3	16.0	17.7	19.8	23.6	28.3	33.7	39.1	42.6	45.7	49.6	52.3
30	13.8	15.0	16.8	18.5	20.6	24.5	29.3	34.8	40.3	43.8	47.0	50.9	53.7
	.995	.990	.975	.950	.900	.750	.500	.250	.100	.050	.025	.010	.005

CUMULATIVE PROBABILITY BETWEEN TABULATED VALUE OF χ^2 AND ∞

Significance probabilities for χ^2 tests

Appendix Table III is adapted from Table 8 of *Biometrika Tables for Statisticians* (Volume I) by E. S. Pearson and H. O. Hartley, published by the Cambridge University Press for the Biometrika Trustees. We are grateful to the authors and publishers for permission to use this material.

APPENDIX TABLE IV. Cumulative F distribution. The body of the table contains values of F; m is the number of degrees of freedom in the numerator and n the number in the denominator. Both "1-sided" and "2-sided" significance probabilities are given.

n	"1-SIDED" TESTS	\multicolumn{12}{c	}{m}	"2-SIDED" TESTS
		1	2	3	4	5	6	7	8	9	10	12	15	20	30	60	120	∞	
1	.10	39.9	49.5	53.6	55.8	57.2	58.2	58.9	59.4	59.9	60.2	60.7	61.2	61.7	62.3	62.8	63.1	63.3	.20
	.05	161	200	216	225	230	234	237	239	241	242	244	246	248	250	252	253	254	.10
	.025	648	800	864	900	922	937	948	957	963	969	977	985	993	1000	1010	1010	1020	.05
	.01	4,050	5,000	5,400	5,620	5,760	5,860	5,930	5,980	6,020	6,060	6,110	6,160	6,210	6,260	6,310	6,340	6,370	.02
	.005	16,200	20,000	21,600	22,500	23,100	23,400	23,700	23,900	24,100	24,200	24,400	24,600	24,800	25,000	25,300	25,400	25,400	.01
2	.10	8.53	9.00	9.16	9.24	9.29	9.33	9.35	9.37	9.38	9.39	9.41	9.42	9.44	9.46	9.47	9.48	9.49	.20
	.05	18.5	19.0	19.2	19.2	19.3	19.3	19.4	19.4	19.4	19.4	19.4	19.4	19.5	19.5	19.5	19.5	19.5	.10
	.025	38.5	39.0	39.2	39.2	39.3	39.3	39.4	39.4	39.4	39.4	39.4	39.4	39.4	39.5	39.5	39.5	39.5	.05
	.01	98.5	99.0	99.2	99.2	99.3	99.3	99.4	99.4	99.4	99.4	99.4	99.4	99.4	99.5	99.5	99.5	99.5	.02
	.005	199	199	199	199	199	199	199	199	199	199	199	199	199	199	199	199	200	.01
3	.10	5.54	5.46	5.39	5.34	5.31	5.28	5.27	5.25	5.24	5.23	5.22	5.20	5.18	5.17	5.15	5.14	5.13	.20
	.05	10.1	9.55	9.28	9.12	9.01	8.94	8.89	8.85	8.81	8.79	8.74	8.70	8.66	8.62	8.57	8.55	8.53	.10
	.025	17.4	16.0	15.4	15.1	14.9	14.7	14.6	14.5	14.5	14.4	14.3	14.3	14.2	14.1	14.0	13.9	13.9	.05
	.01	34.1	30.8	29.5	28.7	28.2	27.9	27.7	27.5	27.3	27.2	27.1	26.9	26.7	26.5	26.3	26.2	26.1	.02
	.005	55.6	49.8	47.5	46.2	45.4	44.8	44.4	44.1	43.9	43.7	43.4	43.1	42.8	42.5	42.1	42.0	41.8	.01
4	.10	4.54	4.32	4.19	4.11	4.05	4.01	3.98	3.95	3.94	3.92	3.90	3.87	3.84	3.82	3.79	3.78	3.76	.20
	.05	7.71	6.94	6.59	6.39	6.26	6.16	6.09	6.04	6.00	5.96	5.91	5.86	5.80	5.75	5.69	5.66	5.63	.10
	.025	12.2	10.6	9.98	9.60	9.36	9.20	9.07	8.98	8.90	8.84	8.75	8.66	8.56	8.46	8.36	8.31	8.26	.05
	.01	21.2	18.0	16.7	16.0	15.5	15.2	15.0	14.8	14.7	14.5	14.4	14.2	14.0	13.8	13.7	13.6	13.5	.02
	.005	31.3	26.3	24.3	23.2	22.5	22.0	21.6	21.4	21.1	21.0	20.7	20.4	20.2	19.9	19.6	19.5	19.3	.01
5	.10	4.06	3.78	3.62	3.52	3.45	3.40	3.37	3.34	3.32	3.30	3.27	3.24	3.21	3.17	3.14	3.12	3.10	.20
	.05	6.61	5.79	5.41	5.19	5.05	4.95	4.88	4.82	4.77	4.74	4.68	4.62	4.56	4.50	4.43	4.40	4.37	.10
	.025	10.0	8.43	7.76	7.39	7.15	6.98	6.85	6.76	6.68	6.62	6.52	6.43	6.33	6.23	6.12	6.07	6.02	.05
	.01	16.3	13.3	12.1	11.4	11.0	10.7	10.5	10.3	10.2	10.1	9.89	9.72	9.55	9.38	9.20	9.11	9.02	.02
	.005	22.8	18.3	16.5	15.6	14.9	14.5	14.2	14.0	13.8	13.6	13.4	13.1	12.9	12.7	12.4	12.3	12.1	.01
6	.10	3.78	3.46	3.29	3.18	3.11	3.05	3.01	2.98	2.96	2.94	2.90	2.87	2.84	2.80	2.76	2.74	2.72	.20
	.05	5.99	5.14	4.76	4.53	4.39	4.28	4.21	4.15	4.10	4.06	4.00	3.94	3.87	3.81	3.74	3.70	3.67	.10
	.025	8.81	7.26	6.60	6.23	5.99	5.82	5.70	5.60	5.52	5.46	5.37	5.27	5.17	5.07	4.96	4.90	4.85	.05
	.01	13.7	10.9	9.78	9.15	8.75	8.47	8.26	8.10	7.98	7.87	7.72	7.56	7.40	7.23	7.06	6.97	6.88	.02
	.005	18.6	14.5	12.9	12.0	11.5	11.1	10.8	10.6	10.4	10.2	10.0	9.81	9.59	9.36	9.12	9.00	8.88	.01
7	.10	3.59	3.26	3.07	2.96	2.88	2.83	2.78	2.75	2.72	2.70	2.67	2.63	2.59	2.56	2.51	2.49	2.47	.20
	.05	5.59	4.74	4.35	4.12	3.97	3.87	3.79	3.73	3.68	3.64	3.57	3.51	3.44	3.38	3.30	3.27	3.23	.10
	.025	8.07	6.55	5.89	5.52	5.29	5.12	4.99	4.90	4.82	4.76	4.67	4.57	4.47	4.36	4.25	4.20	4.14	.05
	.01	12.2	9.55	8.45	7.85	7.46	7.19	6.99	6.84	6.72	6.62	6.47	6.31	6.16	5.99	5.82	5.74	5.65	.02
	.005	16.2	12.4	10.9	10.1	9.52	9.16	8.89	8.68	8.51	8.38	8.18	7.97	7.75	7.53	7.31	7.19	7.08	.01
8	.10	3.46	3.11	2.92	2.81	2.73	2.67	2.62	2.59	2.56	2.54	2.50	2.46	2.42	2.38	2.34	2.32	2.29	.20
	.05	5.32	4.46	4.07	3.84	3.69	3.58	3.50	3.44	3.39	3.35	3.28	3.22	3.15	3.08	3.01	2.97	2.93	.10
	.025	7.57	6.06	5.42	5.05	4.82	4.65	4.53	4.43	4.36	4.30	4.20	4.10	4.00	3.89	3.78	3.73	3.67	.05
	.01	11.3	8.65	7.59	7.01	6.63	6.37	6.18	6.03	5.91	5.81	5.67	5.52	5.36	5.20	5.03	4.95	4.86	.02

v_2	α																			
9	.10	3.36	3.01	2.81	2.69	2.61	2.55	2.51	2.47	2.44	2.42	2.38	2.34	2.30	2.25	2.21	2.18	2.16		
	.05	5.12	4.26	3.86	3.63	3.48	3.37	3.29	3.23	3.18	3.14	3.07	3.01	2.94	2.86	2.79	2.75	2.71		
	.025	7.21	5.71	5.08	4.72	4.48	4.32	4.20	4.10	4.03	3.96	3.87	3.77	3.67	3.56	3.45	3.39	3.33		
	.01	10.6	8.02	6.99	6.42	6.06	5.80	5.61	5.47	5.35	5.26	5.11	4.96	4.81	4.65	4.48	4.40	4.31		
	.005	13.6	10.1	8.72	7.96	7.47	7.13	6.88	6.69	6.54	6.42	6.23	6.03	5.83	5.62	5.41	5.30	5.19		
10	.10	3.28	2.92	2.73	2.61	2.52	2.46	2.41	2.38	2.35	2.32	2.28	2.24	2.20	2.16	2.11	2.08	2.06		
	.05	4.96	4.10	3.71	3.48	3.33	3.22	3.14	3.07	3.02	2.98	2.91	2.84	2.77	2.70	2.62	2.58	2.54		
	.025	6.94	5.46	4.83	4.47	4.24	4.07	3.95	3.85	3.78	3.72	3.62	3.52	3.42	3.31	3.20	3.14	3.08		
	.01	10.0	7.56	6.55	5.99	5.64	5.39	5.20	5.06	4.94	4.85	4.71	4.56	4.41	4.25	4.08	4.00	3.91		
	.005	12.8	9.43	8.08	7.34	6.87	6.54	6.30	6.12	5.97	5.85	5.66	5.47	5.27	5.07	4.86	4.75	4.64		
12	.10	3.18	2.81	2.61	2.48	2.39	2.33	2.28	2.24	2.21	2.19	2.15	2.10	2.06	2.01	1.96	1.93	1.90		
	.05	4.75	3.89	3.49	3.26	3.11	3.00	2.91	2.85	2.80	2.75	2.69	2.62	2.54	2.46	2.38	2.34	2.30		
	.025	6.55	5.10	4.47	4.12	3.89	3.73	3.61	3.51	3.44	3.37	3.28	3.18	3.07	2.96	2.85	2.79	2.72		
	.01	9.33	6.93	5.95	5.41	5.06	4.82	4.64	4.50	4.39	4.30	4.16	4.01	3.86	3.70	3.54	3.45	3.36		
	.005	11.8	8.51	7.23	6.52	6.07	5.76	5.52	5.35	5.20	5.09	4.91	4.72	4.53	4.33	4.12	4.01	3.90		
15	.10	3.07	2.70	2.49	2.36	2.27	2.21	2.16	2.12	2.09	2.06	2.02	1.97	1.92	1.87	1.82	1.79	1.76		
	.05	4.54	3.68	3.29	3.06	2.90	2.79	2.71	2.64	2.59	2.54	2.48	2.40	2.33	2.25	2.16	2.11	2.07		
	.025	6.20	4.77	4.15	3.80	3.58	3.41	3.29	3.20	3.12	3.06	2.96	2.86	2.76	2.64	2.52	2.46	2.40		
	.01	8.68	6.36	5.42	4.89	4.56	4.32	4.14	4.00	3.89	3.80	3.67	3.52	3.37	3.21	3.05	2.96	2.87		
	.005	10.8	7.70	6.48	5.80	5.37	5.07	4.85	4.67	4.54	4.42	4.25	4.07	3.88	3.69	3.48	3.37	3.26		
20	.10	2.97	2.59	2.38	2.25	2.16	2.09	2.04	2.00	1.96	1.94	1.89	1.84	1.79	1.74	1.68	1.64	1.61		
	.05	4.35	3.49	3.10	2.87	2.71	2.60	2.51	2.45	2.39	2.35	2.28	2.20	2.12	2.04	1.95	1.90	1.84		
	.025	5.87	4.46	3.86	3.51	3.29	3.13	3.01	2.91	2.84	2.77	2.68	2.57	2.46	2.35	2.22	2.16	2.09		
	.01	8.10	5.85	4.94	4.43	4.10	3.87	3.70	3.56	3.46	3.37	3.23	3.09	2.94	2.78	2.61	2.52	2.42		
	.005	9.94	6.99	5.82	5.17	4.76	4.47	4.26	4.09	3.96	3.85	3.68	3.50	3.32	3.12	2.92	2.81	2.69		
30	.10	2.88	2.49	2.28	2.14	2.05	1.98	1.93	1.88	1.85	1.82	1.77	1.72	1.67	1.61	1.54	1.50	1.46		
	.05	4.17	3.32	2.92	2.69	2.53	2.42	2.33	2.27	2.21	2.16	2.09	2.01	1.93	1.84	1.74	1.68	1.62		
	.025	5.57	4.18	3.59	3.25	3.03	2.87	2.75	2.65	2.57	2.51	2.41	2.31	2.20	2.07	1.94	1.87	1.79		
	.01	7.56	5.39	4.51	4.02	3.70	3.47	3.30	3.17	3.07	2.98	2.84	2.70	2.55	2.39	2.21	2.11	2.01		
	.005	9.18	6.35	5.24	4.62	4.23	3.95	3.74	3.58	3.45	3.34	3.18	3.01	2.82	2.63	2.42	2.30	2.18		
60	.10	2.79	2.39	2.18	2.04	1.95	1.87	1.82	1.77	1.74	1.71	1.66	1.60	1.54	1.48	1.40	1.35	1.29		
	.05	4.00	3.15	2.76	2.53	2.37	2.25	2.17	2.10	2.04	1.99	1.92	1.84	1.75	1.65	1.53	1.47	1.39		
	.025	5.29	3.93	3.34	3.01	2.79	2.63	2.51	2.41	2.33	2.27	2.17	2.06	1.94	1.82	1.67	1.58	1.48		
	.01	7.08	4.98	4.13	3.65	3.34	3.12	2.95	2.82	2.72	2.63	2.50	2.35	2.20	2.03	1.84	1.73	1.60		
	.005	8.49	5.80	4.73	4.14	3.76	3.49	3.29	3.13	3.01	2.90	2.74	2.57	2.39	2.19	1.96	1.83	1.69		
120	.10	2.75	2.35	2.13	1.99	1.90	1.82	1.77	1.72	1.68	1.65	1.60	1.55	1.48	1.41	1.32	1.26	1.19		
	.05	3.92	3.07	2.68	2.45	2.29	2.18	2.09	2.02	1.96	1.91	1.83	1.75	1.66	1.55	1.43	1.35	1.25		
	.025	5.15	3.80	3.23	2.89	2.67	2.52	2.39	2.30	2.22	2.16	2.05	1.94	1.82	1.69	1.53	1.43	1.31		
	.01	6.85	4.79	3.95	3.48	3.17	2.96	2.79	2.66	2.56	2.47	2.34	2.19	2.03	1.86	1.66	1.53	1.38		
	.005	8.18	5.54	4.50	3.92	3.55	3.28	3.09	2.93	2.81	2.71	2.54	2.37	2.19	1.98	1.75	1.61	1.43		
∞	.10	2.71	2.30	2.08	1.94	1.85	1.77	1.72	1.67	1.63	1.60	1.55	1.49	1.42	1.34	1.24	1.17	1.00		
	.05	3.84	3.00	2.60	2.37	2.21	2.10	2.01	1.94	1.88	1.83	1.75	1.67	1.57	1.46	1.32	1.22	1.00		
	.025	5.02	3.69	3.12	2.79	2.57	2.41	2.29	2.19	2.11	2.05	1.94	1.83	1.71	1.57	1.39	1.27	1.00		
	.01	6.63	4.61	3.78	3.32	3.02	2.80	2.64	2.51	2.41	2.32	2.18	2.04	1.88	1.70	1.47	1.32	1.00		
	.005	7.88	5.30	4.28	3.72	3.35	3.09	2.90	2.74	2.62	2.52	2.36	2.19	2.00	1.79	1.53	1.36	1.00		

Appendix Table IV is adapted from Table 18 of *Biometrika Tables for Statisticians* (Volume I) by E. S. Pearson and H. O. Hartley, published by the Cambridge University Press for the Biometrika Trustees. We are grateful to the authors and publishers for permission to use this material.

APPENDIX TABLE **V**. Significance probabilities for the correlation coefficient (r). The body of the table contains values of the correlation coefficient; n is the number of degrees of freedom.

n	.1	.05	.02	.01	.001	n	.1	.05	.02	.01	.001
1	.98769	.99692	.999507	.999877	.9999988	16	.4000	.4683	.5425	.5897	.7084
2	.90000	.95000	.98000	.990000	.99900	17	.3887	.4555	.5285	.5751	.6932
3	.8054	.8783	.93433	.95873	.99116	18	.3783	.4438	.5155	.5614	.6787
4	.7293	.8114	.8822	.91720	.97406	19	.3687	.4329	.5034	.5487	.6652
5	.6694	.7545	.8329	.8745	.95074	20	.3598	.4227	.4921	.5368	.6524
6	.6215	.7067	.7887	.8343	.92493	25	.3233	.3809	.4451	.4869	.5974
7	.5822	.6664	.7498	.7977	.8982	30	.2960	.3494	.4093	.4487	.5541
8	.5494	.6319	.7155	.7646	.8721	35	.2746	.3246	.3810	.4182	.5189
9	.5214	.6021	.6851	.7348	.8471	40	.2573	.3044	.3578	.3932	.4896
10	.4973	.5760	.6581	.7079	.8233	45	.2428	.2875	.3384	.3721	.4648
11	.4762	.5529	.6339	.6835	.8010	50	.2306	.2732	.3218	.3541	.4433
12	.4575	.5324	.6120	.6614	.7800	60	.2108	.2500	.2948	.3248	.4078
13	.4409	.5139	.5923	.6411	.7603	70	.1954	.2319	.2737	.3017	.3799
14	.4259	.4973	.5742	.6226	.7420	80	.1829	.2172	.2565	.2830	.3568
15	.4124	.4821	.5577	.6055	.7246	90	.1726	.2050	.2422	.2673	.3375
						100	.1638	.1946	.2301	.2540	.3211

Appendix Table V is taken from Table VI of Fisher and Yates, *Statistical Tables for Biological, Agricultural, and Medical Research*, published by Oliver & Boyd, Ltd., Edinburgh, by permission of the authors and publishers.

Symbols

A — estimate of the Y-intercept in Bartlett's "best fit" line.
a — any sample constant to be inserted in an equation, especially the sample estimate of the intercept in a regression equation: a_X is the X-intercept; a_Y is the Y-intercept.
α — (1) parameter estimated by a, especially the intercept parameter in a regression equation: α_X is the X-intercept; α_Y is the Y-intercept.
(2) Coefficient of allometry.
antilog — antilogarithm.
B — estimate of the slope of Bartlett's "best fit" line.
b — usually the estimate of the slope of the rectilinear regression equation, the sample coefficient of regression: b_{YX} is the regression of Y on X; b_{XY} is the regression of X on Y.
β — the true value of the regression coefficient estimated by b. If no subscript appears, it is the parameter estimated by B. Otherwise, β_{YX} is the true regression of Y on X and β_{XY} is the true regression of X on Y.
C — Spearman's coefficient of rank correlation.
cov. — covariance of two variates.
D — a theoretical range of logarithms of observed values as estimated from a single specimen.
d — any deviation from a given value—usually, if not otherwise stated, the deviation of a single observation from its sample mean.
d.f. — degrees of freedom.
E — expected or theoretical frequency of a given class in a frequency distribution or contingency table.
e — the mathematical constant, base of natural logarithms; limit of the series $1 + \dfrac{1}{1} + \dfrac{1}{(2)(1)} + \dfrac{1}{(3)(2)(1)} + \cdots = 2.7183\ldots$.
F — the ratio of two mean squares or two variances. Often written as $F_{n,d}$ where n and d are the numbers of degrees of freedom in numerator and denominator.

f —absolute frequency within a single class of a frequency distribution.
H —harmonic mean.
i —the class interval.
i, j, k —literal subscripts distinguishing a value such as X_i as belonging to the ith class.
K —additive growth rate.
K_g —geometric growth rate.
K_s —measure of kurtosis.
L —sometimes used for class limits; L_l is the lower and L_u the upper limit.
log —logarithm to the base 10.
$\log_e$ —natural logarithm—logarithm to the base e.
M.D. —mean deviation.
μ —true mean of a population; the value of the parameter of which the sample mean is an estimate.
N —total sample size or total frequency.
n —general symbol for any integer.
O —observed frequency in any class of a frequency distribution or contingency table, usually used in χ^2 analysis.
O.R. —observed range.
P —tabulated probability from a standard probability distribution such as Student's t-distribution, the standardized normal distribution, etc.
P.E. —probable error.
p —in a binomial distribution, the probability of occurrence or success.
π —the mathematical constant 3.14159....
Q_1, Q_3 —first and third quartiles.
Q.D. —quartile deviation or semi-interquartile range.
q —in a binomial distribution, the probability of nonoccurrence or failure.
r —the sample estimate of the correlation coefficient, the product-moment correlation coefficient.
ρ —the true value of the correlation coefficient, the parameter of which r is an estimate.
S_k —a measure of skewness.
S.S. —sum of squares.
s —sample estimate of the standard deviation of a population.
$s_{\bar{X}}$ —standard error of the mean. Also standard error of other estimates when appropriate subscripts are used.
s^2 —sample estimate of the variance of a population.
$\sum$ —the summation of all quantities following the symbol, e.g., $\sum X$ is the sum of all the values of X.
σ —population standard deviation estimated by s.

σ^2	–population variance estimated by s^2.
t	–(1) Student's t-statistic, tabulated in Appendix Table II, used for testing the difference between means of two samples. (2) Symbol for time in the study of growth.
τ	–standardized normal deviate tabulated in Appendix Table I.
V	–coefficient of variation.
X	–any individual value of a variate. If two variates are involved, X is usually taken as the independent variate.
χ^2	–a statistic used in testing frequency data, tabulated in Appendix Table III.
Y	–any value of a variate, usually a dependent one, but in general any alternative variable to X.
z	–a measure of correlation transformed from r.
!	–the factorial operator. When preceded by a variate such as X or by a digit, it is shorthand notation for $X(X-1)(X-2)\cdots(3)(2)(1)$.

Bibliography

ALDOUS, C. M. (1937) "Notes on the Life History of the Snow-Shoe Hare." *Journal of Mammalogy*, 18: 46-57.

ANDERSON, E. (1954) "Efficient and Inefficient Methods of Measuring Specific Difference," pp. 93-106 in *Statistics and Mathematics in Biology*, edited by O. Kempthorne, T. A. Bancroft, J. W. Gowen, and J. L. Lush. Ames, Iowa, The Iowa State College Press.

ANTHONY, H. E. (1925) "The Capture and Preservation of Small Animals for Study." *American Museum Guide Leaflet No. 61.*

ATWOOD, K. C., L. L. SCHNEIDER, and F. J. RYAN (1951) "Selective Mechanisms in Bacteria." *Cold Spring Harbor Symposia on Quantitative Biology*, 16: 344-55.

BALDWIN, S. P., H. C. OBERHOLSER, and L. G. WORLEY (1931) "Measurements of Birds." *Scientific Publications of the Cleveland Museum of Natural History*, 2: 1-165.

BARTLETT, M. S. (1949) "Fitting a Straight Line When Both Variables Are Subject to Error." *Biometrics* 5: 207-12.

BLANCHARD, F. N. (1921) "A Revision of the King Snakes: Genus *Lampropeltis*." *United States National Museum*, Bulletin 114.

BLANCHARD, F. N., and F. C. BLANCHARD (1931) "Size Groups and Their Characteristics in the Salamander *Hemidactylium scutatum* (Schlegel)." *American Naturalist*, 65: 149-64.

BOURLIÈRE, F. (1951) *Vie et moeurs des mammifères*. Paris, Payot.

BRUUN, A. FR. (1935) "Flying Fishes (*Exocoetidae*) of the Atlantic." Dana Report, No. 6. Copenhagen and London, the Carlsberg Foundation.

CHAPMAN, J. A. (1954) "Studies on Summit Frequenting Insects in Western Montana." *Ecology*, 35: 41-49.

CLARK, L. B., and A. H. HERSH (1939) "A Study of Relative Growth in *Notonecta undulata*." *Growth*, 3: 347-72.

COCHRAN, W. G. (1954) "Some Methods for Strengthening the Common χ^2 Tests." *Biometrics*, 10: 417-51.

COCHRAN, W. G., and G. M. COX (1957) *Experimental Designs*, New York, John Wiley & Sons.

COLLINS, H. H. (1923) "Studies of the Pelage Phases and of the Nature of Color Variations in Mice of the Genus *Peromyscus*." *Journal of Experimental Zoology*, 38: 45-107.

Computation Laboratory, Harvard University (1955) *Tables of the Cumulative Binomial Probability Distribution*. Cambridge, Massachusetts, Harvard University Press.

C. R. C. *Standard Mathematical Tables*, 11th Edition (formerly *Mathematical Tables from the Handbook of Chemistry and Physics*) (1957) Cleveland, Ohio, Chemical Rubber Publishing Company.

DAVIS, D. E. (1951) "The Relation Between Level of Population and Size and Sex of Norway Rats." *Ecology*, 32: 462-64.

DEEVEY, E. S., JR. (1947) "Life Tables for Natural Populations of Animals." *Quarterly Review of Biology*, 22: 283-314.

DICE, L. R. (1931) "Methods of Indicating the Abundance of Mammals." *Journal of Mammalogy*, 12: 376-81.

DICE, L. R. (1932) "Variation in the Geographic Race of the Deermouse, *Peromyscus maniculatus bairdii*." *Occasional Papers of the Museum of Zoology, University of Michigan*, No. 239.

DICE, L. R. (1933) "Variation in *Peromyscus maniculatus rufinus* from Colorado and Mexico." *Occasional Papers of the Museum of Zoology, University of Michigan*, No. 271: 1-32.

DICE, L. R. (1937) "Additional Data on Variation in the Prairie Deermouse, *Peromyscus maniculatus bairdii*." *Occasional Papers of the Museum of Zoology, University of Michigan*, No. 351.

DICE, L. R., and H. J. LERAAS (1936) "A Graphic Method for Comparing Several Sets of Measurements." *Contributions from the Laboratory of Vertebrate Genetics, University of Michigan*, No. 3: 1-3.

DIXON, J. (1925) "Food Predilections of Predatory and Fur-bearing Mammals." *Journal of Mammalogy*, 6: 34-46.

DUERST, V. V. (1926) "Vergleichende Untersuchungsmethoden am Skelett bei Säugern," *Handbuch der biologischen Arbeitsmethoden*, Abteil 7, Heft 2: 125-530.

EATON, R. J. (1924) "The Migratory Movements of Certain Colonies of Herring Gulls in Eastern North America." *Bird-Banding*, 5: 70-84.

FELLER, W. (1950) *An Introduction to Probability Theory and Its Applications*. New York, John Wiley & Sons.

FISHER, R. A. (1950) *Statistical Methods for Research Workers*, 11th Edition. Edinburgh, Oliver & Boyd.

FISHER, R. A., and F. YATES (1948) *Statistical Tables for Biological and Medical Research*, 3rd Edition. Edinburgh, Oliver & Boyd.

FULTON, B. B. (1952) "Speciation in the Field Cricket." *Evolution*, 6: 283-95.

GREGORY, W. K. (1933) "Fish Skulls: A Study of the Evolution of Natural Mechanisms." *Transactions of the American Philosophical Society*, 23, Article 2.

HALDANE, J. B. S. (1955) "The Measurement of Variation." *Evolution*, 9: 484.

HAMMETT, F. S., and D. W. HAMMETT (1939) "Proportional Length and Growth of Gar *Lepisosteus platyrhincus* (de Kay)." *Growth*, 3: 197-210

HATT, R. T. (1936) "Hyraxes Collected by the American Museum Congo Expedition." *Bulletin of the American Museum of Natural History*, 72: 117-41.

HICKS, L. E. (1934) "Individual and Sexual Variations in the European Starling." *Bird-Banding*, 5: 103-18.

HILDEBRAND, S. F., and W. C. SCHROEDER (1927) "Fishes of Chesapeake Bay." *Bulletin of the United States Bureau of Fisheries*, 43: Part 1.

HOWARD, H. (1927) "A Review of the Fossil Bird, *Parapavo californicus* (Miller), from the Pleistocene Asphalt Beds of Rancho La Brea." *University of California Publication, Bulletin of the Department of Geological Sciences*, 17: Number 1.

HUBBS, C. L., and L. P. SCHULTZ (1932) "*Cottus tubulatus*, a New Sculpin from Idaho." *Occasional Papers of the Museum of Zoology, University of Michigan*, 242: 1-9.

HUBBS, C. L., and K. F. LAGLER (1947) "Fish of the Great Lakes Region." *Cranbrook Institute of Science, Bulletin No. 26*. Bloomfield Hills, Michigan, Cranbrook Press.

HUESTIS, R. R. (1931) "Seasonal Pelagè Differences in *Peromyscus*." *Journal of Mammalogy*, 12: 372-75.

HUXLEY, J. S. (1932) *Problems of Relative Growth*, London, Methuen & Company.

JENKINSON, J. W. (1912) "Growth, Variability and Correlation in Young Trout." *Biometrika*, 8: 444-55.

KÄLIN, J. A. (1933) "Beiträge zur vergleichenden Osteologie des Crocodilidenschädels," *Zoologische Jahrbücher: Abteil für Anatomie und Ontogenie der Tiere*, 57: 535-714.

KERMACK. K. A., and J. B. S. HALDANE (1950) "Organic Correlation and Allometry." *Biometrika*, 37: 30-41.

KURTÉN, B. (1953) "On the Variation and Population Dynamics of Fossil and Mammal Populations," *Acta Zoologica Fennica*, 76: 1-122.

LAMOTTE, M. (1951) "Recherches sur la structure génétique des populations naturelles de *Cepaea nemoralis*." *Bulletin Biologique de France et Belgique*, Supp. 35.

LOW, S. H. (1933) "Further Notes on the Nesting of the Tree Swallows," *Bird-Banding*, 4: 76-87.

MACDONNELL, W. R. (1902) "On Criminal Anthropometry and the Identification of Criminals." *Biometrika*, 1: 177-227.

MARIEN, D. (1950) "Notes on Some Asiatic Sturnidae." *Journal of the Bombay Natural History Society*, 49: 471-87.

MATTHEW, W. D. (1909) "The Carnivora and Insectivora of the Bridger Basin, Middle Eocene." *Memoirs of the American Museum of Natural History*, 9, Part 6: 291-567.

MEDAWAR, P. B. (1945) "Size, Shape and Age," pp. 157-87 in *Essays on Growth and Form*, edited by W. E. LeGros Clark and P. B. Medawar. Oxford, Clarendon Press.

MEEK, S. W., and S. F. HILDEBRAND (1923) "Marine Fishes of Panama." *Field Museum Publications in Zoology*, 15: Part 1.

MILLER, M. A., and E. A. HOY (1939) "Differential Growth and Evolution in a Subterranean Isopod." *American Naturalist*, 73: 347-64.

NICE, M. M. (1933) "Nesting Success During Three Seasons in a Song Sparrow Population." *Bird-Banding*, 4: 119-31.

NICHOLS, J. T. (1935) "The Hawaiian 'Ulva'." *Copeia*, 192-93.

OSBORN, H. F. (1912) "Craniometry of the Equidae." *Memoirs of the American Museum of Natural History*, New Series, 1: 57-100.

OSBORN, H. F. (1929) "The Titanotheres of Ancient Wyoming, Dakota and Nebraska." *United States Geological Survey, Monograph* 55.

PARKER, G. H. (1922) "The Breathing of the Florida Manatee (*Trichechus latirostris*)," *Journal of Mammalogy*, 3: 127-35.

PAYNTER, R. A., JR. (1951) "Clutch Size and Egg Mortality of Kent Island Eiders." *Ecology*, 32: 487-507.

PEARSON, E. S. (1932) "The Percentage Limits for the Distribution of Range in Samples from a Normal Population." *Biometrika*, 24: 406-13.

PEARSON, E. S., and H. O. HARTLEY (1954) *Biometrika Tables for Statisticians*. Cambridge, Cambridge University Press.

PEARSON, H. S. (1928) "Chinese Fossil Suidae." *Palaeontologica Sinica*, 5, Fascicle 5, Series C.

PHILLIPS, J. C. (1920) "Skull Measurements of the Virginia Deer." *Journal of Mammalogy*, 1: 130-33.

RICHARDS, O. W., and A. J. KAVANAUGH (1945) "The Analysis of Growing Form," pp. 180-230 in *Essays on Growth and Form*, edited by W. E. Le Gros Clark and P. B. Medawar. Oxford, Clarendon Press.

RIDGWAY, R. (1901) "The Birds of North and Middle America: Part 1, Family Fringillidae." *United States National Museum, Bulletin* 50: i-xxxii, 1-716.

RIDGWAY, R. (1912) "Color Standards and Color Nomenclature." Washington, D.C. Published by the author.

ROMIG, H. G. (1953) *Binomial Tables.* New York, John Wiley & Sons.

RUTHVEN, A. (1908) "Variations and Genetic Relationships of the Garter Snakes." *United States National Museum, Bulletin* 61.

SCHOOLEY, J. P. (1934) "A Summer Breeding Season in the Eastern Chipmunk, *Tamias striatus*," *Journal of Mammalogy*, 15: 194-96.

SHAW, W. T. (1926) "Age of the Animal and Slope of the Ground Surface, Factors Modifying the Structure of Hibernation Dens of Ground Squirrels." *Journal of Mammalogy*, 7: 91-96.

SIBLEY, C. G. (1954) "Hybridization in the Red-Eyed Towhees of Mexico." *Evolution*, Vol. 8: 252-90.

SIMPSON, G. G. (1941) "Large Pleistocene Felines of North America." *American Museum Novitates*, 1136: 1-27.

SNEDECOR, G. W. (1956) *Statistical Methods Applied to Experiments in Agriculture and Biology*, 5th Edition. Ames, Iowa, Iowa State College Press.

SNYDER, L., and H. YINGLING (1935) "Studies in Inheritance, 12." *Human Biology*, 7: 608-15.

SOERGEL, W. (1925) *Die Fährten der Chirotheria—Eine paläobiologische Studie.* Jena, Gustav Fischer.

SPERRY, C. (1934) "Winter Food Habits of Coyotes: A Report of Progress, 1933." *Journal of Mammalogy*, 15: 287-90.

SUMNER, F. B. (1927) "Linear and Colorimetric Measurements of Small Mammals," *Journal of Mammalogy*, 8: 177-206.

SVIHLA, A. (1934) "Development and Growth of Deer Mice (*Peromyscus maniculatus artemisiae*)." *Journal of Mammalogy*, 15: 99-104.

TEISSIER, G. (1948) "La Relation d'allométrie: sa significance statistique et biologique." *Biometrics*, 4: 14-48.

THOMPSON, D'ARCY (1917) *Growth and Form.* Cambridge, England, Cambridge University Press.

TIPPET, L. H. C. (1925) "On Extreme Individuals and the Range of Samples Taken from a Normal Population." *Biometrika*, 17: 364-87.

ZEUNER, F. (1934) "Die Beziehungen zwischen Schädelform und Lebensweise bei rezenten und fossilen Nashörnern." *Bericht der Naturforschenden Gesellschaft zu Freiburg im Breisgau*, 34: 21-80.

Index

Absolute size, 90
Acheta assimilis, 363
Acropithecus rigidus, 49, 51, 250
Age groups, 384-90
   factors affecting, 385
   fossils, 388-90
   living animals, 385-88
   methods of separating, 385-90
Aldous, C. M., 59, 367
Allen's rule, 178, 216
Allometric constant, 397-415
   changes in, 409-15
   negative, 404
Allometric equation, 397-98, 401, 406
   fitting, 399-405
Allometry, 396-415
   coefficient of, 396-98
   deviations from, 398, 409-15
   negative, 398
   size, 406-09
Allopatric species, 298
Analysis of chi-square, 326-38
Analysis of variance, 258-305
   complex analyses, 305
   fixed model, 266-89
   hierarchical model, 298-304
   meaning, 259-60
   mixed model, 296-97
   one factor, 266-71, 290-92
   random model, 289-96
   in regression, 271-77
   table, 269
   three factor, 287, 293-95
   two factor, 277-87, 292-93
Anderson, E., 346
Angles, 11
Anisonchus sectorius, 345
Anthony, H. E., 25
Aquatic insects, 284
Areas, 13
Armadillo, 116
Asellus californicus, 411, 412, 413
Association, 306, 312-24
Attributes, 2, 34
Atwood, K. C., 379
Austroicetes cruciata, 148, 149
Average, meaning of, 75-76

Baldwin, S. P., 25
Bartlett, M. S., 232-37
Bergmann's rule, 216
"Best fit" line, 232-37
Bias, 27-29, 100
   elimination of, 28
   in paleontological samples, 115-16
   in qualitative samples, 105-06
Binomial distribution tables, 155
"Biometrics," 305
Birds, measurement of, 25
Blanchard, F. C., 387
Blanchard, F. N., 24, 35, 44, 243, 246, 387
Bourlière, F., 351
Bridger fauna, 33
Bruun, A. F., 57, 319

Capture-mark-recapture method, 112-13
Caranx melampygus, 44, 59, 60
Carapace, 24
Centiles, 89
Central tendency measures, 65-77
Cepaea nemoralis, 113, 114, 188, 191, 328, 332
Chapman, J. A., 32
Chipmunk, 33, 367
Chi-square, distribution, 161-63
   table, 423
Chi-square test, 187-91, 306-07
   small samples, 189-90
   Yates' correction, 190-91
Chleuastochoerus, 183
Citellus columbianus columbianus, 322
Clark, L. B., 394, 405, 410, 411
Classes, non-numerical, 3
Clines, 354-56
Cochran, W. G., 305
Coefficient of variation, 89-95
   assumed, 206
Colbert, E., 372
Collins, H. H., 27
Color, quantifying, 3
   standards, 26-27
Color top, 27
Comparison of samples, 172-212
Confidence, measure of, 151
Confidence intervals, 148-71
   "best fit" intercept, 236
   "best fit" slope, 233-36
   binomial proportion, 154-58
   coefficient of variation, 168-71
   construction of, 152-54
   intercept, 224-29
   mean, 168-71
   mean deviation, 168-70
   median, 168-70
   quantiles, 168-70
   ratios, 163-65
   slope, 224-29

436 INDEX

Confidence intervals, *cont.*
  standard deviation, 161-63, 168-71
  variance, 161-63, 168-71
Confidence limits, 155
Consistency, 29-30, 99-100
Contingency classification, 314
Contingency table, 314-18
  higher order, 316-18
  two-by-two, 314-16
Contingency test
  meaning, 323-24, 325
  partitioning, 329-38
  small samples, 322-23
Continuous probability distribution, 135
Coordinates, 340-44
  angular, 342-43
  arithmetic, 340-42
  deformed, 369-72
  logarithmic, 340-43
  polar, 342-43, 366
  rectangular, 340-43
  semilogarithmic, 340-42
Correlation, 213-57, 312, 313
  cause and effect, 246-48
  definition, 214
  multiple, 248
  partial, 248-53
  rank, 253-57
  scope of, 216-17
  spurious, 246-48
Correlation coefficient
  partial, 248, 249
  product moment, 238-41
  Spearman's rank, 253-57
  table of distribution, 426
  transformation to z, 244-46
Correlation tables, 241, 242, 243
Cox, G. M., 305
Curve fitting
  "best fit" line, 402-05
  major axis, 401-02
  reduced major axis, 401-03
Cottus, 81, 82
Counts, 4
Covariance, 221
Coyote, 350, 364
Crocodile, 25

Data from observation, 10-13
Davis, D. E., 193, 194
Deciles, 89
Deer mouse, 69. *See also Peromyscus*
Deevey, E. S., 351
Degrees of freedom
  association tests, 317-18
  chi-square distribution, 161-62, 187
  F tests, 185
  goodness-of-fit test, 309-10
  partial correlation, 249
  partitioning in chi-square test, 336
  Student's t, 159, 183
Dendrohyrax dorsalia emini, 92
Dentition, use in age grouping, 389-90
Deposition of fossils, 388-90
Deviations, in the analysis of variance, 260-63
Deviations from normality, 142-47
Diagrams
  bar, 348, 349, 350, 360-61, 366
  deformed coordinate, 369-72
  Dice-Leraas, 351-55
  frequency, 339, 344
  frequency surface, 345-46
  genetic descent, 369
  gradient, 354-56
  histogram, 340, 344-45, 356
  line, 339, 343, 344
  phylogeny, 368
  pictorial, 366-72
  pie, 364-65
  point, 367
  population composition, 362-63
  proportions, 358-65
  pyramid, 348, 351
  ratio, 357-58
  scatter, 346-48, 366
Dice, L. R., 3, 69, 97, 177, 204, 351, 362
Dicerorhinus sumatrensis, 372
Diodon, 371
Diptera, 32
Discriminant function, 14
Dispersion, 78-95
  absolute, 90
  relative, 89-95
Distribution-free tests, 158, 191
Distribution patterns
  types, 53-61
  meaning of, 52-53
Distributions
  bimodal, 102
  binomial, 124-29
  cumulative, 61-62, 121-24
  discontinuous, 34
  F, 185-86
  F', 197-98
  J-shaped, 57
  noncentral t, 195
  normal, 133-39, 306
  Pearsonian, 158
  Poisson, 129-33, 306
  Student's t, 159-61
  U-shaped, 58
Dixon, J., 350
Drosophila, 14, 113
Drosophila persimilis, 301-02
Duerst, V. V., 26

Eaton, R. J., 342
Effect, in analysis of variance, 260-63
Efficiency, 100, 101
Escherichia coli, 379
Estimates, properties of, 99-101
Estimation
 and calculation, 97-99
 of population size, 112-13
Events in probability, 118
Exact test for association, 323
Exocoetus obtusirostris, 319
Expected mean square, 290

F-distribution, table, 424-25
Factorial, 125
Factorial design, 277
Factors in analysis of variance, 260-66
 fixed, 263-66
 random, 263-66
Faunal sampling, 133
Feller, W., 133
Fisher, R. A., 14, 97, 127, 323
Fishes, linear dimensions, 24
Frequencies, 10, 12
 absolute, 119
 definition, 31
 relative, 62-64, 119
Frequency distribution, 31-47
Frequency polygon, 49
Fulton, B. B., 363

Gaindatherium browni, 372
Gauss, K. F., 133
Gaussian curve, 133
Geissler, 127
Genetic relationship, measure of, 2
Geographical races, 301
Geometrical growth increment, 378
Geometric mean, 145, 146
Gloger's rule, 216
Goodness of fit, 306, 307-12
 normal distribution, 307-08
 Poisson distribution, 309-10
Gosset, W. S., 159
Graphic methods, 48-52, 339-72
Grouping
 attributes, 32-34
 discontinuous variates, 35, 36
 numerical qualitative, 43-44
 qualitative, 10, 31, 32
 secondary, 38-43, 45-47
Growth, 373-420
 additive, 375-76
 allometric, 374, 376-415
 allomorphic, *see* allometric
 of area, 376
 geometric, 376-82
 gradients, 405, 406, 408

heterogonic, *see* allometric
 of mass, 376
 shape changes, 415-20
 in time, 373, 375-96
Growth curves
 additive, 375, 391, 395
 complex, 378, 396
 fitting, 391-96
 geometric, 376-77, 392-95
 idealized, 380-81
 in systematics, 419-20
Growth rate (growth constant)
 additive, 375
 changes in additive rate, 391-92, 393
 changes in geometric rate, 393-94
 geometric, 377-80, 396-98

Haldane, J. B. S., 101, 402
Hammett, D. W., 399, 400
Hammett, F. S., 399, 400
Hassler, W., 284
Hatt, R. T., 92
Hemidactylium scutatum, 385, 386, 387-88
Herring gull, 342
Hersh, A. H., 394, 405, 410, 411
Heustis, R. R., 327
Hicks, L. E., 33
Hildebrand, S. F., 55, 109, 349
Histogram, 51, 52
Homogeneity, 102-04
Homogeneity of error variance, 286-87
Homogeneity tests, 324-37
 partitioning, 331-38
 several variates, 326-34
Howard, H., 177
Hoy, E. A., 412, 413
Hubbs, C. L., 24, 81
Huxley, J., 406
Hybridization, 346-48
Hypotheses, 149-50

Ictitherium wongii, 389
Index
 hybrid, 14
 meaning of, 13, 14
 speed, 17
Inference
 basic form of, 150-51
 principles of, 148-50
Integers, 2
Interaction, 260-63
 first order, 263
 second order, 263
Intercept
 "best fit," 232-33
 confidence interval, 224-29
 estimate of, 221

International rules of zoological nomenclature, 104
Iridoprocne bicolor, 34
Isometric growth, *see* Allometric growth

Jenkinson, J. W., 392, 393

Kälin, J. A., 24, 25
Kavanaugh, A. J., 411, 413
Kermack, K. A., 402
King snake, 35, 44. *See also Lampropeltis*
Kurtén, B., 351, 389, 390
Kurtosis, coefficient of, 146-47

Lagler, K. F., 24
Lamotte, M., 114, 188, 191, 328, 332
Lampropeltis, 236, 239, 240
Lampropeltis elapsoides elapsoides, 44
Lampropeltis getulus getulus, 35, 47
Lampropeltis polyzona, 218, 219, 222, 243, 245, 246
La Place, P. S., 133
Least squares, 225-27
Lepisosteus platyrhincus, 399, 400, 403, 404, 408
Leptokurtosis, 146
Leraas, H. J., 351
Level, in analysis of variance, 263
Line diagram, 339, 343, 344
Litolestes notissimus, 93, 130, 132, 145
Lizards, linear dimensions, 24
Low, S. H., 34
Lowry, E. M., 278
Lumping of classes, 309, 311
  in contingency tests, 323

Macdonnell, W. R., 144
Maia squinado, 414, 415
Manatee, 34
Matthew, W. D., 33
Mean
  arithmetic, 65-68
  calculation, 66-68
  geometric, 74-75
  graphic comparison, 350-56
  harmonic, 75
  of means, 68-69
  quadratic, 75
  sample, 101
Mean deviation, 82-83
Mean square, 269
Measurement
  accuracy of, 5-7, 9
  adequacy of, 21-22
  condition of material, 23-24
  delimitation of, 22
  of mammals, 25, 26

paleontological, 20
systems, 24-27
Medawar, P. B., 416-19
Median, 70-72, 101
Meek, S. W., 109
Meleagris gallopavo, 177
Melospiza melodia beata, 52
Mesokurtosis, 146
Miller, M. A., 412, 413
Mode, 72-74
  in skewed distributions, 72
Modules, 13, 19
Moulting, effect on allometry, 415
Mus musculus, 172

Natural logarithms, 244
Nichols, J. T., 44, 59
Normal, meaning of, 77
Normal distribution
  approximation to binomial, 137-39
  areas under, 136
  fitting, 136-37
  mean, 134-35
  special properties, 139-42
  table, 421
  variance, 134-35
Notonecta undulata, 394, 395, 404, 405, 409, 410, 411
Null hypothesis, 174-76
  test of, 175-76
Numbers, meaning in zoology, 4-5

Oberholser, H. C., 25
Observations
  primary, 10
  quantitative, 105
  true values, 7
Observed range, 139
Odocoileus virginianus borealis, 91
Orthagoristus, 371
Osborn, H. F., 26, 368, 370
Ovis dalli dalli, 351

Paleontological sampling, 113-16
Parameter, defined, 97
Parapavo californicus, 177
Parexocoetus brachypterus hillianus, 56, 57, 58, 76
Parker, G. H., 34
Paynter, R. A., 155
Pearson, E. S., 141
Pearson, H. S., 183
Pearson, K., 90
Perissodactyls, osteometry, 26
Peromyscus, 3, 142
Peromyscus maniculatus, 119, 159, 197, 204, 309

INDEX **439**

Peromyscus maniculatus artemisiae, 359
Peromyscus maniculatus bairdii, 69, 97, 177, 353, 362
Peromyscus maniculatus gambelii, 23, 29
Peromyscus maniculatus rubidus, 327
Phenacodus primaevus, 180, 181, 182, 224, 243
Phillips, J. C., 91
Photometric spectroscopic analysis, 27
Pipilo erythrophthalmus, 14, 347
Pipilo ocai, 14, 347
Plastron, 24
Platykurtosis, 146
Plesiadapis gidleyi, 207
Plesiadapis tricuspidens, 207, 208
Plesiaddax depereti, 389-90
Poisson distribution
  fitting, 130, 131
  mean, 129, 132
Polymorphism in snails, 328-34
Pomolobus aestivalis, 55, 63
Population, defined, 96
Population size estimate, 112-13
Predation pressure on snails, 328-32
Prediction
  confidence limits, 238
  equation, 231-32
Probability, definition, 117-18
Probability density function, 121
Probability distribution
  defined, 119
  continuous, 120-21
Ptilodontidae, 94, 203
Ptilodus montanus, 34, 41, 66, 94, 162, 204, 317, 326

Quadrat, 110-11
Qualitative sampling, 109
Quantitative sampling, 109-11
Quartile deviation, 88, 89
Quartiles, 88
Quételet, L. A. J., 54, 139, 142
Quintiles, 89

Rainbow trout, 392, 393
Range, 78-82
  estimate, 140-42
  graphic comparison, 350-56
  inferred, 206
  observed, 78-80
  real, 80
  relation to standard deviation, 139
  standard, 142
Range midpoint, 74
Ratios, 13-19
  confidence interval, 163-65
  of continuous variates, 14, 15
  of discontinuous variates, 15, 16
  frequencies, 15, 16
Rattus norvegicus, 193-94
Rectilinearity, 276
Reed's wing index, 14
Regression, 213-57
  definition, 215-16
  scope, 216-17
Regression coefficient 221-24
  choice of, 230-32
Rhea, 60
Rhinoceros unicornis, 372
Richards, O. W., 411, 413
Ridgway, R., 25, 27
Romig, H. G., 126, 155, 199
Rounding numbers, 9
Rupicapra rupicapra, 351
Ruthven, A., 350, 352
Ryan, F. J., 379

Sample mean, 120
  distribution, 160
Samples
  characteristics of good samples, 102-07
  heterogeneous, 201-05
  paleontological, 93
  random, 106-07
Sample variance, 120
  distribution, 161-62
Sampling
  faunal, 108-12
  station method, 111-12
  without replacement, 107-08
Sampling distributions, 119-20
Sampling for growth study, 382-90
  age groups, 384-90
  methods, 382-84
Sampling limits, 199-201
Scatter diagram, 217-18, 219
Schneider, L. L., 379
Schooley, J. P., 33, 367
Schroeder, W. C., 55, 349
Schultz, L. P., 81
Scores, 3
Sculpin, 81
Semi-interquartile range, 88-89
Sex ratio in man, 127-29
Shape constant, 418-19
Shaw, W. T., 322
Short-cut contingency tests
  multiple categories, 320-21
  two-by-two, 318-19
Sibley, C. G., 14, 336, 337, 355, 356
Significance of a difference
  biological, 173-75, 312
  statistical, 173-74
Significant figures, 5-9

Simpson, G. G., 357-58
Skewness, 54-55
  coefficient of, 143-46
Skimming, 109
Slope of a line, 217-20
  "best fit," 232-37
  confidence interval, 224-29
  estimate of, 221
Smallmouth bass, 278
Snedecor, G. W., 305, 416
Snyder, L., 369
Soergel, W., 26
Somateria mollissima dresseri, 154, 155
Sperry, C., 364
Standard deviation, 84-88
  estimate of, 142
Standard error
  of coefficient of variation, 166, 167, 168
  of mean, 160-61, 166, 168
  of mean deviation, 166, 168
  of median, 166, 168
  of quartiles, 166, 168
  of standard deviation, 167, 168
Standard population, 141
Standardized normal deviate, 135-36
"Student," 159
Student's t distribution, 422
Student's t test, 176-83
  assumptions, 183-84
  in regression, 226, 227, 229
Sturnus contra, 271, 272
Sturnus fuscus, 271, 272
Sturnus ginginiamus, 271, 272
Sturnus vulgaris vulgaris, 33
Sum of squares
  deviation from regression, 273, 274
  due to regression, 273, 274
  error, 269
  interaction, 277, 281, 287-88
  main effect, 268, 269
  total, 268, 269
Subhyracodon occidentalis, 372
Sumner, F. B., 23, 25, 27, 29, 30, 361
Svihla, A., 359

Tamias striatus, 33
Teissier, G., 401, 414, 415
Temperature, 11
Tests of differences
  attributes, 312-24
  "best fit" intercept and slope, 237
  binomial proportions, 186-91
  chi-square, 306-07
  chi test, 330-31
  correlation coefficient, 242-46
  frequencies, 306-38
  homogeneity, 324-27
  hypothetical value, 192-93
  means, 176-83
  one-sided, 178-80
  paired comparisons, 180-82
  partial correlations, 249-50
  regression coefficients, 229-30, 276
  required sample sizes, 193-201
  single specimen, 205-12
  single specimen and a sample, 182-83
  variance ratio test, 310-12
  variances, 184-86
Thamniophus megalops, 352
Theory formation, 148
Thompson, D'Arcy W., 369, 371, 374, 415
Times series, 366, 367
Tint photometer, 27
Tippett, L. H. C., 141
Tooth wear, 251-53
Traverse method of sampling, 111
Tree swallow, 34
Turtles, linear dimensions of, 24-25
Typical, meaning of, 76-77

Variables
  dependent, 217
  independent, 217
Variables in zoology, 1-4
Variability, 90-93
Variance, 83-88
  binomial distribution, 311
  calculation of, 85-88
  error, 290
  of estimate in regression, 225
  interaction, 292-93, 294
  main effect, 290
  Poisson distribution, 311
  sample, 83, 101
  of z, 244
Variate, defined, 2
  continuous, 1-4, 8, 11-13, 36-38, 53
  discontinuous (discrete), 1-5, 8, 11-13, 34-36, 53, 57-58
  measurement of, 7
  time period, 11
  true value of, 4
Volume, 11

Width, 24
Worley, L. G., 25

X intercept, 217-20

Y intercept, 217-20
Yingling, H., 369

Zeuner, F., 26

A CATALOG OF SELECTED
DOVER BOOKS
IN SCIENCE AND MATHEMATICS

A CATALOG OF SELECTED

DOVER BOOKS

IN SCIENCE AND MATHEMATICS

Astronomy

BURNHAM'S CELESTIAL HANDBOOK, Robert Burnham, Jr. Thorough guide to the stars beyond our solar system. Exhaustive treatment. Alphabetical by constellation: Andromeda to Cetus in Vol. 1; Chamaeleon to Orion in Vol. 2; and Pavo to Vulpecula in Vol. 3. Hundreds of illustrations. Index in Vol. 3. 2,000pp. 6⅛ x 9¼.
23567-X, 23568-8, 23673-0 Three-vol. set

THE EXTRATERRESTRIAL LIFE DEBATE, 1750–1900, Michael J. Crowe. First detailed, scholarly study in English of the many ideas that developed from 1750 to 1900 regarding the existence of intelligent extraterrestrial life. Examines ideas of Kant, Herschel, Voltaire, Percival Lowell, many other scientists and thinkers. 16 illustrations. 704pp. 5⅜ x 8½. 40675-X

A HISTORY OF ASTRONOMY, A. Pannekoek. Well-balanced, carefully reasoned study covers such topics as Ptolemaic theory, work of Copernicus, Kepler, Newton, Eddington's work on stars, much more. Illustrated. References. 521pp. 5⅜ x 8½.
65994-1

AMATEUR ASTRONOMER'S HANDBOOK, J. B. Sidgwick. Timeless, comprehensive coverage of telescopes, mirrors, lenses, mountings, telescope drives, micrometers, spectroscopes, more. 189 illustrations. 576pp. 5⅝ x 8¼. (Available in U.S. only.)
24034-7

STARS AND RELATIVITY, Ya. B. Zel'dovich and I. D. Novikov. Vol. 1 of *Relativistic Astrophysics* by famed Russian scientists. General relativity, properties of matter under astrophysical conditions, stars, and stellar systems. Deep physical insights, clear presentation. 1971 edition. References. 544pp. 5⅝ x 8¼. 69424-0

Chemistry

CHEMICAL MAGIC, Leonard A. Ford. Second Edition, Revised by E. Winston Grundmeier. Over 100 unusual stunts demonstrating cold fire, dust explosions, much more. Text explains scientific principles and stresses safety precautions. 128pp. 5⅜ x 8½. 67628-5

THE DEVELOPMENT OF MODERN CHEMISTRY, Aaron J. Ihde. Authoritative history of chemistry from ancient Greek theory to 20th-century innovation. Covers major chemists and their discoveries. 209 illustrations. 14 tables. Bibliographies. Indices. Appendices. 851pp. 5⅜ x 8½. 64235-6

CATALYSIS IN CHEMISTRY AND ENZYMOLOGY, William P. Jencks. Exceptionally clear coverage of mechanisms for catalysis, forces in aqueous solution, carbonyl- and acyl-group reactions, practical kinetics, more. 864pp. 5⅜ x 8½.
65460-5

CATALOG OF DOVER BOOKS

THE HISTORICAL BACKGROUND OF CHEMISTRY, Henry M. Leicester. Evolution of ideas, not individual biography. Concentrates on formulation of a coherent set of chemical laws. 260pp. 5⅜ x 8½. 61053-5

A SHORT HISTORY OF CHEMISTRY, J. R. Partington. Classic exposition explores origins of chemistry, alchemy, early medical chemistry, nature of atmosphere, theory of valency, laws and structure of atomic theory, much more. 428pp. 5⅜ x 8½. (Available in U.S. only.) 65977-1

GENERAL CHEMISTRY, Linus Pauling. Revised 3rd edition of classic first-year text by Nobel laureate. Atomic and molecular structure, quantum mechanics, statistical mechanics, thermodynamics correlated with descriptive chemistry. Problems. 992pp. 5⅜ x 8½. 65622-5

Engineering

DE RE METALLICA, Georgius Agricola. The famous Hoover translation of greatest treatise on technological chemistry, engineering, geology, mining of early modern times (1556). All 289 original woodcuts. 638pp. 6¾ x 11. 60006-8

FUNDAMENTALS OF ASTRODYNAMICS, Roger Bate et al. Modern approach developed by U.S. Air Force Academy. Designed as a first course. Problems, exercises. Numerous illustrations. 455pp. 5⅜ x 8½. 60061-0

DYNAMICS OF FLUIDS IN POROUS MEDIA, Jacob Bear. For advanced students of ground water hydrology, soil mechanics and physics, drainage and irrigation engineering and more. 335 illustrations. Exercises, with answers. 784pp. 6⅛ x 9¼. 65675-6

ANALYTICAL MECHANICS OF GEARS, Earle Buckingham. Indispensable reference for modern gear manufacture covers conjugate gear-tooth action, gear-tooth profiles of various gears, many other topics. 263 figures. 102 tables. 546pp. 5⅜ x 8½. 65712-4

MECHANICS, J. P. Den Hartog. A classic introductory text or refresher. Hundreds of applications and design problems illuminate fundamentals of trusses, loaded beams and cables, etc. 334 answered problems. 462pp. 5⅜ x 8½. 60754-2

MECHANICAL VIBRATIONS, J. P. Den Hartog. Classic textbook offers lucid explanations and illustrative models, applying theories of vibrations to a variety of practical industrial engineering problems. Numerous figures. 233 problems, solutions. Appendix. Index. Preface. 436pp. 5⅜ x 8½. 64785-4

STRENGTH OF MATERIALS, J. P. Den Hartog. Full, clear treatment of basic material (tension, torsion, bending, etc.) plus advanced material on engineering methods, applications. 350 answered problems. 323pp. 5⅜ x 8½. 60755-0

A HISTORY OF MECHANICS, René Dugas. Monumental study of mechanical principles from antiquity to quantum mechanics. Contributions of ancient Greeks, Galileo, Leonardo, Kepler, Lagrange, many others. 671pp. 5⅜ x 8½. 65632-2

Math–Geometry and Topology

ELEMENTARY CONCEPTS OF TOPOLOGY, Paul Alexandroff. Elegant, intuitive approach to topology from set-theoretic topology to Betti groups; how concepts of topology are useful in math and physics. 25 figures. 57pp. 5⅜ x 8½. 60747-X

COMBINATORIAL TOPOLOGY, P. S. Alexandrov. Clearly written, well-organized, three-part text begins by dealing with certain classic problems without using the formal techniques of homology theory and advances to the central concept, the Betti groups. Numerous detailed examples. 654pp. 5⅜ x 8½. 40179-0

EXPERIMENTS IN TOPOLOGY, Stephen Barr. Classic, lively explanation of one of the byways of mathematics. Klein bottles, Moebius strips, projective planes, map coloring, problem of the Koenigsberg bridges, much more, described with clarity and wit. 43 figures. 210pp. 5⅜ x 8½. 25933-1

CONFORMAL MAPPING ON RIEMANN SURFACES, Harvey Cohn. Lucid, insightful book presents ideal coverage of subject. 334 exercises make book perfect for self-study. 55 figures. 352pp. 5⅝ x 8¼. 64025-6

THE GEOMETRY OF RENÉ DESCARTES, René Descartes. The great work founded analytical geometry. Original French text, Descartes's own diagrams, together with definitive Smith-Latham translation. 244pp. 5⅜ x 8½. 60068-8

THE THIRTEEN BOOKS OF EUCLID'S ELEMENTS, translated with introduction and commentary by Sir Thomas L. Heath. Definitive edition. Textual and linguistic notes, mathematical analysis. 2,500 years of critical commentary. Unabridged. 1,414pp. 5⅜ x 8½. Three-vol. set.
Vol. I: 60088-2 Vol. II: 60089-0 Vol. III: 60090-4

GEOMETRY OF COMPLEX NUMBERS, Hans Schwerdtfeger. Illuminating, widely praised book on analytic geometry of circles, the Moebius transformation, and two-dimensional non-Euclidean geometries. 200pp. 5⅝ x 8¼. 63830-8

DIFFERENTIAL GEOMETRY, Heinrich W. Guggenheimer. Local differential geometry as an application of advanced calculus and linear algebra. Curvature, transformation groups, surfaces, more. Exercises. 62 figures. 378pp. 5⅜ x 8½. 63433-7

CURVATURE AND HOMOLOGY: Enlarged Edition, Samuel I. Goldberg. Revised edition examines topology of differentiable manifolds; curvature, homology of Riemannian manifolds; compact Lie groups; complex manifolds; curvature, homology of Kaehler manifolds. New Preface. Four new appendixes. 416pp. 5⅜ x 8½. 40207-X

TOPOLOGY, John G. Hocking and Gail S. Young. Superb one-year course in classical topology. Topological spaces and functions, point-set topology, much more. Examples and problems. Bibliography. Index. 384pp. 5⅝ x 8¼. 65676-4

CATALOG OF DOVER BOOKS

Physics

OPTICAL RESONANCE AND TWO-LEVEL ATOMS, L. Allen and J. H. Eberly. Clear, comprehensive introduction to basic principles behind all quantum optical resonance phenomena. 53 illustrations. Preface. Index. 256pp. 5⅜ x 8½. 65533-4

ULTRASONIC ABSORPTION: An Introduction to the Theory of Sound Absorption and Dispersion in Gases, Liquids and Solids, A. B. Bhatia. Standard reference in the field provides a clear, systematically organized introductory review of fundamental concepts for advanced graduate students, research workers. Numerous diagrams. Bibliography. 440pp. 5⅜ x 8½. 64917-2

QUANTUM THEORY, David Bohm. This advanced undergraduate-level text presents the quantum theory in terms of qualitative and imaginative concepts, followed by specific applications worked out in mathematical detail. Preface. Index. 655pp. 5⅜ x 8½. 65969-0

ATOMIC PHYSICS (8th edition), Max Born. Nobel laureate's lucid treatment of kinetic theory of gases, elementary particles, nuclear atom, wave-corpuscles, atomic structure and spectral lines, much more. Over 40 appendices, bibliography. 495pp. 5⅜ x 8½. 65984-4

AN INTRODUCTION TO HAMILTONIAN OPTICS, H. A. Buchdahl. Detailed account of the Hamiltonian treatment of aberration theory in geometrical optics. Many classes of optical systems defined in terms of the symmetries they possess. Problems with detailed solutions. 1970 edition. xv + 360pp. 5⅜ x 8½. 67597-1

THIRTY YEARS THAT SHOOK PHYSICS: The Story of Quantum Theory, George Gamow. Lucid, accessible introduction to influential theory of energy and matter. Careful explanations of Dirac's anti-particles, Bohr's model of the atom, much more. 12 plates. Numerous drawings. 240pp. 5⅜ x 8½. 24895-X

ELECTRONIC STRUCTURE AND THE PROPERTIES OF SOLIDS: The Physics of the Chemical Bond, Walter A. Harrison. Innovative text offers basic understanding of the electronic structure of covalent and ionic solids, simple metals, transition metals and their compounds. Problems. 1980 edition. 582pp. 6⅛ x 9¼. 66021-4

HYDRODYNAMIC AND HYDROMAGNETIC STABILITY, S. Chandrasekhar. Lucid examination of the Rayleigh-Benard problem; clear coverage of the theory of instabilities causing convection. 704pp. 5⅜ x 8¼. 64071-X

INVESTIGATIONS ON THE THEORY OF THE BROWNIAN MOVEMENT, Albert Einstein. Five papers (1905–8) investigating dynamics of Brownian motion and evolving elementary theory. Notes by R. Fürth. 122pp. 5⅜ x 8½. 60304-0

THE PHYSICS OF WAVES, William C. Elmore and Mark A. Heald. Unique overview of classical wave theory. Acoustics, optics, electromagnetic radiation, more. Ideal as classroom text or for self-study. Problems. 477pp. 5⅜ x 8½. 64926-1

CATALOG OF DOVER BOOKS

PHYSICAL PRINCIPLES OF THE QUANTUM THEORY, Werner Heisenberg. Nobel Laureate discusses quantum theory, uncertainty, wave mechanics, work of Dirac, Schroedinger, Compton, Wilson, Einstein, etc. 184pp. 5⅜ x 8½.   60113-7

ATOMIC SPECTRA AND ATOMIC STRUCTURE, Gerhard Herzberg. One of best introductions; especially for specialist in other fields. Treatment is physical rather than mathematical. 80 illustrations. 257pp. 5⅜ x 8½.   60115-3

AN INTRODUCTION TO STATISTICAL THERMODYNAMICS, Terrell L. Hill. Excellent basic text offers wide-ranging coverage of quantum statistical mechanics, systems of interacting molecules, quantum statistics, more. 523pp. 5⅜ x 8½.   65242-4

THEORETICAL PHYSICS, Georg Joos, with Ira M. Freeman. Classic overview covers essential math, mechanics, electromagnetic theory, thermodynamics, quantum mechanics, nuclear physics, other topics. First paperback edition. xxiii + 885pp. 5⅜ x 8½.   65227-0

PROBLEMS AND SOLUTIONS IN QUANTUM CHEMISTRY AND PHYSICS, Charles S. Johnson, Jr. and Lee G. Pedersen. Unusually varied problems, detailed solutions in coverage of quantum mechanics, wave mechanics, angular momentum, molecular spectroscopy, more. 280 problems plus 139 supplementary exercises. 430pp. 6½ x 9¼.   65236-X

THEORETICAL SOLID STATE PHYSICS, Vol. 1: Perfect Lattices in Equilibrium; Vol. II: Non-Equilibrium and Disorder, William Jones and Norman H. March. Monumental reference work covers fundamental theory of equilibrium properties of perfect crystalline solids, non-equilibrium properties, defects and disordered systems. Appendices. Problems. Preface. Diagrams. Index. Bibliography. Total of 1,301pp. 5⅜ x 8½. Two volumes.   Vol. I: 65015-4   Vol. II: 65016-2

A TREATISE ON ELECTRICITY AND MAGNETISM, James Clerk Maxwell. Important foundation work of modern physics. Brings to final form Maxwell's theory of electromagnetism and rigorously derives his general equations of field theory. 1,084pp. 5⅜ x 8½. Two-vol. set.   Vol. I: 60636-8   Vol. II: 60637-6

OPTICKS, Sir Isaac Newton. Newton's own experiments with spectroscopy, colors, lenses, reflection, refraction, etc., in language the layman can follow. Foreword by Albert Einstein. 532pp. 5⅜ x 8½.   60205-2

THEORY OF ELECTROMAGNETIC WAVE PROPAGATION, Charles Herach Papas. Graduate-level study discusses the Maxwell field equations, radiation from wire antennas, the Doppler effect and more. xiii + 244pp. 5⅜ x 8½.   65678-5

INTRODUCTION TO QUANTUM MECHANICS With Applications to Chemistry, Linus Pauling & E. Bright Wilson, Jr. Classic undergraduate text by Nobel Prize winner applies quantum mechanics to chemical and physical problems. Numerous tables and figures enhance the text. Chapter bibliographies. Appendices. Index. 468pp. 5⅜ x 8½.   64871-0

METHODS OF THERMODYNAMICS, Howard Reiss. Outstanding text focuses on physical technique of thermodynamics, typical problem areas of understanding, and significance and use of thermodynamic potential. 1965 edition. 238pp. 5⅜ x 8½. 69445-3

TENSOR ANALYSIS FOR PHYSICISTS, J. A. Schouten. Concise exposition of the mathematical basis of tensor analysis, integrated with well-chosen physical examples of the theory. Exercises. Index. Bibliography. 289pp. 5⅜ x 8½. 65582-2

RELATIVITY IN ILLUSTRATIONS, Jacob T. Schwartz. Clear nontechnical treatment makes relativity more accessible than ever before. Over 60 drawings illustrate concepts more clearly than text alone. Only high school geometry needed. Bibliography. 128pp. 6⅛ x 9¼. 25965-X

THE ELECTROMAGNETIC FIELD, Albert Shadowitz. Comprehensive undergraduate text covers basics of electric and magnetic fields, builds up to electromagnetic theory. Also related topics, including relativity. Over 900 problems. 768pp. 5⅝ x 8¼. 65660-8

GREAT EXPERIMENTS IN PHYSICS: Firsthand Accounts from Galileo to Einstein, edited by Morris H. Shamos. 25 crucial discoveries: Newton's laws of motion, Chadwick's study of the neutron, Hertz on electromagnetic waves, more. Original accounts clearly annotated. 370pp. 5⅜ x 8½. 25346-5

RELATIVITY, THERMODYNAMICS AND COSMOLOGY, Richard C. Tolman. Landmark study extends thermodynamics to special, general relativity; also applications of relativistic mechanics, thermodynamics to cosmological models. 501pp. 5⅜ x 8½. 65383-8

LIGHT SCATTERING BY SMALL PARTICLES, H. C. van de Hulst. Comprehensive treatment including full range of useful approximation methods for researchers in chemistry, meteorology and astronomy. 44 illustrations. 470pp. 5⅜ x 8½. 64228-3

STATISTICAL PHYSICS, Gregory H. Wannier. Classic text combines thermodynamics, statistical mechanics and kinetic theory in one unified presentation of thermal physics. Problems with solutions. Bibliography. 532pp. 5⅜ x 8½. 65401-X

Paperbound unless otherwise indicated. Available at your book dealer, online at **www.doverpublications.com**, or by writing to Dept. GI, Dover Publications, Inc., 31 East 2nd Street, Mineola, NY 11501. For current price information or for free catalogues (please indicate field of interest), write to Dover Publications or log on to **www.doverpublications.com** and see every Dover book in print. Dover publishes more than 500 books each year on science, elementary and advanced mathematics, biology, music, art, literary history, social sciences, and other areas.